Biodegradation of Toxic and Hazardous Chemicals

This timely reference book discusses the biotechnological applications of microorganisms as a crucial solution for the sustainable management of different types of toxic pollutants. It reviews the sustainable biodegradation approach and resource recovery for different kinds of pollutants like plastic wastes, pharmaceutical wastes, pesticides, and textile industry wastes. The book provides an understanding of biotechnology-based interventions toward a zero-waste route.

KEY FEATURES

- Provides a deep understanding of biodegradation of toxic pollutants from industries ranging from textiles to pharmaceuticals
- Presents novel technologies for the sustainable treatment of environmental pollution
- Reviews crucial considerations like energy requirements and cost analysis
- Brings out the latest themes such as the eco-design study of bio-nanomaterials, bio-nanofilters, and assessment for the treatment of emerging pollutants and IoT-based technology
- Covers the latest research developments in the biodegradation of antibiotics, pesticides, and electronic wastes

This book is meant for scientists, ecologists, microbiologists, industry experts, researchers, students, innovators involved in biotechnology research, and policymakers focused on bioremediation.

SUSTAINABLE INDUSTRIAL AND ENVIRONMENTAL BIOPROCESSES

This book series aims to provide a comprehensive collection of books focusing on bioprocesses in industrial and environmental biotechnology. The multidisciplinary content encompasses the chemical and biochemical engineering, industrial microbiology, and energy biosciences, all with a central theme of sustainable development and circular economy principles. The books align with the Sustainable Development Goals (SDGs) and offer state-of-the-art information and in-depth knowledge on the subject matter. The book in these series also emphasizes on the application of emerging tools, such as machine learning and artificial intelligence, for the advancement of bioprocesses. While primarily targeting academicians and researchers, the series is valuable for policy planners and industry professionals, with carefully tailored contents to cater to their specific needs and interests.

Series Editor: Dr Ashok Pandey

Professor Ashok Pandey is currently Distinguished Scientist at the Centre for Innovation and Translational Research, CSIR-Indian Institute of Toxicology Research, Lucknow, India. His major research and technological development interests are industrial & environmental biotechnology and energy biosciences, focusing on biomass to biofuels & chemicals, waste to wealth & energy, etc. He has 16 patents, 120 books, >1000 papers and book chapters, etc with h index of 131 and >74,000 citations. Professor Pandey is the recipient of many national and international awards and honours, which include Fellow, The World Academy of Sciences (TWAS), Highly Cited Researcher (Top 1% in the world), Clarivate Analytics (since 2018–till date), Rank #1 in India in Biology in India and Biochemistry and #417 in the world, Research.Com (2023), Rank #1 in India in Microbiology under Enabling and Strategic Technologies sector, Elsevier (2021), Rank #1 in India in Biotechnology and #8 in the world, Stanford University Report (2020–2021–2022), etc.

Diatoms Biotechnology
Archana Tiwari, Ashok Pandey, Pau-Loke Show, Binod Parameswaran

Sustainable Treatment and Management of Sewage Sludge
Mukesh Kumar Awasthi, Zengqiang Zhang, Ashok Pandey

Pharmaceuticals in Aquatic Environment: Remediation Technologies and Future Challenges
Vinod Kumar Garg, Ashok Pandey, Navish Kataria, Caterina Faggio

Biomass Hydrolysing Enzymes: Basics, Advancements, and Applications
Reeta Rani Singhania, Anil Kumar Patel, Héctor A. Ruiz and Ashok Pandey

Biodegradation of Toxic and Hazardous Chemicals: Detection and Mineralization
Kashyap K. Dubey, Kamal K. Pant, Ashok Pandey and María Ángeles Sanromán

Biodegradation of Toxic and Hazardous Chemicals: Remediation and Resource Recovery
Kashyap K. Dubey, Kamal K. Pant, Ashok Pandey and María Ángeles Sanromán

For more information about this series, please visit: https://www.routledge.com/Sustainable-Industrial-and-Environmental-Bioprocesses/book-series/SIEB.

Biodegradation of Toxic and Hazardous Chemicals

Remediation and Resource Recovery

Edited by

Kashyap K. Dubey

School of Biotechnology, Jawaharlal Nehru University, New Delhi, India

Kamal K. Pant

Department of Chemical Engineering, Indian Institute of Technology, Delhi, India

Indian Institute of Technology, Roorkee, India

Ashok Pandey

Centre for Innovation and Translational Research, CSIR-Indian Institute of Toxicology Research, Lucknow, India

Sustainability Cluster, School of Engineering, University of Petroleum and Energy Studies, Dehradun, India

Centre for Energy and Environmental Sustainability, Lucknow, India

María Ángeles Sanromán

Department of Chemical Engineering, Universidade de Vigo, Vigo, Spain

CRC Press

Taylor & Francis Group

Boca Raton London New York

CRC Press is an imprint of the Taylor & Francis Group, an **informa** business

Design cover image: Shutterstock, image numbers 1325050973 and 2092866652

First edition published 2024
by CRC Press
2385 Executive Center Drive, Suite 320, Boca Raton, FL 33431

and by CRC Press
4 Park Square, Milton Park, Abingdon, Oxon, OX14 4RN

CRC Press is an imprint of Taylor & Francis Group, LLC

ISBN: 978-1-032-45369-9 (hbk)
ISBN: 978-1-032-48932-2 (pbk)
ISBN: 978-1-003-39148-7 (ebk)

DOI: 10.1201/9781003391487

Typeset in Times LT Std
by Apex CoVantage, LLC

Contents

Preface

The increasing pace of industrialization and modern way of living has brought a lot of comforts for mankind; however, it also has resulted in several kinds of problems, in particular impacting the environment negatively. One of the major concerns in this regard has been the generation of toxic and hazardous chemicals as wastes or residues from the manufacturing of various products of a chemical nature and from the biomedical industries. Disposal of these wastes in the environment poses several health and environmental problems and has been a serious challenge to tackle. Several methods have been in development in the past couple of decades on the removal and/or treatment of such wastes, in particular from wastewater and solid wastes, among which microbial-based processes have gained specific attention. Use of genetically modified microorganisms, in particular, has been of great significance for the treatment of the toxic and hazardous chemicals, mineralizing them to non-toxic forms.

Biodegradation of Toxic and Hazardous Chemicals: Remediation and Resource Recovery highlights the characteristics, aims and applications of microorganisms as a crucial solution for sustainable management of the toxic pollutants enriched in the ecosystem/aquatic ecosystem, soil, and air. It facilitates biotechnology toward the development of more sustainable biological systems by minimizing the level of harmful toxic substances and reducing the toxic effects of current chemical processes. It also provides a comprehensive understanding of the subject knowledge with contemporary environmental challenges using biotechnology-based interventions. Readers will learn all about the recent, sustainable progress in theoretical, practical aspects and future potential applications of bio-systems in the environment.

In this book, a multidisciplinary team of researchers contributed on the biodegradation of toxic pollutants present in the different environmental matrices with their advancement and highlighting the technical, scientific, regulatory, safety, and societal impacts. The chapters focus on the different roles of biotechnology as remediation of emerging toxic pollutants/contaminants from industrial wastewater, eco-design, and modification study of bio-nanoparticles, life-cycle assessment, nano-filtration, bionanomaterials-based sensors for monitoring air and water pollution, resource recovery from wastewater using biotechnology, and limitations for a healthy and economically sustainable environment. Integration of biochemical principles, procedures, concepts, and attitudes into the current research and curriculum can form a robust platform and a powerful pedagogical foundation.

The book comprises 11 chapters. 1. Biodegradation of Petrochemical Refinery Wastewater. 2. Biodegradation of Anti-Inflammatory, Non-Steroidal, Pharmaceutical Active Compounds. 3. Biodegradation of Catechol in Sequential Batch Reactor. 4. Biodegradation and Decolorization of Textile Wastewater by Sequential Aerobic/Microaerophilic, Aerobic and Microaerophilic Processes. 5. Bio-Adsorbent-Based Integrated System. 6. Carbon-Based Hybrid Metal Composites for the Degradation of Toxic Chemicals. 7. Microbial Biofilms in Biodegradation of Toxic Chemicals. 8. Microbial Degradation of Endocrine-Disrupting Chemicals. 9. Bioremediation of Heavy Metals Mediated by Cyanobacteria. 10. Metal Recovery from Lithium-Ion Batteries Using Bioleaching: A Greener Approach. 11. Microplastics Pollution: A Perspective on the Source, Fate, Impact, Identification and Extraction from the Environment.

We hope that this book will be of immense usefulness as reference literature for a diverse community of professionals, scientists, ecologists, environment microbiologists, industrialists, researchers, students (undergraduate and postgraduate), innovators involved in biotechnology research, and policymakers who are engaged in the development of strategies to deal with the challenges of the current environmental issues and working in the bioremediation field.

We are grateful to the reviewers for their time and efforts in critically reviewing the manuscripts of the chapters. We sincerely acknowledge the Biotech Research Society, India for giving us the opportunity and support to prepare this book under the agreement with Taylor & Francis. We thank the team of Taylor & Francis Group, comprising Dr Bhavik Sawhney, Senior Publisher and Ms. Neha Bhatt, Editorial Assistant, CRS Press, and the entire team of CRC Press related to the production of this book for their consistent support during the publication process.

Editors
Kashyap K. Dubey
Kamal K. Pant
Ashok Pandey
María Ángeles Sanromán

Editors

Professor Kashyap Kumar Dubey is currently Professor and Dean School of Biotechnology at Jawaharlal Nehru University, New Delhi (India). His R& interests are in the area of bioprocess development of microbial secondary metabolites, with focus on developing low cost decentralised technology for onsite treatment of hospital aqueous water.

Professor Kamla K Pant is currently Director, Indian Institute of Technology Roorkee, and has been Petrotech Chair Professor and Dean Faculty and Head of the Chemical Engineering Department at IIT Delhi. He is working on cutting-edge and futuristic technologies for coal-to-methanol conversion, e-waste treatment, waste plastic conversion, biomass valorization, and so on using catalysis and reaction engineering.

Professor Ashok Pandey is currently Distinguished Scientist at Centre for Innovation and Translational Research, CSIR-Indian Institute of Toxicology Research, Lucknow, India; Distinguished Professor, Sustainability Cluster, University of Petroleum and Energy Studies, Dehradun, India and Executive Director (Hon), Centre for Energy and Environmental Sustainability, Lucknow, India. His major R&D interests are in industrial/environmental biotechnology and energy biosciences.

Prof. María Ángeles Sanromán is the Head of the Bioengineering & Sustainable Processes Group at the University of Vigo, Spain. Currently, main research lines carried out are environmental technology applied to soil and water remediation; advanced oxidation processes; reuse, recycling, and valorization of wastes; green chemistry; and catalysts.

Contributors

Mark A. T. Blaskovich
Centre for Superbug Solutions, Institute for Molecular Bioscience
The University of Queensland
St Lucia, Australia
Indian Institute of Technology Delhi Academy of Research (UQIDAR)

Vallari R. Chourasia
Department of Chemical Engineering
Indian Institute of Technology
Delhi, India
University of Queensland
St Lucia, Australia

Kashyap Kumar Dubey
School of Biotechnology
Jawaharlal Nehru University
New Delhi, India

Silvia Escudero-Curiel
Department of Chemical Engineering, CINTECX
Universidade de Vigo
Vigo, España

Shally Gupta
Chemical Engineering Department
Indian Institute of Technology
Delhi, India

Gopinath Halder
Chemical Engineering Department
National Institute of Technology
Durgapur, India

Prashant Ram Jadhao
Chemical Engineering Department
Indian Institute of Technology
Delhi, India

Marut Jain
Department of Chemical Engineering
Indian Institute of Technology
New Delhi, India

Neethu Thoppil Joshikumar
Department of Chemical Engineering
Indian Institute of Technology
Delhi, India
University of Queensland
St Lucia, Australia

Pradeep Kumar
Department of Vocational Studies and Skill Development,
Central University of Haryana,
Mahendergarh, India

Punit Kumar
Department of Morphology
Karaganda Medical University
Karaganda, Kazakhstan

Ravindra Kumar
Department of Biotechnology
SGRR University Dehradun
Uttarakhand, India

Nidhi Kushwaha
Department of Chemical Engineering
Indian Institute of Technology
Delhi, India

Naseeb
Department of Vocational Studies and Skill Development,
Central University of Haryana,
Mahendergarh, India

K. D. P. Nigam
Chemical Engineering School
The University of Adelaide
Australia

Nikita
Laboratory of AIDS Research and Immunology, School of Biotechnology
Jawaharlal Nehru University
New Delhi, India

Antonina Nikolayeva
Medical University of Karaganda
Kazakhstan

Aisha Noor
Department of Chemical Engineering
Indian Institute of Technology
New Delhi, India

Ashok Pandey
Centre for Innovation and Translational Research
CSIR-Indian Institute of Toxicology Research, Lucknow, India
Sustainability Cluster, School of Engineering
University of Petroleum and Energy Studies Dehradun, India
Centre for Energy and Environmental Sustainability
Lucknow, India

Kamal K. Pant
Department of Chemical Engineering
Indian Institute of Technology New Delhi, India
Indian Institute of Technology
Roorkee, India

Marta Pazos
Department of Chemical Engineering, CINTECX
Universidade de Vigo
España

Pinky
Department of Vocational Studies and Skill Development,
Central University of Haryana,
Mahendergarh, India

Abhisek Sahoo
Department of Chemical Engineering
Indian Institute of Technology
Delhi, India

María Ángeles Sanromán
Department of Chemical Engineering, CINTECX
Universidade de Vigo
Vigo, España

Shalini Sharma
Department of Entomology
SGRR University Dehradun
Uttarakhand, India

Uma Singh
Department of Botany
University of Allahabad
Prayagraj

Ravi Tandon
Laboratory of AIDS Research and Immunology, School of Biotechnology
Jawaharlal Nehru University
New Delhi, India

Richa Tandon
Department of Botany
University of Allahabad
Prayagraj

G. L. Tiwari
Department of Botany
University of Allahabad
Prayagraj

Ankush Yadav
Department of Botany
Zakir Husain Delhi College
University of Delhi
New Delhi, India

Zyta M. Ziora
Centre for Superbug Solutions
Institute for Molecular Bioscience The University of Queensland
St Lucia, Australia

1 Biodegradation of Petrochemical Refinery Wastewater

Marut Jain, Aisha Noor, Abhisek Sahoo, Kamal K. Pant, Zyta M. Ziora and Mark A. T. Blaskovich

1.1 INTRODUCTION

The petroleum industry products and their derived compounds mainly comprise hydrocarbons, which are an essential raw source for different industries and various energy resources. Increased demand for petroleum products in daily life has heightened the cost of availability and simultaneously depleted crude oil resources (Varjani et al. 2015; Varjani and Upasani 2017). In addition, petrochemical refineries and industries produce a massive amount of wastewater that contains a series of toxic and hazardous pollutants such as hydrocarbons and other inorganic and organic impurities. These hydrocarbon pollutants are termed "recalcitrant" compounds and are categorized in the priority pollutant list (Costa et al. 2012). Petroleum hydrocarbon products cause pollution in the form of offshore/onshore industrial activities, accidental spills, effluent release, and municipal runoffs. The caused pollution is equally threatening to the environment as well as living organisms on the earth (Sajna et al. 2015). These pollutants affect the marine environment rigorously; therefore, it critically demands sustainable development to reduce the pollution (Varjani et al. 2020). The two highlighted issues in the context of the petroleum industry are the remediation of hydrocarbon pollutants while expediting oil recovery (Varjani et al. 2020; Varjani 2017). Previously, a lot of established technologies such as oxidation, incineration, and solidification were employed to treat wastewater generated from petrochemical refineries and industries (Jain et al. 2022), though the aforementioned methods were efficient but were energy intensive and also produced toxic by-products during remediation. Notably, bioremediation is found to be one of the effective yet sustainable technologies to treat petrochemical refinery wastewater. Thus, proper to understanding related to the scope and efficient processes of bioremediation, we need to understand crude oil properties, environmental concerns, consequences of oil recovery, biodegradation mechanisms, and factors affecting biodegradation rate and controlling rate (Boopathy 2000; Varjani et al. 2015).

The petroleum industry's discharges various complex compounds into water bodies; for instance, resins, paraffins, aromatics, and asphaltenes and other miscellaneous hydrocarbons. To degrade such lethal compounds, bioremediation is one of the foremost approaches due to its facile and eco-friendly nature where degradation occurs using various microorganisms, and they are capable enough of breaking down complex moieties (Zanaroli et al. 2010). The bioremediation technique utilizes microorganisms like bacteria and fungi for degrading and removing toxic pollutants due to their metabolic enzyme capability. They are a well-known source for removal and degradation purposes of pollutants involving petroleum industry effluents (Medina-Bellver et al. 2005). Bioremediation using a natural source of microorganisms is the primary mechanism for the removal of pollutants (petroleum and hydrocarbon products) from the environment (Amirabdollahian and Datta 2013). Additionally, bioremediation is cost-effective and non-invasive compared to other present techniques (Jain et al. 2021).

DOI: 10.1201/9781003391487-1

One significant petrochemical industry pollutant is phenol, mostly found in effluents released from industrial wastewater. It has many applications, such as the production of fertilizers, textiles, explosives, dyes, and plastics (Jain et al. 2022). Because of its extensive use, the concentration of phenol itself and its derivatives can be quite high (50 to 2000 mg/L) in industrial effluents. These phenolic compounds are toxic to aquatic flora and fauna, even at low concentrations (above 0.001 mg/L) (Jain et al. 2021). Phenol and its vapors are harsh to the eyes, the skin, and the respiratory tract (Cahan et al. 2013).

Dimethyl phthalate (DMP) is used in different manufacturing applications, including plastics, insect repellents, safety glass, and lacquer coatings, and it is another top priority pollutant in various countries, such as China, European countries, United States of America, India, and others. It is a plasticizer used as an additive during the synthesis of products derived from polymers; for example, automobile applications, PVC sheets, plastic toys, and paints (Zhang et al. 2015). Polychlorinated biphenyls (PCBs) are a family of related compounds containing approximately 209 different congeners. Traces are present in sediments and soil and are also found to be accumulated in the food chain, mainly in animal fatty tissues (Sadañoski et al. 2020). Their production was banned in the USA in 1978, and globally, by the Stockholm Convention on Persistent Organic Pollutants in 2001, which also regulated the use of other organic pollutants (Porta and Zumeta 2002).

The purpose of this current chapter is to focus on expanding the scope of the bioremediation approach and technique for petrochemical refinery wastewater treatment. It considers the toxicity rate of hydrocarbons and its consequences on the environment. Moreover, various factors involved in the bioremediation pathways are also discussed. It provides insights about persistent hydrocarbons and their degradation pathways, biodegradation rate, microbial metabolism, enzymes, bioremediation technologies, and related processes.

1.2 TOXIC BEHAVIOR OF PETROLEUM HYDROCARBON

The most persistent organic pollutants present in the petrochemical industry wastewater are polyaromatic hydrocarbons (PAHs), components which are carcinogenic and hemotoxic. Hydrocarbons create extensive damage to ecosystems due to their biological magnification property (Chandra et al. 2012). The spillages produced and leakages found in underground tanks, oil wells, and refinery sites release an abundant amount of hydrocarbon pollutants, which cause contamination of water bodies as well as the soil surface (Mani et al. 2019; Meckenstock et al. 2016).

The direct/indirect effects and acute/chronic effects of petroleum pollutants are described by (Zhang et al. 2011; Costa et al. 2012; Mani et al. 2019; Meckenstock et al. 2016). It directly or indirectly affects hormonal imbalance, disturbance in the pathway of metabolic reactions, suffocation, and anoxia in living organisms (Walker et al. 2005). The short-term impacts comprise acute necrosis mortality, drowning, ingestion, and hypothermia (Varjani et al. 2020; Desforges et al. 2016). In contrast, the long-term effects of pollutants include abnormalities related to sea animals, mainly discoloration, jaw reductions, and skull defects (Varjani 2017). These effects are responsible for population or community changes, which create further imbalances in the ecosystem (Walker et al. 2005). The volatile organic compounds (VOCs) present in crude oil pose a risk to human health. The PAHs in crude oil have the potential to alter or damage DNA, RNA, and protein, which, in turn, generate tumors affecting the skin and epithelial tissues (Desforges et al. 2016).

For assessing health risk problems of humans and the environment, equivalent carbon numbers or equivalent fractions are divided into a range of petroleum hydrocarbons. These carbon numbers/fractions are subdivided according to different characteristics based on physical, chemical, and toxicological aspects. Fraction 1 includes carbon numbers from C6 to C10, representing a volatile fraction of mixtures present in hydrocarbon. Fraction 2 contains equivalent carbon numbers C10 to C16, representing a semi-volatile fraction. Fraction 3 includes carbon numbers from C16 to C34, representing non-volatile fractions. Fraction 4 consists of the carbon compounds whose carbon number is equivalent to C35+, with the lowest volatility and less solubility (Varjani et al. 2015).

1.3 MICROBIAL DEGRADATION OF PETROCHEMICAL HYDROCARBON POLLUTANTS

Petroleum hydrocarbon biodegradation is tedious and complex because of the different complex hydrocarbon chains. The quantity of hydrocarbons also determines the intricacy of the biodegradation route. Hydrocarbons in petroleum by-products are classified into four classes: Saturated hydrocarbons, aromatics, resins, and asphaltenes (Colwell et al. 2008). Hydrocarbon degradation depends on many factors, such as the ease of availability of microorganisms, temperature, and other climatic factors. The major influential problem inhibiting biodegradation in the environment is the poor availability of oil pollutants to the source microorganisms. Removal of petroleum hydrocarbons is even more difficult, as these hydrocarbons get firmly attached to the soil component (Barathi and Vasudevan 2001). Different natures of hydrocarbons are differentially prone to microbial attack; for instance, linear alkanes are most susceptible to microbial degradation, followed by branched alkanes and then more complex hydrocarbons, such as small aromatics to cyclic alkane. On the contrary, various hydrocarbons of higher molecular weight compounds, like polycyclic aromatic hydrocarbons, are difficult to degrade or cannot be eliminated (Atlas and Bragg 2009).

Johnsen et al. (2005) defined the different functions of PAHs' microbial degradation (Johnsen et al. 2005). These functions can be described as follows: a) The microorganism degradation by assimilative biodegradation produces carbon and energy for it and mineralization of some parts of compounds. b) The purpose of increasing solubility of PAHs is done by a process named intracellular detoxification. c) Cometabolism – a process of degradation related to PAHs where there is no requirement of energy and process outcome – is the generation of carbon for cell metabolism. The oxidoreductase that integrates atoms of oxygen in substrates is known as an oxygenase.

The ultimate and eco-friendly route to remove petroleum hydrocarbon from the environment can be achieved by microbial degradation (Atlas 1984). It is a natural mechanism through which the environment gets clean and becomes less polluted; for example, Jones et al. (Jones et al. 1983) reported the biodegradation of petroleum hydrocarbons from marine sediments. The extensive study reveals the microorganisms involved in the biodegradation of alkyl aromatic and n-alkanes of oil are *Pseudomonas, Arthrobacter, Mycobacterium, and Rhodococcus.* Another evidence of microbial degradation of petroleum hydrocarbon was reported from Nigeria, where bacterial strains were isolated from the polluted stream. *Pseudomonas fluorescence, P. aeruginosa, Bacillus spp., B. subtilis,* and some other micrococcus and *Corynebacterium spp.* were found having immense potential to biodegrade crude oil from the environment.

The leading organisms responsible for the biodegradation of the environment's pollutants are bacteria and yeast, followed by fungi. The efficiency of reported organisms is 6–82% (Jones et al. 1970) for fungi; 0.13–50% for bacteria; and for marine bacteria, it ranges from 0.003–100% (Hollaway et al. 1980; Mulkins Phillips and Stewart 2011). Studies have supported that a mixed consortium of microorganisms provides a wide array of enzymatic degradation actions on petroleum hydrocarbon pollutants to get rid of them from the soil, freshwater, and other water bodies (Atlas 1984). For instance, laccase enzymes of basidiomycetes help in degradation of polyaromatic hydrocarbons. Similarly, bacteria such as *Pseudomonas spp.* and acenitobacteria expedite the degradation rate of naphthalene, aromatic, and linear chain hydrocarbons with the help of enzymes. Beside this, other fungal and bacterial enzymes are also involved in the elimination of toxic pollutants from the environment, as shown in the Table 1.1.

Among all the microorganisms, bacteria are primarily responsible for the active biodegradation of petroleum hydrocarbons like spilled oil in the soil and environment (Rahman et al. 2003; Brooijmans et al. 2009). Some specific bacterial strains exclusively utilize hydrocarbons as their feedstock (Yakimov et al. 2007). It is reported that 25 genera of bacteria, especially feeding on hydrocarbons, and 25 genera of fungi degrading hydrocarbons are extracted from the marine area. Bartha and Bossert (Atlas 1984) have documented a compilation of 22 bacterial genera and 31 fungal genera, whereas, in the initial period, a minimal study was done on biodegradation of the

TABLE 1.1
Microorganisms and Their Enzymes for Their Degradation

S No.	Microorganism	Enzymes	Substrates	References
1.	*Methylococcus spp.*	Monooxygenase	C1-C10 linear chain hydrocarbons (HC), fatty acids (FA)	(McDonald et al. 2006)
2.	*Candida spp.*	Cytochrome P450	C10-C16 HC, FA	(Das and Chandran 2011)
3.	*Acinetobacter spp.*	Dioxygenase	C10-C30 HC	(Al-Jobouri 2020)
4.	*Basidiomycetes spp.*	Laccase	Polyaromatic hydrocarbons (PAH)	(Abdel-Hamid et al. 2013)
5.	*Phanerochaete spp.*	Lignin peroxidase	PAH	(Falade et al. 2017)
6.	*Pseudomonas petudia*	Naphthalene dioxygenase	Naphthalene and other aromatic compounds	(Pozdnyakova-Filatova et al. 2020)
7.	*Gordonia spp.*	Lys1-3	n-hexadecane	(Liu et al. 2022)

oil hydrocarbons by yeast, fungi, and bacteria. Further, Das and Mukherjee (2007) have extracted crude oil from the petroleum hydrocarbon-contaminated soil in Northeast India. *Acinobacter spp.* has the potential to utilize hydrocarbons of chain length ranging from C10-C40 as their food source (Throne-Holst et al. 2007). Various other bacteria isolated from hydrocarbon-contaminated soil, such as *Aeromicrobium, Brevibacterium, Burkholderia, Dietzia, Gordonia*, and mycobacterium, are found to be capable of degradation and removal of petroleum hydrocarbons from the soil. Similarly, Daugulis and McCracke (Daugulis and McCracken 2003) show *Sphingomonas* bacteria have promising potential to biodegrade polyaromatic hydrocarbons from the environment.

Furthermore, many fungal strains such as *Amorphoteca, Candida, Talaromyces, Graphium, Pichia, and Yarrowia* are found to be effective degrading agents for removal of hydrocarbons (Chaillan et al. 2004). They are isolated from the oil-spilled, hydrocarbon-contaminated soil site. Another terrestrial fungus group, which includes *Aspergillus, Cephalosporium,* and *Penicillium*, also have the capability to degrade petroleum hydrocarbons, as reported by Singh. Another fungal group – i.e., yeast spp. like *Candida lipolytica, Geotrichum sp., and Trichosporon mucoides* – are also extracted from oil-contaminated soil and noted to have potential to degrade petroleum hydrocarbons (Boguslawska-Wąs and Dąbrowski 2001).

Algae and protozoa are a dominant and abundant group among the microbial community in aquatic ecosystems as well as terrestrial ecosystems, but their potential to degrade hydrocarbon is low and rarely found. A group of researchers shows that alga – namely, *Prototheca zopfi* – have the ability to use crude oil (Colwell et al. 2008). It also demonstrates the capability of utilizing mixed hydrocarbons and other n-chain alkanes, isoalkanes, and aromatic hydrocarbons too.

1.4 FACTORS AFFECTING DEGRADATION

1.4.1 Effect of Temperature

Temperature is a primary factor in bioremediation, as it regulates the physical composition of organic pollutant hydrocarbons present at the contaminated site and also affects the microbial population acting on them (Ansari et al. 2017; Jain et al. 2022). Temperature also plays a vital role in the rate of growth of microorganisms, solubility of gases, metabolism of microbes, soil composition, as well as physical and chemical properties of pollutants. Shreds of evidence suggest the hydrophobic pollutants' solubility is directly related to an increase in temperature and an enhanced diffusion rate and also aids in transferring of long chain hydrocarbon alkane from solid state to water liquid state

(Aislabie et al. 2006). On the other hand, viscosity is inversely proportional to increase in temperature. Furthermore, at lower temperatures, oil's viscosity increases, and the reduction observed in volatilization of shorter n-alkane chains depleted their solubility in water (Leahy and Colwell 1990). These factors, together, obstruct the rate of biodegradation of pollutants. Biosurfactants/bioemulsifiers are an important tool employed for the degradation of oil contaminated soil in colder areas like polar regions of very low temperature (Kauppi et al. 2011). Bacterium *B. thuringenesis* is found to biodegrade oil polluted soil by 80% in just 27 days, as reported by Thamer et al. (2013). This biodegradation rate has supported the presence of environmental factors, availability of bacterial enzymes, or production of biosurfactants, and the presence of ambient temperature has increased the efficiency and potential of bacteria to degrade the toxic oil hydrocarbon compounds.

1.4.2 Effect of Electron Acceptors

Permeability of soil air is an important factor that directly and indirectly affects the activity of microbes. Petroleum pollutants deplete the soil air permeability, which retards the growth of microbes and thereby degradation rate (Meckenstock et al. 2016). To uplift the biodegradation rate, electrons get captured by oxygen, which, in turn, enhances the bioremediation process (Grace Pavithra et al. 2021). On the other hand, ensuring enough oxygen to the soil for complete mineralization of oil hydrocarbon in soil is costly and problematic (Jain et al. 2022). However, in anaerobic digestion of oil hydrocarbon pollutants, microbes use other substrate as an electron acceptor such as sulphate, manganese (II), iron (III), nitrate, or carbon dioxide. Electron acceptors other than oxygen found in oil-polluted soil are often more economical and affordable than oxygen. However, only little success is gained by anaerobic degradation, which can be utilized for the effective degradation of hydrocarbons (Saavedra et al. 2018).

1.4.3 Effect of Salinity and Pressure

Like temperature, high salinity helps in increasing microorganism growth and catalyzes their activity and products (Leahy and Colwell 1990; Das et al. 2011). Several studies are designed to analyze the salinity and pressure effect on the bioremediation rate of crude oil hydrocarbon contaminants (Gopinathan et al. 2017). Pressure and salinity play crucial role in lakes, deep sea, and other water ecosystems that have specific environments and are contaminated by crude oil hydrocarbon compounds (Atlas and Bragg 2009). A notable study has observed the effect of NaCl presence and absence in the soil on biodegradation and found that around 41% degradation of oil pollutant in a period of four months occured in the absence of NaCl, whereas, upon addition of 50 g/L NaCl, the degradation was only 12% in the same time period and same soil site. An experiment was conducted by mixing deep sea sedimented bacteria to degrade n-alkane such as hexadecane, tetradecane, and other various hydrocarbon at high pressure ~500 atm as well as atmospheric pressure (Mishra et al. 2020). It was observed that, when an oil pollutant touches the deep ocean zone, few hydrocarbons had contaminated the deep benthic oceanic environment. Contaminants at benthic zones are not easily degraded because of the recalcitrant amount of crude oil persisting for many years or even decades.

1.4.4 Effect of Substrate and Properties

The mineralization is also affected by the quantity and concentration of crude oil hydrocarbon (Leahy and Colwell 1990). Herein, when the lag phase of microbial metabolism decreases, the crude oil concentration increases and the maximum degrading rate collectively increases mineralization (Varjani et al. 2015). Moreover, the chain length of crude oil hydrocarbons also determines its degradation rate such that the simpler chain degrades rapidly, whereas the complex hydrocarbon chain degrades at a slower rate. Linear alkanes are degraded faster, as compared to branched alkanes, followed by low-molecular-weight alkyl aromatics (Grace Pavithra et al. 2021). Concentrated petroleum hydrocarbons are found to be lethal to the microorganism degrading it, thus limiting the potential degradation activity of the microbes. Moreover, a low amount of hydrocarbon will also inhibit the microbial activity because of a limited supply of the nutrient and carbon source for the growth of microbes. In addition to it, the simple structure and low molecular weight of hydrocarbon supports a good biodegradation rate (Varjani et al. 2015). However, the same hydrocarbons obtained from different crude oils are observed to degrade at a different rate with different microorganisms or even with the same microbes. This variability is due to the abundance and bioavailability of the compound aiding in the degradation process. Physicochemical properties of the contaminated soil and type of hydrocarbon present also govern the bioremediation rate, as these features directly affect the type, population, and metabolic activities of the microorganism of the ecosystem (Ghazali et al. 2004; Grace Pavithra et al. 2021). Therefore, the composition of pollutants at the site influences the degradation rate, like microbes degrading hydrocarbons prefer simpler chains. Under optimum conditions, medium-hydrocarbon-chain simpler alkenes – for instance, kerosene – are completely degradable (Atlas 1984). However, the pollutants consisting of aromatic rings are degradable, but high molecular weight compounds such as asphaltic-naphthenic complexes are negligibly degradable. In this regard, around 11% degradation rate is reported for heavy petroleum hydrocarbons in ambient conditions for microbes' growth in a particular period of time (Ghazali et al. 2004).

1.5 MECHANISM OF DEGRADATION

Complete degradation at a faster rate of most of the organic pollutants can be observed in an oxygenated environment – i.e., favorable in aerobic conditions. A principal reaction involved in the degradation of hydrocarbons under aerobic degradation is the attack of oxygen. Initial hydrocarbon degradation is an oxidative process where oxygenase and peroxidases enzymatic reactions happen by the incorporation of oxygen. The enzyme-catalyzed reaction by oxygen is a key step for the degradation of organic pollutants. A peripheral pathway follows a series of step-by-step reactions where organic contaminants get converted into intermediate products, which act as a substrate for the Kreb's cycle, as shown in Figure 1.1. Succinate, pyruvate, and acetyl-CoA, along with other salts, aid in the synthesis of cell biomass from the intermediate product of the cycle. The gluconeogenesis

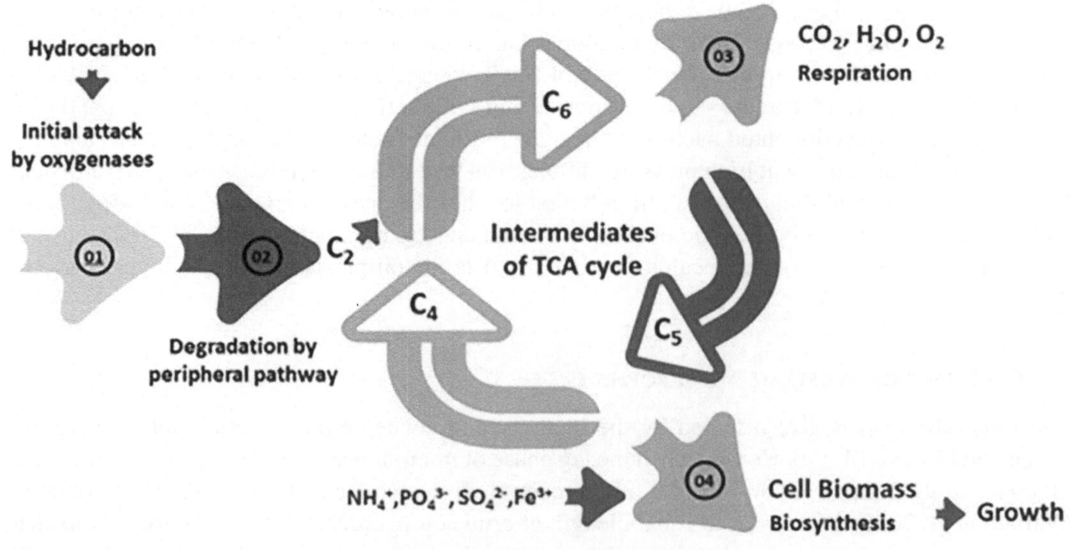

FIGURE 1.1 Pathway of degradation of hydrocarbons.

process and sugar help in the biosynthesis, which assists in the growth of cell biomass. Specific enzymes such as oxygenase and peroxidase assist in the degradation of organic pollutant hydrocarbons. Notably, the first enzymatic degradation was noted by oxygenase and was on xenobiotics (Fritsche and Hofrichter 2008). Various other steps such as microbial cell adherence on the substrate and the biosurfactants availability are further involved in regulation of pollutant degradation.

1.6 ENZYME'S ROLE IN HYDROCARBON DEGRADATION

As aforementioned enzymes' activity in hydrocarbon degradation, heme-thiolate monooxygenase is a well-known, super-ubiquitous enzyme consisting of cytochrome P450 alkane hydroxylases, which play a pivotal role in degradation of long-chain hydrocarbons, oil, chlorinated hydrocarbons, fuel additives, and other organic pollutants (Van Beilen and Funhoff 2007). Enzymes like cytochrome P450, soluble di iron methane monooxygenase, integrated di iron alkane hydroxylases (alkB), and membrane-bound copper containing monooxygenase are involved. Enzymes-mediated oxygenated degradation of the organic hydrocarbons is directly dependent on the chain length of the hydrocarbons. Higher eukaryotes with well-organized systems carry various P450 cytochromes, which assemble to contribute to the metabolic conversion of intermediate substrate. Biodegradation of organic oil pollutants is directly related and dependent on the availability of cytochrome P450, and such kinds of P450 additivity is limited to few species of the microbe's genera (Zimmer et al. 1996). Bundles of cytochrome P450 availability help fungi and other yeast species to utilize and degrade n-alkanes and other aliphatic chains as their energy and carbon source. Many yeast species such as *Candida apicola, C. maltose, and C. tropicalis* are studied for the isolation of cytochrome P450 and were found to have plenty of cytochrome P450 (Scheuer et al. 1998). On the other hand, Van Beilen and Funhoff (2005) have evidently supported the presence of diverse and multiple alkane oxygenase systems' availability in eukaryotes as well as prokaryotes. It was noticed that they were actively involved in crude oil hydrocarbon degradation in aerobic condition.

1.7 MICROALGAE APPROACH

In a particular aquaculture system, a huge amount of freshwater is utilized in producing the rapid growth of food. Approximately 20,000 L of freshwater is consumed in the production of 1 kg of penaeid shrimp (Timmons and Losordo 1994). This results in the generation of a huge amount of wastewater that is rich in nutrients, particularly nitrogen and phosphorous (Ansari et al. 2017). In general, this wastewater contains nutrients COD in the range of 100–150 mg/L, 3–7 mg/L of ammonia, 2–110 mg/L of nitrates, and 2–50 mg/L of phosphates (Viegas et al. 2021). Around 60 kg N_2 and 12 kg P are discharged in wastewater from the rearing of 1 ton of live channel catfish (Hu et al. 2017). The direct release of this nutrient-rich effluent into the environment without any remediation may lead to eutrophication. Many researchers have applied chemical precipitation with salts like ferrous chloride to remove phosphorus. Anaerobic bacteria could be used in sophisticated, expensive systems to obtain the direct conversion of ammonia into nitrogen and biomass. These conventional treatment methodologies are not only expensive but may also release noxious by-products into the environment as secondary pollutants (Nasir et al. 2015). Phycoremediation – bioremediation with the help of microalgae – is an emerging solution, in this regard, wherein the algal population grows and produces biomass. Microalgae are single-celled photosynthetic organisms that possess the ability to utilize energy from sunlight, inorganic nutrients, and carbon dioxide from the environment to generate beneficial biomass (Gupta et al. 2016).

The major interests of cultivation of microalgae in wastewater effluent from the fish industry are as follows:

(1) Microalgae could help in remediation of aquaculture wastewater.
(2) Reutilization of this treated water could be done for the rearing of fishes.

(3) Biomass from microalgal cultivation is a potential high nutritional value aqua feed (Li et al. 2021).

Malibari et al. (2018) studied the cultivation of six different species – *Chlorella sp., Dunaliella sp., Nannochloropsis sp., Navicula sp.* (two strains), and *Tetraselmis sp.* – in nitrogen and phosphate-rich shrimp wastewater (Malibari et al. 2018).

1.8 WATER REUTILIZATION

Wastewater arising from uses such as washing and bathing does not contain concentrated waste and is known as greywater. The pathogens' concentration, trace constituents, organic nutrients, and organic matter are lower than present in wastewater (Metcalf & Eddy Inc. et al. 2007). The evolution of using wastewater in agricultural irrigation arises from ancient times, when untreated sewage wastewater was added to farmlands and crops directly (Leong and Chang 2020). Untreated greywater is used in subsurface irrigation, but it should be treated before any application. Treatment technologies include disinfection, biological treatment, and removal of suspended particles (Ludwig 1997). The parameters considered for use in agricultural irrigation comprises infrastructure facilities, groundwater characteristics, the drainage through sewage and runoff water quality, dissolved solids in water, properties related to topography and soil, properties of crops, and water quality.

The majority of treated water is being used by various industrial users, which include refineries and power plants related to manufacturing units where water is utilized for chilling purposes (Nagarajan et al. 2020). Water quality is highly important for reuse of water in industrial processes, and treatment is being used to improve the quality before use. The various parameters that give the estimation of the quality of water – dissolved oxygen, the total dissolved solids, and chlorides – are important issues because they create potential scaling and corrosion issues in the heat exchanger and piping systems. The other important factors that affect it are system reliability, system supply, and discarding of accumulated waste from cooling towers (Metcalf & Eddy Inc. et al. 2007).

1.9 CHALLENGES AND FUTURE PERSPECTIVES

Further investigation and research are required on various biochemical pathways; crucial and limiting factors for pollutant degradation related to microorganisms are required before utilization of microbes; thus, it can contribute positively to bioremediation processes. The area of aerobic and anaerobic degradation related to polyaromatic hydrocarbons is yet to be explored. Economic awards should be rewarded for clean-up and finding out new ways for disposing certain wastes, if results turn out to be achievable. There are many research areas yet to be explored before focusing only on bioremediation for clean-up methods. Such types of feasible research areas are as follows:

- The solitude identification and ex situ as well as in situ application of various microorganisms that identifies pollutants to be their independent energy and carbon sources.
- Finding out the unknown microorganisms and their cytochrome P450, their mechanism, and pathway of hydrocarbon degradation mechanisms.
- Describing the relationship among various microorganisms in associated systems for degrading hydrocarbon pollutants.
- Explanation of both aerobic and anaerobic biodegradation mechanisms pathways related to pollutants and intermediates found in the degradation process.
- Authenticating the involvement of important microorganisms for various biochemical processes.
- Additionally, finding out factors and process of improving those factors which affects the pollutants' availability and their degradation sources.

1.10 CONCLUSIONS

The priority pollutants are mainly petroleum hydrocarbon products containing polyaromatic hydrocarbons (PAHs), which are carcinogenic (damages DNA, RNA) and hemotoxic. They show low reactivity and are more stable and resistant to degradation. The persistent organic pollutants (POPs) in the environment seriously threaten the entire ecosystem and environment. They directly or indirectly affect hormonal imbalance, disturbance in the pathway of metabolic reactions, suffocation, as well as anoxia.

The ultimate and eco-friendly route to remove petroleum hydrocarbon from the environment can be achieved by microbial degradation, which is a natural mechanism through which the environment gets clean and becomes less polluted. The microorganisms primarily involved in the biodegradation of alkyl aromatic and n-alkanes of oil are *Pseudomonas, Arthrobacter,* mycobacterium, *and Rhodococcus.* There have been 25 genera of bacteria feeding on hydrocarbons and 25 genera of fungi degrading hydrocarbons extracted from the marine area. Fungal strains such as *Amorphoteca, Candida, Talaromyces, Graphium, Pichia,* and *Yarrowia* have been found to be effective degrading agents for the removal of hydrocarbons. Bioremediation is a very efficient, versatile, and highly economical alternative, compared to various available physicochemical methods of hydrocarbon degradation. The microorganisms, both in individual form and consortium form, contribute to biodegradation, thus controlling pollution in the environment. The process of biodegradation (aerobic/anaerobic) involves various catabolic pathways, which signify the efficient strategies for designing bioremediation pathways for various petroleum hydrocarbon pollutants, which can make the environment clean and green. Factors such as temperature, electron acceptors, salinity, pressure, and substrates control and regulate the biodegradation route and require further investigation and research.

BIBLIOGRAPHY

Abdel-Hamid, Ahmed M., Jose O. Solbiati, and Isaac K. O. Cann. 2013. "Insights into Lignin Degradation and Its Potential Industrial Applications." *Advances in Applied Microbiology* 82 (January). Academic Press: 1–28. doi:10.1016/B978-0-12-407679-2.00001-6.

Aislabie, Jackie, David J. Saul, and Julia M. Foght. 2006. "Bioremediation of Hydrocarbon-Contaminated Polar Soils." *Extremophiles* 10 (3). Springer: 171–179. doi:10.1007/S00792-005-0498-4/FIGURES/1.

Al-Jobouri, Ahmed H. A. 2020. "Studying Some the Functional Properties of Tamarind Tamarindus Indica L. Mucilage." *Journal For Agriculture Sciences (QJAS)* 10 (2): 304–307. https://jouagr.qu.edu.iq/.

Alonso-Alvarez, Carlos, Ignacio Munilla, Marta López-Alonso, and Alberto Velando. 2007. "Sublethal Toxicity of the Prestige Oil Spill on Yellow-Legged Gulls." *Environment International* 33 (6). Pergamon: 773–781. doi:10.1016/J.ENVINT.2007.02.012.

Amirabdollahian, Mahsa, and Bithin Datta. 2013. "Identification of Contaminant Source Characteristics and Monitoring Network Design in Groundwater Aquifers: An Overview." *Journal of Environmental Protection* 4 (5). Scientific Research Publishing, Inc.: 26–41. doi:10.4236/JEP.2013.45A004.

Amund, O. O., and N. Nwokoye. 1993. *Hydrocarbon Degradation Potentials of Yeast Isolates from a Polluted Lagoon.* http://localhost:8080/xmlui/handle/123456789/788.

Ansari, Faiz Ahmad, Poonam Singh, Abhishek Guldhe, and Faizal Bux. 2017. "Microalgal Cultivation Using Aquaculture Wastewater: Integrated Biomass Generation and Nutrient Remediation." *Algal Research* 21 (January). Elsevier: 169–177. doi:10.1016/J.ALGAL.2016.11.015.

Atlas, Ronald, and James Bragg. 2009. "Bioremediation of Marine Oil Spills: When and When Not—the Exxon Valdez Experience." *Microbial Biotechnology* 2 (2). Wiley-Blackwell: 213. doi:10.1111/J.1751-7915.2008.00079.X.

Atlas, Ronald M. 1984. *Petroleum Microbiology.* Macmillan, 692.

Barathi, S., and N. Vasudevan. 2001. "Utilization of Petroleum Hydrocarbons by *Pseudomonas Fluorescens* Isolated from a Petroleum-Contaminated Soil." *Environment International* 26 (5–6). Environ Int: 413–416. doi:10.1016/S0160-4120(01)00021-6.

Boguslawska-Wąs, Elżbieta, and Waldemar Dąbrowski. 2001. "The Seasonal Variability of Yeasts and Yeast-like Organisms in Water and Bottom Sediment of the Szczecin Lagoon." *International Journal of Hygiene and Environmental Health* 203 (5–6). Urban & Fischer: 451–458. doi:10.1078/1438-4639-00056.

Boopathy, R. 2000. "Factors Limiting Bioremediation Technologies." *Bioresource Technology* 74 (1). Elsevier: 63–67. doi:10.1016/S0960-8524(99)00144-3.

Brooijmans, Rob J. W., Margreet I. Pastink, and Roland J. Siezen. 2009. "Hydrocarbon-Degrading Bacteria: The Oil-Spill Clean-up Crew." *Microbial Biotechnology* 2 (6). John Wiley & Sons, Ltd: 587–594. doi:10.1111/J.1751-7915.2009.00151.X.

Cahan, Rivka, Matanya Stein, Yaakov Anker, Yakov Langzam, and Yeshayahu Nitzan. 2013. "Innovative Utilization of Coal Bottom Ash for Bioremediation of Toxic Organic Pollutants." *International Biodeterioration & Biodegradation* 85 (November). Elsevier: 421–428. doi:10.1016/J.IBIOD.2013.08.010.

Chaillan, Frédéric, Anne Le Flèche, Edith Bury, Y. Hui Phantavong, Patrick Grimont, Alain Saliot, and Jean Oudot. 2004. "Identification and Biodegradation Potential of Tropical Aerobic Hydrocarbon-Degrading Microorganisms." *Research in Microbiology* 155 (7). Res Microbiol: 587–595. doi:10.1016/J.RESMIC.2004.04.006.

Chandra, Subhash, Richa Sharma, Kriti Singh, and Anima Sharma. 2012. "Application of Bioremediation Technology in the Environment Contaminated with Petroleum Hydrocarbon." *Annals of Microbiology* 63 (2). BioMed Central: 417–431. doi:10.1007/S13213-012-0543-3.

Colwell, Rita R., John D. Walker, and Joseph J. Cooney. 2008. "Ecological Aspects of Microbial Degradation of Petroleum in the Marine Environment." *Critical Reviews in Microbiology* 5 (4). Taylor & Francis: 423–445. doi:10.3109/10408417709102813.

Costa, A. S., L. P. C. Romão, B. R. Araújo, S. C. O. Lucas, S. T. A. Maciel, A. Wisniewski, and M. R. Alexandre. 2012. "Environmental Strategies to Remove Volatile Aromatic Fractions (BTEX) from Petroleum Industry Wastewater Using Biomass." *Bioresource Technology* 105 (February). Elsevier: 31–39. doi:10.1016/J.BIORTECH.2011.11.096.

Das, Kishore, and Ashis K. Mukherjee. 2007. "Crude Petroleum-Oil Biodegradation Efficiency of Bacillus Subtilis and Pseudomonas Aeruginosa Strains Isolated from a Petroleum-Oil Contaminated Soil from North-East India." *Bioresource Technology* 98 (7). Bioresour Technol: 1339–1345. doi:10.1016/J.BIORTECH.2006.05.032.

Das, Nilanjana, and Preethy Chandran. 2011a. "Microbial Degradation of Petroleum Hydrocarbon Contaminants: An Overview." *Biotechnology Research International* 2011. Hindawi Limited. doi:10.4061/2011/941810.

Das, Nilanjana, and Preethy Chandran. 2011b. "Microbial Degradation of Petroleum Hydrocarbon Contaminants: An Overview." *Biotechnology Research International* 2011 (September). Hindawi Limited: 1–13. doi:10.4061/2011/941810.

Daugulis, Andrew J., and Colleen M. McCracken. 2003. "Microbial Degradation of High and Low Molecular Weight Polyaromatic Hydrocarbons in a Two-Phase Partitioning Bioreactor by Two Strains of Sphingomonas Sp." *Biotechnology Letters* 25 (17). Biotechnol Lett: 1441–1444. doi:10.1023/A:1025007729355.

Deppe, Uta, Hans Hermann Richnow, Walter Michaelis, and Garabed Antranikian. 2005. "Degradation of Crude Oil by an Arctic Microbial Consortium." *Extremophiles* 9 (6). Springer: 461–470. doi:10.1007/S00792-005-0463-2/FIGURES/5.

Desforges, Jean Pierre W., Christian Sonne, Milton Levin, Ursula Siebert, Sylvain De Guise, and Rune Dietz. 2016. "Immunotoxic Effects of Environmental Pollutants in Marine Mammals." *Environment International* 86 (January). Pergamon: 126–139. doi:10.1016/J.ENVINT.2015.10.007.

Falade, Ayodeji O., Uchechukwu U. Nwodo, Benson C. Iweriebor, Ezekiel Green, Leonard V. Mabinya, and Anthony I. Okoh. 2017. "Lignin Peroxidase Functionalities and Prospective Applications." *Microbiology Open* 6 (1). John Wiley & Sons, Ltd: e00394. doi:10.1002/MBO3.394.

Fritsche, Wolfgang, and Martin Hofrichter. 2008. "Aerobic Degradation by Microorganisms." *Biotechnology: Second, Completely Revised Edition* 11–12 (May). John Wiley & Sons, Ltd: 144–167. doi:10.1002/9783527620999.CH6M.

Fuchs, Georg, Matthias Boll, and Johann Heider. 2011. "Microbial Degradation of Aromatic Compounds—from One Strategy to Four." *Nature Reviews Microbiology* 9 (11). Nature Publishing Group: 803–816. doi:10.1038/nrmicro2652.

Ghazali, Farinazleen Mohamad, Raja Noor Zaliha Abdul Rahman, Abu Bakar Salleh, and Mahiran Basri. 2004. "Biodegradation of Hydrocarbons in Soil by Microbial Consortium." *International Biodeterioration & Biodegradation* 54 (1). Elsevier: 61–67. doi:10.1016/J.IBIOD.2004.02.002.

Gopinathan, R., Avijit Bhowal, and Chandrasekhar Garlapati. 2017. "Adsorption Characteristics of Activated Carbon for the Reclamation of Colored Effluents Containing Orange G and New Solid-Liquid Phase Equilibrium Model." *Journal of Chemical and Engineering Data* 62 (1). American Chemical Society: 558–567. doi:10.1021/ACS.JCED.6B00845/ASSET/IMAGES/LARGE/JE-2016-008452_0012.JPEG.

Grace Pavithra, Kirubanandam, P. Senthil Kumar, and Sunita Varjani. 2021. "Bioelectrochemical Systems for Remediation and Recovery of Nutrients From Industrial Wastewater." *Biomass, Biofuels, Biochemicals*, January. Elsevier, 445–474. doi:10.1016/B978-0-12-821878-5.00018-0.

Gupta, Sanjay Kumar, Faiz Ahmad Ansari, Amritanshu Shriwastav, Narendra Kumar Sahoo, Ismail Rawat, and Faizal Bux. 2016. "Dual Role of Chlorella Sorokiniana and Scenedesmus Obliquus for Comprehensive Wastewater Treatment and Biomass Production for Bio-Fuels." *Journal of Cleaner Production* 115 (March). Elsevier: 255–264. doi:10.1016/J.JCLEPRO.2015.12.040.

Hollaway, Stephen L., Gary M. Faw, and Ronald K. Sizemore. 1980. "The Bacterial Community Composition of an Active Oil Field in the Northwestern Gulf of Mexico." *Marine Pollution Bulletin* 11 (6). Pergamon: 153–156. doi:10.1016/0025-326X(80)90141-1.

Hu, Jinzhao, Rui Hu, Dan Qi, and Xujie Lu. 2017. "Study on Treatment of Aquaculture Wastewater Using a Hybrid Constructed Wetland." *IOP Conference Series: Earth and Environmental Science* 61 (1). IOP Publishing: 012015. doi:10.1088/1755-1315/61/1/012015.

Jain, Marut, Sadaf Aiman Khan, Ashish Pandey, Kamal Kishore Pant, Zyta Maria Ziora, and Mark A. T. Blaskovich. 2021. "Instructive Analysis of Engineered Carbon Materials for Potential Application in Water and Wastewater Treatment." *Science of The Total Environment* 793 (November). Elsevier: 148583. doi:10.1016/j.scitotenv.2021.148583.

Jain, Marut, Sadaf Aiman Khan, Abhisek Sahoo, Prashant Dubey, Kamal Kishore Pant, Zyta Maria Ziora, and Mark A. T. Blaskovich. 2022. "Statistical Evaluation of Cow-Dung Derived Activated Biochar for Phenol Adsorption: Adsorption Isotherms, Kinetics, and Thermodynamic Studies." *Bioresource Technology* 352 (May). Elsevier: 127030. doi:10.1016/J.BIORTECH.2022.127030.

Jain, Marut, Sadaf Aiman Khan, Komal Sharma, Prashant Ram Jadhao, Kamal Kishore Pant, Zyta Maria Ziora, and Mark A. T. Blaskovich. 2022. "Current Perspective of Innovative Strategies for Bioremediation of Organic Pollutants from Wastewater." *Bioresource Technology* 344 (January). Elsevier: 126305. doi:10.1016/J.BIORTECH.2021.126305.

Johnsen, Anders R., Lukas Y. Wick, and Hauke Harms. 2005. "Principles of Microbial PAH-Degradation in Soil." *Environmental Pollution* 133 (1). Elsevier: 71–84. doi:10.1016/J.ENVPOL.2004.04.015.

Jones, D. M., A. G. Douglas, R. J. Parkes, J. Taylor, W. Giger, and C. Schaffner. 1983. "The Recognition of Biodegraded Petroleum-Derived Aromatic Hydrocarbons in Recent Marine Sediments." *Marine Pollution Bulletin* 14 (3). Pergamon: 103–108. doi:10.1016/0025-326X(83)90310-7.

Jones, J. G., M. Knight, and J. A. Byrom. 1970. "Effect of Gross Pollution by Kerosine Hydrocarbons on the Microflora of a Moorland Soil." *Nature* 227 (5263). Nature: 1166. doi:10.1038/2271166A0.

Kauppi, Sari, Aki Sinkkonen, and Martin Romantschuk. 2011. "Enhancing Bioremediation of Diesel-Fuel-Contaminated Soil in a Boreal Climate: Comparison of Biostimulation and Bioaugmentation." *International Biodeterioration & Biodegradation* 65 (2). Elsevier: 359–368. doi:10.1016/J.IBIOD.2010.10.011.

Lal, B., and S. Khanna. 1996. "Degradation of Crude Oil by Acinetobacter Calcoaceticus and Alcaligenes Odorans." *Journal of Applied Bacteriology* 81 (4). John Wiley & Sons, Ltd: 355–362. doi:10.1111/J.1365-2672.1996.TB03519.X.

Leahy, Joseph G., and Rita R. Colwell. 1990. "Microbial Degradation of Hydrocarbons in the Environment." *Microbiological Reviews* 54 (3). American Society for Microbiology (ASM): 305. doi:10.1128/MR.54.3.305-315.1990.

Leong, Yoong Kit, and Jo Shu Chang. 2020. "Bioremediation of Heavy Metals Using Microalgae: Recent Advances and Mechanisms." *Bioresource Technology* 303 (May). Elsevier: 122886. doi:10.1016/J.BIORTECH.2020.122886.

Li, Huankai, Sihua Chen, Kang Liao, Qian Lu, and Wenguang Zhou. 2021. "Microalgae Biotechnology as a Promising Pathway to Ecofriendly Aquaculture: A State-of-the-Art Review." *Journal of Chemical Technology & Biotechnology* 96 (4). John Wiley & Sons, Ltd: 837–852. doi:10.1002/JCTB.6624.

Liu, Yinsong, Jingchun Wu, Yikun Liu, and Xiaolin Wu. 2022. "Biological Process of Alkane Degradation by Gordonia Sihwaniensis." *ACS Omega* 7 (1). American Chemical Society: 55–63. doi:10.1021/ACSOMEGA.1C01708/ASSET/IMAGES/LARGE/AO1C01708_0010.JPEG.

Ludwig, A. 1997. *Create an Oasis with Greywater. Distributed by Oasis Design*. Santa Barbara. https://oasis-design.net/greywater/createanoasis/.

Malibari, Raghdah, Fotoon Sayegh, Ahmed M. Elazzazy, Mohamed N. Baeshen, Marianna Dourou, and George Aggelis. 2018. "Reuse of Shrimp Farm Wastewater as Growth Medium for Marine Microalgae Isolated from Red Sea—Jeddah." *Journal of Cleaner Production* 198 (October). Elsevier: 160–169. doi:10.1016/J.JCLEPRO.2018.07.037.

Mani, Priyadharshini, V. T. Fidal, Kyle Bowman, Mark Breheny, T. S. Chandra, Taj Keshavarz, and Godfrey Kyazze. 2019. "Degradation of Azo Dye (Acid Orange 7) in a Microbial Fuel Cell: Comparison Between Anodic Microbial-Mediated Reduction and Cathodic Laccase-Mediated Oxidation." *Frontiers in Energy Research* 0 (September). Frontiers: 101. doi:10.3389/FENRG.2019.00101.

McDonald, Ian R., Carlos B. Miguez, Gerlinde Rogge, Denis Bourque, Karin D. Wendlandt, Denis Groleau, and Jcolin Murrell. 2006. "Diversity of Soluble Methane Monooxygenase-Containing Methanotrophs Isolated from Polluted Environments." *FEMS Microbiology Letters* 255 (2). Oxford Academic: 225–232. doi:10.1111/J.1574-6968.2005.00090.X.

Meckenstock, Rainer U., Matthias Boll, Housna Mouttaki, Janina S. Koelschbach, Paola Cunha Tarouco, Philip Weyrauch, Xiyang Dong, and Anne M. Himmelberg. 2016. "Anaerobic Degradation of Benzene and Polycyclic Aromatic Hydrocarbons." *Microbial Physiology* 26 (1–3). Karger Publishers: 92–118. doi:10.1159/000441358.

Medina-Bellver, Javier I., Patricia Marín, Antonio Delgado, Alicia Rodríguez-Sánchez, Emilio Reyes, Juan L. Ramos, and Silvia Marqués. 2005. "Evidence for in Situ Crude Oil Biodegradation after the Prestige Oil Spill." *Environmental Microbiology* 7 (6). Environ Microbiol: 773–779. doi:10.1111/J.1462-2920.2005.00742.X.

Metcalf & Eddy, Inc. an AECOM Company, Takashi Asano, Franklin Burton, and Harold Leverenz. 2007. *Water Reuse: Issues, Technologies, and Applications*. McGrawHill, New York. McGraw-Hill Education, 1570. www.accessengineeringlibrary.com/content/book/9780071459273.

Mishra, Bishwambhar, Sunita Varjani, Dinesh Chand Agrawal, Sanjeeb Kumar Mandal, Huu Hao Ngo, Mohammad J. Taherzadeh, Jo Shu Chang, Siming You, and Wenshan Guo. 2020. "Engineering Biocatalytic Material for the Remediation of Pollutants: A Comprehensive Review." *Environmental Technology & Innovation* 20 (November). Elsevier: 101063. doi:10.1016/J.ETI.2020.101063.

Mulkins Phillips, G. J., and J. E. Stewart. 2011. "Distribution of Hydrocarbon-Utilizing Bacteria in Northwestern Atlantic Waters and Coastal Sediments." *Canadian Journal of Microbiology* 20 (7). NRC Research Press Ottawa, Canada: 955–962. doi:10.1139/M74-147.

Nagarajan, Dillirani, Duu Jong Lee, Chun Yen Chen, and Jo Shu Chang. 2020. "Resource Recovery from Wastewaters Using Microalgae-Based Approaches: A Circular Bioeconomy Perspective." *Bioresource Technology* 302 (April). Elsevier: 122817. doi:10.1016/J.BIORTECH.2020.122817.

Nasir, Nurfarahana Mohd, Nur Syuhada Abu Bakar, Fathurrahman Lananan, Siti Hajar Abdul Hamid, Su Shiung Lam, and Ahmad Jusoh. 2015. "Treatment of African Catfish, Clarias Gariepinus Wastewater Utilizing Phytoremediation of Microalgae, Chlorella Sp. with Aspergillus Niger Bio-Harvesting." *Bioresource Technology* 190 (August). Elsevier: 492–498. doi:10.1016/J.BIORTECH.2015.03.023.

Pérez-Cadahía, Beatriz, Anunciación Lafuente, Teresa Cabaleiro, Eduardo Pásaro, Josefina Méndez, and Blanca Laffon. 2007. "Initial Study on the Effects of Prestige Oil on Human Health." *Environment International* 33 (2). Pergamon: 176–185. doi:10.1016/J.ENVINT.2006.09.006.

Porta, Miguel, and E Zumeta. 2002. "Implementing the Stockholm Treaty on Persistent Organic Pollutants." *Occupational and Environmental Medicine*. doi:10.1136/oem.59.10.651.

Pozdnyakova-Filatova, Irina, Kirill Petrikov, Anna Vetrova, Alina Frolova, Rostislav Streletskii, and Marina Zakharova. 2020. "The Naphthalene Catabolic Genes of Pseudomonas Putida BS3701: Additional Regulatory Control." *Frontiers in Microbiology* 11 (June). Frontiers Media S.A.: 1217. doi:10.3389/FMICB.2020.01217/BIBTEX.

Rahman, K. S. M., Thahira J. Rahman, Y. Kourkoutas, I. Petsas, R. Marchant, and I. M. Banat. 2003. "Enhanced Bioremediation of N-Alkane in Petroleum Sludge Using Bacterial Consortium Amended with Rhamnolipid and Micronutrients." *Bioresource Technology* 90 (2). Elsevier Ltd: 159–168. doi:10.1016/S0960-8524(03)00114-7.

Ron, Eliora Z., and Eugene Rosenberg. 2014. "Enhanced Bioremediation of Oil Spills in the Sea." *Current Opinion in Biotechnology* 27 (June). Elsevier Current Trends: 191–194. doi:10.1016/J.COPBIO.2014.02.004.

Saavedra, Ricardo, Raúl Muñoz, María Elisa Taboada, Marisol Vega, and Silvia Bolado. 2018. "Comparative Uptake Study of Arsenic, Boron, Copper, Manganese and Zinc from Water by Different Green Microalgae." *Bioresource Technology* 263 (September). Elsevier: 49–57. doi:10.1016/J.BIORTECH.2018.04.101.

Sadañoski, Marcela Alejandra, Ana Silvia Tatarin, Mónica Lucrecia Barchuk, Mariana Gonzalez, César Nicolás Pegoraro, María Isabel Fonseca, Laura Noemí Levin, and Laura Lidia Villalba. 2020. "Evaluation of Bioremediation Strategies for Treating Recalcitrant Halo-Organic Pollutants in Soil Environments." *Ecotoxicology and Environmental Safety* 202 (October). Academic Press. doi:10.1016/J.ECOENV.2020.110929.

Sajna, Kuttuvan Valappil, Rajeev Kumar Sukumaran, Lalitha Devi Gottumukkala, and Ashok Pandey. 2015. "Crude Oil Biodegradation Aided by Biosurfactants from Pseudozyma Sp. NII 08165 or Its Culture Broth." *Bioresource Technology* 191 (September). Elsevier: 133–139. doi:10.1016/J.BIORTECH.2015.04.126.

Scheuer, Ulrich, Thomas Zimmer, Dörte Becher, Frieder Schauer, and Wolf Hagen Schunck. 1998. "Oxygenation Cascade in Conversion of N-Alkanes to α,ω-Dioic Acids Catalyzed by Cytochrome P450 52A3." *Journal of Biological Chemistry* 273 (49). Elsevier: 32528–32534. doi:10.1074/JBC.273.49.32528.

Souza, Ellen Cristina, Thereza Christina Vessoni-Penna, and Ricardo Pinheiro De Souza Oliveira. 2014. "Biosurfactant-Enhanced Hydrocarbon Bioremediation: An Overview." *International Biodeterioration & Biodegradation* 89 (April). Elsevier: 88–94. doi:10.1016/J.IBIOD.2014.01.007.

Thamer, Marwah, Abdul Rahman Al-Kubaisi, Zahraa Zahraw, Hussam Aldin Abdullah, Iman Hindy, Ammar Abd Khadium, Marwah Thamer, et al. 2013. "Biodegradation of Kirkuk Light Crude Oil by Bacillus Thuringiensis, Northern of Iraq." *Natural Science* 5 (7). Scientific Research Publishing: 865–873. doi:10.4236/NS.2013.57104.

Throne-Holst, Mimmi, Alexander Wentzel, Trond E. Ellingsen, Hans Kristian Kotlar, and Sergey B. Zotchev. 2007. "Identification of Novel Genes Involved in Long-Chain n-Alkane Degradation by Acinetobacter Sp. Strain DSM 17874." *Applied and Environmental Microbiology* 73 (10). American Society for Microbiology (ASM): 3327. doi:10.1128/AEM.00064-07.

Timmons, Michael B. (Michael Ben), and Thomas Losordo. 1994. *Aquaculture Water Reuse Systems: Engineering Design and Management*. Elsevier, 333.

Van Beilen, Jan B., and Enrico G. Funhoff. 2005. "Expanding the Alkane Oxygenase Toolbox: New Enzymes and Applications." *Current Opinion in Biotechnology* 16 (3). Elsevier Current Trends: 308–314. doi:10.1016/J.COPBIO.2005.04.005.

Van Beilen, Jan B., and Enrico G. Funhoff. 2007. "Alkane Hydroxylases Involved in Microbial Alkane Degradation." *Applied Microbiology and Biotechnology* 74 (1). Appl Microbiol Biotechnol: 13–21. doi:10.1007/S00253-006-0748-0.

Van Meter, Robin J., James R. Spotila, and Harold W. Avery. 2006. "Polycyclic Aromatic Hydrocarbons Affect Survival and Development of Common Snapping Turtle (Chelydra Serpentina) Embryos and Hatchlings." *Environmental Pollution* 142 (3). Elsevier: 466–475. doi:10.1016/J.ENVPOL.2005.10.018.

Varjani, Sunita J. 2017. "Microbial Degradation of Petroleum Hydrocarbons." *Bioresource Technology* 223 (January). Elsevier: 277–286. doi:10.1016/J.BIORTECH.2016.10.037.

Varjani, Sunita J., Parita Rakholiya, How Yong Ng, Siming You, and Jose A. Teixeira. 2020. "Microbial Degradation of Dyes: An Overview." *Bioresource Technology* 314 (October). Elsevier: 123728. doi:10.1016/J.BIORTECH.2020.123728.

Varjani, Sunita J., Dolly P. Rana, Ajay K. Jain, Surendra Bateja, and Vivek N. Upasani. 2015. "Synergistic Ex-Situ Biodegradation of Crude Oil by Halotolerant Bacterial Consortium of Indigenous Strains Isolated from on Shore Sites of Gujarat, India." *International Biodeterioration & Biodegradation* 103 (September). Elsevier: 116–124. doi:10.1016/J.IBIOD.2015.03.030.

Varjani, Sunita J., and Vivek N. Upasani. 2017. "A New Look on Factors Affecting Microbial Degradation of Petroleum Hydrocarbon Pollutants." *International Biodeterioration & Biodegradation* 120 (May). Elsevier: 71–83. doi:10.1016/J.IBIOD.2017.02.006.

Viegas, Catarina, Luísa Gouveia, and Margarida Gonçalves. 2021. "Aquaculture Wastewater Treatment through Microalgal. Biomass Potential Applications on Animal Feed, Agriculture, and Energy." *Journal of Environmental Management* 286 (May). J Environ Manage. doi:10.1016/J.JENVMAN.2021.112187.

Walker, C. H., S. P. Hopkin, R. M. Sibly, and D. B. Peakall. 2005. "Principles of Ecotoxicology, Third Edition." In *Principles of Ecotoxicology*, Third Edition, January. CRC Press, 1–305. doi:10.1201/9781439894224/PRINCIPLES-ECOTOXICOLOGY-WALKER-SIBLY-SIBLY-PEAKALL.

Yakimov, Michail M., Kenneth N. Timmis, and Peter N. Golyshin. 2007. "Obligate Oil-Degrading Marine Bacteria." *Current Opinion in Biotechnology* 18 (3). Curr Opin Biotechnol: 257–266. doi:10.1016/J.COPBIO.2007.04.006.

Zanaroli, Giulio, Sara Di Toro, Daniela Todaro, Giovanna C. Varese, Antonio Bertolotto, and Fabio Fava. 2010. "Characterization of Two Diesel Fuel Degrading Microbial Consortia Enriched from a Non Acclimated, Complex Source of Microorganisms." *Microbial Cell Factories* 9 (1). BioMed Central: 1–13. doi:10.1186/1475-2859-9-10/FIGURES/5.

Zhang, Lulu, Jingling Liu, Huayong Liu, Guisheng Wan, and Shaowei Zhang. 2015. "The Occurrence and Ecological Risk Assessment of Phthalate Esters (PAEs) in Urban Aquatic Environments of China." *Ecotoxicology* 24 (5). Springer: 967–984. doi:10.1007/S10646-015-1446-4.

Zhang, Zhengzhi, Zhaowei Hou, Chunyu Yang, Cuiqing Ma, Fei Tao, and Ping Xu. 2011. "Degradation of N-Alkanes and Polycyclic Aromatic Hydrocarbons in Petroleum by a Newly Isolated Pseudomonas Aeruginosa DQ8." *Bioresource Technology* 102 (5). Elsevier: 4111–4116. doi:10.1016/J.BIORTECH.2010.12.064.

Zimmer, Thomas, Moriya Ohkuma, Akinori Ohta, Masamichi Takagi, and Wolf Hagen Schunck. 1996. "The CYP52 Multigene Family of Candida Maltosa Encodes Functionally Diverse N-Alkane-Inducible Cytochromes P450." *Biochemical and Biophysical Research Communications* 224 (3). Biochem Biophys Res Commun: 784–789. doi:10.1006/BBRC.1996.1100.

2 Biodegradation of Anti-Inflammatory, Non-Steroidal, Pharmaceutical Active Compounds

Sumona Show and Gopinath Halder

2.1 INTRODUCTION

Water is vital to all living creatures and should never be disregarded. Unfortunately, hazardous chemicals and untreated wastewater discharged by fast-developing industries and municipalities have contaminated a substantial amount of potable water, impacting the quality of the water. As per UNESCO, approximately 80% of untreated wastewater discharged into water bodies around the world is not adequately treated. Due to these reasons, it is clear that this wastewater discharge is a major concern because both humans as well as microbes are directly and indirectly affected. Furthermore, according to the World Health Organization (WHO), polluted water is responsible for 40% of all fatalities and 30% of all diseases. In these waters, bacteria are more likely to evolve drug resistance. In actuality, humans will be directly exposed through the consumption of already-irrigated plants or contaminated fish when pharmaceutical residues from untreated wastewater containing numerous dangerous pharmaceuticals are dumped into the environment (Figure 2.1) (Ramezani et al. 2021).

Pharmaceuticals play a crucial role in the basic care and living standards of people, and their deterioration is turning into an environmental issue in the form of contaminants. The key concern for world health today is how to manage pharmaceutical pollutants, such as discarded and expired medicines, personal home care items, and over-the-counter pharmaceuticals. They can be divided into several groups according to their pharmacological characteristics, depending on their mode of action, chemical composition, administration route, therapeutic impact, and biological systems they affect (Woolridge and Hoboy 2019). Typical pharmaceutical drugs include antipyretics, analgesics, anti-malarial, antibiotics, and others. Non-steroidal, anti-inflammatory drugs (NSAIDs), which are classified as analgesics, are known to persist in the aqueous phase for prolonged periods of time due to their relative stability as well as hydrophilicity. NSAIDs are a group of drugs that are used to treat inflammation, fever, as well as other inflammatory disorders. Inhibiting the enzyme cyclooxygenase is the primary mechanism of NSAIDs. Arachidonic acid must be transformed into thromboxanes, prostaglandins, and prostacyclins via cyclooxygenase (Qureshi and Dua 2022). The rate-limiting process of converting arachidonic acid into prostaglandins, which are inflammatory mediators, requires the cyclooxygenase (COX) enzyme. COX consists of two well-known isoforms: COX-1 and COX-2. In the body, COX-1 is constitutively produced and helps to keep the lining of the gastrointestinal tract and the kidneys functioning properly and helps platelet aggregation. In the case of COX-2, it is not constitutively expressed in the body; rather, it is induced to express during an inflammatory response (Chaiamnuay et al. 2006). However, NSAIDs are known to have harmful effects on the hepatic, cardiovascular, renal, and hematologic systems as well as the gastric mucosa. There are several ways that pharmaceutical pollutants enter the environment. Since NSAIDs are believed to be persistent, they can be detected in drinking water as well as other aquatic environments, such as rivers and lakes, in concentrations ranging from ng/L to μg/L. They are regarded as

 DOI: 10.1201/9781003391487-2

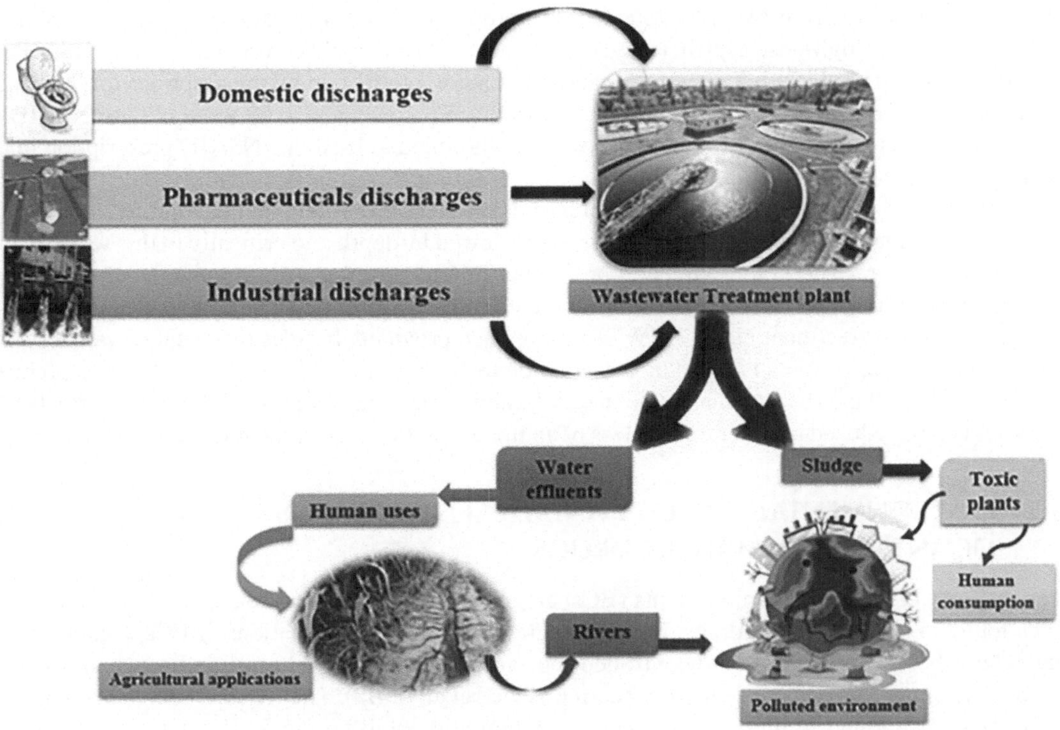

FIGURE 2.1 Sources of pharmaceutical wastewater.

an emerging pollutant of concern owing to their widespread prevalence in the marine environment as well as long-term ecotoxic impacts on the biological components of the environment Rastogi et al. (2021).

As a consequence, developing an efficient and environmentally acceptable approach for eliminating contaminants from the aquatic environment is important. For the removal of pharmaceutical contaminants from wastewater, a number of treatment techniques, including bio-filtration, oxidation, extraction, and so on, are typically used. These treatment methods have a variety of negative aspects that are characterised by their complexity, the development of toxic sludge, expensive operating and capital expenses, and these drawbacks are brought on by the ineffective exclusion of contaminants from aquatic systems.

Presently, bioremediation techniques are important for treating wastewater and are considered eco-friendly. Despite the fact that various research studies have shown that bioremediation has relatively less potential for long-term pollutant clearance, it is nevertheless a feasible way to decrease harmful chemicals. This approach effectively eliminates toxic substances by allowing harmful substances to biodegrade into comparatively harmless substances like carbon dioxide and water Show et al. (2021). As a by-product of the bioremediation process, CO_2 evolution has been utilised as an indicator of bacterial respiration rate (Kao et al. 2001).

2.1.1 Frequently Used Anti-Inflammatory Drugs

The FDA (Food and Drug Administration) has granted approval for the use of NSAIDs as analgesic, anti-inflammatory, and antipyretic agents. These actions make NSAIDs beneficial for treating disorders like gout, migraines, pyrexia, dysmenorrhea, arthritis, and muscle pain. They can also be used as opioid-sparing drugs in some cases of severe trauma. NSAIDs can be broadly classified as follows: Ibuprofen, Naproxen, Diclofenac, Ketoprofen, Celecoxib, Mefenamic acid, Etoricoxib, Indomethacin, and high-dose aspirin (low-dose aspirin is not normally considered to be an NSAID). NSAIDs are often used in clinical practises and account for 5 to 10% of all prescriptions for medications each year. In general practise settings, up to 96% of patients above 65 years old use NSAIDs regularly. The percentage of senior patients over 60 who filled at least one NSAID prescription in a year was about 7.3% Bindu et al. (2020).

The major representative NSAIDs (Diclofenac, Naproxen, Ibuprofen, and Ketoprofen) and their microbial degradation potential and pathways are reviewed in-depth and critically in this work in an effort to fill the gap. Their environmental prevalence and toxicity are also discussed. Additionally, special considerations on the variables influencing circumstances of occurrence as well as remediation in wastewater treatment plants (WWTPs) have been provided, besides increasing awareness of the drawbacks of applying advanced bioremediation techniques in field-scale WWTPs. The overall objective of the chapter is to enhance the field of knowledge regarding how the harmful effects of NSAIDs can be reduced in order to have less of an impact on the environment Banerjee et al. (2019).

2.2 ENVIRONMENTAL TOXICITY AND HEALTH HAZARDS OF ANTI-INFLAMMATORY DRUGS

The majority of emerging contaminants (ECs) are synthetic, organic substances that have recently been found in natural surroundings. Environmentally relevant concentrations of ECs, a significant and relatively recent category of uncontrolled substances that may have harmful impacts on aquatic as well as human life, are a reason for concern Rasheed et al. (2019). They are the chemicals that are most frequently found in municipal sewage, common household products, pharmaceutical manufacturing facilities, wastewater, hospitals, landfills, and the environment around natural aquatic life. The concentration of ECs might be anything between a few ng/L and a few hundred μg/L. As these ECs are not completely removed by conventional wastewater treatment processes, such concentrations in the marine environment may pose an environmental risk such as interference with the endocrine system of high organisms, microbiological resistance, and accumulation in soil, plants, and animals Sibeko et al. (2019).

NSAIDs, endocrine disrupting compounds (EDCs), surfactants, pesticides, flame retardants, and industrial additives are the majority of the substances classified as ECs. Both groundwater and surface waters frequently contain NSAIDs. The quality of the surface water and aquatic life are significantly affected by such a pervasive presence of NSAIDs, which ultimately causes a decrease in the quantity of aquatic creatures. Although NSAID dosages purposefully used by many people for medical purposes and excessive NSAID doses can have an impact on people, concerns about long-term health issues due to the chronic exposure issues cannot be ignored.

2.3 EXCLUSION OF PHARMACEUTICAL COMPOUNDS FROM WASTEWATERS

The removal of NSAIDs from water can be accomplished using a variety of treatment methods. However, the problem of NSAID removal from water has been sparked and clarified by the use of a combination of physical, chemical, and biological processes in the present treatment techniques (Figure 2.2). Each treatment method has its own benefits and drawbacks, both in terms of capital as well as operating expenditures and also in terms of efficacy, the generation of sludge and harmful by-products, pre-treatment requirements, environmental effect, operability, dependability, efficiency, and other factors.

NSAIDs and other micro-pollutants are now thought to be best removed by highly selective and rapid reactions, like advanced oxidation processes (AOPs), or adsorbents, like activated carbon. AOPs are characterised by the generation of tremendously reactive species, like hydroxyl radicals, that can completely mineralise refractory molecules into carbon dioxide, water, and inorganic ions

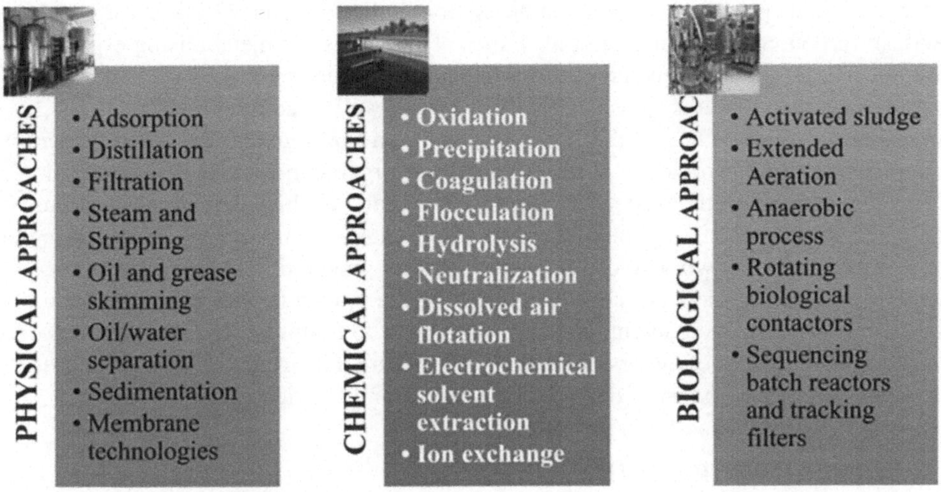

FIGURE 2.2 Different treatment methods.

or break them down into possible biodegradable, intermediate compounds. AOPs use combinations of reactive oxidants, such as ultrasound oxidation, ozonation, and photocatalysis. Fenton oxidation, ion exchange, electrochemical oxidation, soil aquifer treatment, and membrane filtration are some further trendy techniques. To increase removal efficacy, membrane technologies are frequently used in a combination of chemical, physical, as well as biological methods Chaiamnuay et al. (2006). Tetrabutyl ammonium bromide (TBAB) was used as the carrier in an emulsion liquid membrane that Seifollahi and Rahbar-Kelishami (2017) used to extract diclofenac from an aqueous medium. Fenton oxidation describes the procedures that result in the formation of extremely reactive oxygen species. Ibuprofen was found to be degraded very effectively by ZVI (Zero-Valent Iron)-Fenton. After secondary wastewater treatment, the treatment of sewage effluents in soil aquifers is effective for a number of medicines. Diclofenac, Ketoprofen, and Naproxen have all been known to be successfully degraded through soil aquifer treatment.

These techniques could, however, result in the development of several by-products that could be hazardous, and these oxidation technologies are characterised by a big environmental impact due to high energy demand. For instance, AOPs result in less sludge, faster reaction rates, and shorter retention times, but they are expensive to install, run, and maintain. Fenton's reaction has a wide range of applications, is straightforward to carry out, does not produce any bromated by-products, and can degrade both soluble and insoluble colours in industrial effluents. Its requirement for low pH (2.5–4) raises operating costs, which is one of its disadvantages. However, it produces a significant amount of ferrous iron sludge and highly concentrated anions in the treated effluent. Photochemical degradation, another kind of chemical treatment, has little chance of producing by-products that are bromated. However, when adding the catalyst as slurry, there needs to be a separation step.

These factors lead to the conclusion that microbial degradation could be used to get rid of these micro-pollutants. Using various microorganisms or microbial consortia in a laboratory setting, some pharmaceuticals have been successfully degraded. Depending on the medicine and the microorganisms utilised, they show varying rates of removal efficiency Barkin (2015).

2.3.1 Biodegradation Mechanism

Biological oxidation and hydrolysis are the two recognised, primary biodegradation processes. These methods are further subdivided according to the sources employed, such as bacteria or enzymes. Microorganisms use co-metabolic mechanisms and growth to convert xenobiotics.

Organic-contaminant growth processes completely mineralise organic pollutants using only organic pollutants as carbon energy Banerjee et al. (2019). The process of metabolising organic contaminants is known as co-metabolism and occurs when a growth substrate like carbon is present. In the breakdown of active pollutants, yeast, fungi, bacteria, as well as some protozoa are involved. The primary benefit of this method is the improvement of carbon dioxide and water mineralisation, which can happen either aerobically or anaerobically, of organic molecules Samanta et al. (2020).

Three stages of biodegradation occur: 1) Natural attenuation, which involves the decrease of contaminants by naturally occurring microbes. 2) Adding nutrients and oxygen to the biodegradation process is a technique known as biostimulation. 3) Bioaugmentation, which targets specific contaminants by introducing more effective microbes into the system in addition to the natural microflora. In order to use organic contaminants as substrate or to co-metabolise them, microorganisms are influenced by a number of factors, including temperature, nitrogen, pH and phosphorus content, which affects the rate and extent of degradation Fritsche and Hofrichter (2008).

2.4 MICROBIAL DEGRADATION

Due to their advantages over other approaches to treatment, the microbial degradation of NSAIDs has attracted a great deal of attention from the research community worldwide. There are numerous studies examining the degradability of various pure bacterial and fungal cultures. Even though the removal durations and rates for each microbe varied widely, the end result was reduced toxicity in the ultimate products. Research on bacterial breakdown of NSAIDs has been conducted more frequently than studies on fungal degradation. But, Diclofenac, Ibuprofen, and Naproxen have all been successfully degraded and mineralised by a number of fungi, including *Phanerochaetesordida* YK-624 and *Trametesversicolor* (white-rot fungus). *Trametes versicolor* has also been reported to decompose Ketoprofen.

Even though the concentration of NSAIDs that is environmentally relevant is typically between ng/L and μg/L, numerous research studies use starting values much above that range. These strains are particularly useful for bioremediating hospital wastewater effluents or pharmaceutical production sites where the concentrations will be in accordance with these researches Vandraas et al. (2010).

2.4.1 IBUPROFEN

Ibuprofen is a NSAID with the molecular formula$C_{13}H_{18}O_2$. IBP is made up of two functional groups, benzene and carboxylic acid, and it has low water solubility, high consumption, and mobility in water. IBP is also known as 2-(4-isobutyl phenyl) propanoic acid because its name is derived from three functional groups: ibu from isobutyl, pro from propionic acid, and fen from phenyl.

It is frequently used to relieve pain, fever, and inflammation. IBP is very slightly water soluble, but when organic solvents such methanol, acetone, ethanol, and dichloromethane are added, it becomes more soluble. IBP seems to be on the World Health Organization's (WHO) "Essential Drugs List," which lists the medications that are required for basic health care. Daily use can cause serious side effects such as liver failure, kidney problems, mild heart failure, and other complications for human beings. IBP poses a serious threat to fish populations in the aquatic environment, according to a report on the effect of the drug on fish, which indicated that negative impacts on the reproductive axis appear at low doses of up to 0.1 μg/L Mathias et al. (2018).

The exploitation and uncontrolled discharge of specific toxins into water sources has detrimental impacts on aquatic life and, as a result, on human health as well. This problem is made worse by the unintentional discharge of these pollutants from the pharmaceutical sector, hospitals, or cities. In order to remove ibuprofen from wastewater, several researchers employ bioremediation techniques that involve microbes, which has a high percentage of ibuprofen elimination (Table 2.1).

Fortunato et al. (2016) determined through research that batch and continuous reactors used indigenous bacteria collected from various locations to break down IBP. Since the bacterial community

TABLE 2.1
Some Applications of Bioremediation for Elimination of Ibuprofen

Microorganisms	Genera	Parametric conditions	Removal %	References
P. chrysosporium	White-rot fungi	250 mg/L, 4 days	-	Langenhoff et al. 2013
B. thuringiensis	Gram-positive bacteria	25 mg/L, 20 days	46.56%	Marchlewicz et al. 2016
C. aquatica and *Bacillus sp.*	Bacteria	100 mg/L, 33 h	95.9%	Fortunato et al. 2016
P. tricornutum	Living and dead biomass Microalgae	pH 8.2, 2 mg/L & 0.8 mg/L, 200 rpm, 18°C, 6 h	99.9%	Santaeufemia et al. 2018
T. versicolor, I. lacteus, G. lucidum, P. chrysosporium	White-rot fungi	pH 4.5, 10 g/L, 250 mg/L, 135 rpm, 25°C, 7 days	*P. chrysosporium* showed % of degradation: 88% (Low nitrogen medium) and 70% (Malt extract medium). Others degrade IBP to non-detectable levels.	Marco-urrea et al. 2009
C. sorokiniana, Nannochloris sp., Navicula sp., mixed consortium	Microalgae	-	-	Sutherland and Ralph 2019
M. yunnanensis	Bacteria	pH 7, 150 rpm, 25°C, 12 h	83.50%	Sharma et al. 2019
P. chrysosporium	White-rot fungi	30°C	> 90%	Girijan and Kumar 2019
S. yanoikuyae	Bacteria	50 mg/L, 130 days	68%	Balciunas et al. 2020
G. applanatum and *L. sulphurous*	Fungi	15 mg/L, 72 h	66% and 79% correspondingly	Bankole et al. 2020

from the Noria bridge contained *Bacillus* and *Comamonas aquatica* species, it showed the greatest potential for degradation. The efficacy of IBP degradation from an initial level of 100 mg/L was shown to reach 92% of the chemical oxygen demand (COD) in batch reactors as well as continuous reactors. Santaeufemia et al. (2018) tested with *Phaeodactylum tricornutum* microalgae biomass that was taken from Bohlin to test the sorption characteristics of IBP with living or dead microalgae. Following investigation, it was observed that both living as well as dead biomass showed notable efficiency and eradicated roughly 99.9% of IBP. Isolating the most efficient bacterial species from the location of pharmaceutical companies and discussing about IBP degradation were the key points of Sharma et al. (2019)'s study. Out of the six isolated bacterial strains, *Micrococcus yunnanensis*, also known as KGP04 isolate, had the highest potential to degrade IBP, at a rate of about 41.57% before optimisation. The parametric parameters had been optimised using the Taguchi L8 orthogonal array, which led to an 83.50% increase in the rate of degradation.

2.4.2 Naproxen

Naproxen (NAP) is an anti-inflammatory agent with analgesic and antipyretic properties. It is known as 2-arylpropionic acid (profen). Its molecular formula is $C_{14}H_{14}O_3$. An odourless, white to off-white, crystalline material is the free acid. It is essentially insoluble in water but lipid-soluble. It is used to treat fever, menstrual cramps, discomfort, and inflammatory conditions like gout and rheumatoid arthritis. It is consumed orally. It comes in formulations for both immediate and delayed release.

NAP may result in ulcers, bleeding, or perforations in the stomach, intestine, or oesophagus (tube connecting the mouth and stomach). These issues can arise at any point during therapy, without any prior symptoms, and they have the potential to be fatal. Also, those who smoke or drink a lot of alcohol while using naproxen may be at higher risk. According to Picquet (2013), the Albemarle Company manufactures 500 tonnes of naproxen annually. This drug is among those that are frequently found in aquatic environments.

The addition of the cytochrome P450 inhibitor 1-aminobenzotriazole significantly inhibited the degradation of naproxen in both *T. versicolor* and *A. niger*, indicating that both enzymatic systems may be involved in the breakdown of NAP. Since various additional carbon sources triggered various sets of enzymes, multiple degradation pathways for NAP were found during degradation utilising the same strain under various co-metabolic circumstances. Gentisate 1,2-dioxygenase was activated when glucose was added as an additional carbon source, leading to the oxygenolytic cleavage of its product, while *Stenotrophomonas maltophilia* KB2 and *Planococcus sp.* used phenol as a growth substrate. Catechol 2,3-dioxygenase activity has been detected in strain S5. Marco-Urrea et al. (2010), who showed minor and practically total NAP degradation in studies with isolated laccase and purified laccase plus mediator 1-hydroxybenzotriazol, accordingly, are also recognised as the field's pioneers in the study of NAP microbial degradation.

Wojcieszynska et al. (2014) described that a bacterial strain transformed the polycyclic NSAID naproxen through biotransformation. Within 35 days, *Stenotrophomonas maltophilia* KB2 degraded naproxen with a 28% degradation efficiency. Degradation efficiency was 78% and 40%, correspondingly, under cometabolic conditions using glucose or phenol as a carbon source. Cometabolic systems may be used for the bioremediation of degraded sites, according to an increase in NAP degradation efficacy in the presence of extra carbon sources. The immobilised bacteria that can degrade NAP may also be employed as a bio-preparation in sewage treatment plants. In another study, Marchlewicz et al. (2016) studied that a gram-positive bacterium known as strain B1(2015b) was identified in the soil of the chemical plant "Organika-Azot" Poland. The isolated strain was identified as a *Bacillus thuringiensis* species on the basis of investigation of the 16S rRNA gene sequence. Under cometabolic conditions, this strain demonstrated the capacity to metabolise 6 mg/l of NAP over the period of 35 days. These facts suggest that the B1(2015b) strain may be useful for the bioremediation of environments affected by NSAIDs. Similarly, Lu et al. (2019) *Pseudoxanthomonas sp.* DIN-3, a bacterial strain isolated from the biological activated carbon, was capable of simultaneously removing NAP, which were provided as the only carbon source. 39% of NAP (50 mg/L) was bio-removed in 14 days. This study demonstrates the potential of using such a strain to bioenhance contaminated sites or bioreactors; however, a realistic evaluation of bio-removal in the ecosystem should be carried out taking into account the combined impacts of multiple pollutants. Furthermore, these findings might offer a theoretical foundation for the creation of microbiological agents that efficiently remove NAP (Table 2.2) and for the use of NSAIDs to reduce environmental contamination.

2.4.3 Diclofenac

Diclofenac (DCF) is a monocarboxylic acid made of phenylacetic acid with a 2,6-dichlorophenyl amino group at position 2. It functions as a xenobiotic, antipyretic, prostaglandin-endoperoxide synthase inhibitor, drug allergy, non-steroidal, anti-inflammatory medicine, non-narcotic analgesic, environmental pollutant, and prostaglandin-endoperoxide synthase inhibitor. It contains an aromatic amine, an amino acid, a dichlorobenzene, a monocarboxylic acid, and a secondary amino compound. It shares functional similarities with diphenylamines and phenylacetic acids. It is a DCF conjugate acid. Like with most NSAIDs, it is believed that COX-inhibition inhibits prostaglandin synthesis, which is the main mechanism underlying its anti-inflammatory,

TABLE 2.2
Some Applications of Bioremediation for Elimination of Naproxen

Microorganisms	Genera	Parametric Conditions		Removal %	References
S. maltophilia KB2	Bacteria	6 mg/L	35 days	28%	Wojcieszynska et al. 2014
		6 mg/L (supplementary feeding: glucose)		78%	
		6 mg/L (supplementary feeding: phenol)		40%	
B. thuringiensis B1	Bacteria	6 mg/L, 35 days		100%	Marchlewicz et al. 2016
P. chrysosporium	Fungi	20 mg/L, 7 days		86–90%	Li et al. 2015
A. niger	Fungi	50 mg/L, 48 h		98%	Aracagök et al. 2017
Cymbella sp. and *S. Quadricauda*	Algae	1 mg/L, 30 days		97.1% and 58.8% correspondingly	Ding et al. 2017
*Planococcus*sp. S5	Bacteria	6 mg/L, 35 days		30% as sole substrate; 75–86% with supplementary sources of carbon	Domaradzka et al. 2015
*Pseudoxanthomonas*sp. DIN-3	Bacteria	50 µg/L, 14 days		39%	Lu et al. 2019
T. versicolor	Fungi	10 mg/L, 6 h		100%	Marco-Urrea et al. 2010
		55 µg/L, 5 h		95%	

antipyretic, and analgesic effects. Both COX-1 and COX-2 are relatively equally inhibited by diclofenac Alfaro and Davis (2022).

DCF is a drug that has acquired FDA approval for the treatment and management of acute as well as chronic pain carried on by inflammatory disorders, particularly those that affect the musculoskeletal system. Ankylosing spondylitis, rheumatoid arthritis, and osteoarthritis are a few of these. Abdominal pain, gastrointestinal bleeding, nausea, dizziness, headaches, and swelling can all result from taking too much DCF. Heart disease, stroke, kidney issues, and stomach ulcers are among a few of the serious side effects that could occur. Due to its relatively high lipid solubility, diclofenac is one of the few NSAIDs that can pass through the blood-brain barrier and reach the brain. It is believed that it works in the brain as well via inhibiting COX-2. Additionally, the spinal cord may be affected Sandri (2014).

DCF can be broken down by a few microbial strains as a main substrate and as a secondary substrate. In a few experiments, co-metabolism increased the removal efficiency when a carbon substrate, like acetate or toluene, was added. However, the presence of a source of carbon has also led to a drop in elimination efficacy, which is most likely the result of competitive inhibition from secondary substrates. Additionally, distinct metabolites were produced throughout the same strain's breakdown process under various circumstances (mono-substrate and cometabolic conditions). In *Pseudomonas moorei* KB4, 4'-hydroxydiclofenac and diclofenac lactam were detected in both mono- and co-metabolic situations; however, phthalate acid was only present in co-metabolic circumstances. The main process involved in the degradation of diclofenac by the fungal strains was indicated to be hydroxylation catalysed by cytochrome P450 due to the presence of mono- and di-hydroxylated metabolites (CYP). This was further corroborated by the identification of lower reductions in diclofenac in *P. sordida* YK-624 cultured in the presence of the CYP inhibitor 1-aminobenzotriazole. DCF could also be converted to 4-(2,6-dichlorophenylamino)-1,3 benzenedimethanol with the help of purified laccase, but this process did not appear to be carried out by the same enzyme system in *T. versicolor* pellets.

Moreira et al. (2018) used the bacterial strain *Labrysportu calensis* F11; the biodegradation of DCF was evaluated. In 30 days, biotransformation of 70% of DCF (1.7–34 M), provided as the only carbon source, was accomplished. Over the course of 6 days for 1.7 M and 25 days for 34 M of DCF, complete degradation was attained through co-metabolism with acetate. Ultra-high-pressure liquid chromatography/quadrupole time-of-flight, high-resolution, tandem mass-spectrometry was used to find and identify intermediate metabolites. The total ion chromatogram, collected in full-scan mode, was screened in order to find the degradation intermediates. Due to the complexity of the biological degradation materials, this method produced a significant amount of data. In addition, it was important to discriminate between chromatographic peaks originating from DCF biodegradation and those resulting from nutrient metabolism as well as bacterial lysis. The main method of DCF degradation by strain F11 involves hydroxylation events, and the production of benzoquinone imine species appears to be a key step. Furthermore, this is the first publication to identify conjugated metabolites produced by bacteria's sulfation reactions on DCF. Strong indication for strain F11's full breakdown of DCF includes stoichiometric chlorine liberation and the absence of metabolite identification at the completion of the experiments. Another research study examined by Lu et al. (2019) at the ability of biological activated carbon (BAC) filters to remove DCF through the formation of biofilms on a bench-scaled BAC column that was operated for 400 days. The outcomes demonstrated that the BAC column eliminated DCF successfully (> 90%). *Pseudoxanthomonas sp.* DIN-3, a bacterial strain isolated from the BAC, was able to simultaneously eliminate DCF, which was provided as the only carbon source. 23% of DCF (50 g/L) was bioremoved in 14 days. Through co-metabolism with acetate, glucose, or methanol, more effective bioremoval was accomplished concurrently with bacterial growth. These findings indicated that *Pseudoxanthomonas sp.* DIN-3 may act as a DCF degrader. Similarly, Bankole et al. (2020) has done a unique investigation on biodegradation of 30 mg/L of DCF by two wood-rot fungus; *Ganoderma applanatum* (GA) and *Laetiporus sulphureus* (LS) were explored for 72 h. The fungal consortium (GA + LS) had a 96% removal rate for DCF. However, the removal efficiency of GA and LS from NSAIDs was rather modest (61 and 73%, respectively). However, when the fungal consortium (GA + LS) was added to the experimental setup, 99.5% drug degradation was attained. Overall, LS showed greater degradation efficiency than GA, with an 87% rate of breakdown on DCF. Table 2.3 includes a list of additional studies conducted by different researchers for removal of DCF.

2.4.4 Ketoprofen

In comparison to other NSAIDs that are delivered systemically, Ketoprofen (KPF) is a relatively recent drug. In 1990, the FDA authorised its usage in horses. In order to exert Ketoprofen's anti-inflammatory, antipyretic, and analgesic effects, it is given orally, topically, intravenously, or intramuscularly. KPF is frequently used to treat a variety of musculoskeletal conditions in horses, including colic, endotoxemia, and acute injuries as well as more chronic conditions like osteoarthritis. Its chemical formula is $C_{16}H_{14}O_3$. A 3-benzoylphenyl group is substituted for propionic acid at position 2 in the oxomonocarboxylic acid KPF. It functions as an antipyretic, an EC 1.14.99.1 (prostaglandin-endoperoxide synthase) inhibitor, an environmental pollutant, a xenobiotic, a drug allergen, and a non-steroidal anti-inflammatory medication. It is an oxo monocarboxylic acid and a benzophenone. It is similar to a propionic acid in terms of function.

Humans also use KPF to treat pain associated with a variety of diseases. It also decreases pain, swelling, as well as joint stiffness from arthritis. But excess intake of KPF medicine may result in stomach or intestine perforations, bleeding, or ulcers. These issues can arise at any point during

TABLE 2.3
Some Applications of Bioremediation for Elimination of Diclofenac

Microorganisms	Genera	Parametric Conditions		Removal %	References
T. versicolor	White-rot fungus	10 mg/L, 1 h		94%	Marco-Urrea et al. 2010
P. sordida YK-624	White-rot fungus	45 µg/L	0.5 h	100%	Hata et al. 2010
		29.615 µg/L	3 days	90%	
			6 days	100%	
*Klebsiella*sp. KSC	Bacteria	70 mg/L, 72 h		100%	Stylianou et al. 2018
*R. ruber*strain IEGM 346	Bacteria	50 µg/L	6 days	100%	Ivshina et al. 2019
		50 µg/L	60 days	50%	
Labrysportucalensis F11	Bacteria	503.455–10069.1 g/L	30 days	70%	Moreira et al. 2018
		503.455 g/L and 10069.1 g/L (supplementary feeding: acetate)	6 and 25 days	100%	
Brevibacterium sp. D4	Bacteria	10 mg/L	30 days	35%	Bessa et al. 2017
		10 mg/L (periodic supplementary feeding: acetate)	30 days	90%	
E. cloacae D16	Bacteria	10 mg/L, 48 h		67.57%	Aissaoui et al. 2017
Pseudoxanthomonas sp. DIN-3	Bacteria	50 µg/L, 14 days		23%	Lu et al. 2019
G. applanatum and *L. sulphurous*	Wood-rot fungi	15 mg/L, 72 h		80% and 87% correspondingly	Bankole et al. 2020
P. moorei KB4	Bacteria	1 mg/L	12 days	51%	Zur et al. 2020
		1 mg/L (supplementary: glucose)	11 days	100%	
		1 mg/L (supplementary: glucose and acetate)	12 days	100%	

TABLE 2.4
Some Applications of Bioremediation for Elimination of Ketoprofen

Microorganisms	Genera	Parametric Conditions	Removal %	References
Trametesversicolor	Fungi	10 mg/L, 24 h	100%	Marco-Urrea et al. 2010
		40 µg/L, 5 h	100%	
Pleurotusostreatus	Fungi	10 mg/L, 7–15 days	90% batch stage, 70–85% continuous stage	Palli et al. 2017
P. djamor	Fungi	10 mg/L, 48 h	85%	Cruz-Ornelas et al. 2019

therapy, without any prior symptoms, and they have the potential to be fatal. As a result, numerous researchers have examined at this issue and their findings are shown in Table 2.4.

Palli et al. (2017) investigated the ability of the fungus *Pleurotus ostreatus* to degrade hospital wastewater containing 10 mg/L of KPF. In a fluidised bed bioreactor, the degradation test was conducted in batch and continuous modes (hydraulic retention time in the range 1.63–3 days). KPF's degradation in batch mode reached approximately 50% in just 5 days. About 70% of the KPF was removed in continuous mode. Additionally, compared to other fungi, *P. ostreatus* was able to lower the chemical oxygen demand of the hospital wastewater, which is a significant advantage in the development of a wastewater treatment process. The white-rot fungus *Pleurotusdj amor*was utilised by Cruz-Ornelas et al. (2019) to analyse the kinetics of KPF degradation was used in this study. After 48 hours of incubation, KPF was removed in 87% of cases. However, the enzymatic activities of laccases, manganese peroxidases, and lignin peroxidases have been assessed during the degradation kinetics analysis, yielding values of 3700, 270, and 31 U/L, respectively. These outcomes demonstrate the potential use of this fungus for future studies of the degradation of other NSAIDs and other drugs emerging combinations of contaminants, emerging contaminants, or substances with chemical structures comparable to those examined in this study.

Both *Trametesversicolor* and *Pleurotusostreatus* showed inhibition of KPF degradation in response to the presence of the cytochrome P450 inhibitor 1-aminobenzotriazole, indicating that cytochrome P450 is involved in the first oxidation step of KPF degradation. Both of these biotransformations appeared to include the reduction of the keto group as a significant step. In *T. versicolor* and *P. ostreatus*, an aromatic ring underwent hydroxylation, while *T. versicolor* was the only organism to report deoxygenation Kim (2001).

2.5 MICROBIAL DEGRADATION UTILISING MIXED CULTURES

Even though lab tests using pure cultures of algal, fungal, or bacterial organisms to study the microorganism's breakdown of NSAIDs yielded some promising results, maintaining a pure culture on a field scale of remediation is extremely difficult. As a result, other research studies looked into the use of mixed consortiums for a variety of NSAIDs as well as for individuals. The mixed consortium's individual parts, which are listed in Table 2.5, were only partially characterised. With respect to DCF, IBP, and KPF, numerous genera, including *Bacillus pseudomycoides, Nitratireductor, Pseudacidovorax, Asticcacaulis, Vibrio mediterraneus*, and *Rhodococcusruber*, were found to exhibit competitive advantages or higher toxicity resistance, which has been interpreted as a sign that they may contribute to the drugs' correlating biodegradation.

Using two photobioreactors, an anoxic-aerobic photobioreactor (phase A) and an anaerobic-anoxic-aerobic photobioreactor (phase B), IBP and NAP were degraded by an algal-bacterial method. The microalgae population in the photobioreactors has been primarily made up of *Phormidium sp., Chlorella vulgaris*, and *Scenedesmus acuminatus* in phase A and *Pseudonabaenaacicularis, Chlorella vulgaris*, and *Scenedesmus acuminatus* in phase B, correspondingly, even though the bacterial component was not discussed (López-Serna et al. 2019). Another study, Ismail et al. (2016), identified three novel metabolites, (3-ethylphenyl) (phenyl) methanone, (3-hydroxyphenyl) (phenyl) methanone, and (3-hydroxyphenyl) (oxo) acetic acid as a result of Ketoprofen degradation using the K_2 bacterial consortium but not in studies using pure culture. Experiments using mixed cultures were also performed to analyse degradation under various situations. Illumination, nitrification, concentration of ammonium, and dissolved organic matter were some of the criteria evaluated. A K2 bacterial consortium has been shown to biodegrade Ketoprofen more quickly in the dark (within 48 hours) than when exposed to light. It was also stated how a diurnal cycle affected everything. High ammonium concentrations resulted in slower IBP biotransformation rates, but another study found that DCF degradation efficiency increased.

TABLE 2.5
Microbial Degradation of NSAIDs Using Mixed Cultures

	Parameters			
Microorganism	Initial Concentration	Interval	Removal %	References
G. amicalis, A. bouvetii, P. aminophilus, and *P. americanu*	1000 μg/L or 500 μg/L IBP	2–6 h	100%	Almeida et al. (2013)
M. populi, G. hydrophobica, T. spumae, P. aminovorans, R. qingshengii, G. terrae, R. zopfii, and *B. thiooxidans*	250 μg/L or 100 μg/L Ketoprofen	10–25 h	100%	
A. nicotianae. Pseudomonas sp., E. hormaechei, and *C. youngae*	3 mg/L of every single NSAIDs	48 h	23.08% for IBP and 9.12% for DCF	Aissaoui et al. (2017)
	3 mg/L of every single NSAIDs (supplementary feeding: glucose)	48 h	100% for IBP and 56% for DCF	
K_2 bacterial consortium (*R. ornithinolytica* B6, *P. aeruginosa* strain JPP, *Pseudomonas sp.* P16, *Stenotrophomonas sp.* 5LF 19TDLC)	1.27 g/L of Ketoprofen under dark situation	48 h	100%	Ismail et al. (2016)
K_2 bacterial consortium + microalga *Chlorella sp.* Iso 4	508.5 mg/L of Ketoprofen under diurnal situation	10 days	100%	
Stable nitrifying enrichment culture	25–100 μg/L of IBP	24 h	100%	Dawas-Massalha et al. (2014)
	25–100 μg/L of Ketoprofen	~150 h	100%	
Pentane enrichment culture	2.9 mg/L of DCF, 2 mg/L IBP and 2.3 mg/L NAP correspondingly	96 h	0%, 100% and 64% correspondingly	Bragança et al. (2016)

DCF was observed to be persistent in the presence of nitrification, while IBP and NAP (1 mg/L) were entirely eliminated in 35 days. Ibuprofen and Naproxen biodegradation were shown to be enhanced by the addition of dissolved organic matter, which may have been because of increased biomass synthesis and microbial respiration, which indicate a more active microbial community with increased drug biodegradation potential.

2.6 APPLICATIONS AND DRAWBACKS OF BIOREMEDIATION TECHNIQUES

Although numerous laboratory researches have been conducted on the biodegradation of NSAIDs, there are few applications at pilot scale or in significant water treatment facilities. The type of microorganisms used for biodegradation needs to be the main area of focus when scaling up to a WWTP. Based on the kind of wastewater that the WWTP treats, this decision needs to be made. The microorganisms used need to have a very high Minimum Inhibitory Concentration (MIC) for their successful functioning in the case of hospital or pharmaceutical company effluent, where NSAIDs are present in comparatively high concentration. Since most microorganisms degrade NSAIDs at low pH in laboratory experiments, the type of microorganisms utilised also relies on the pH of the contaminated water Vieno and Sillanpää (2014).

The site of the tertiary biological water treatment reactor for the elimination of pharmaceuticals in the WWTP needs to be taken into account when scaling up. A pure culture or a mixed microbial culture must be maintained, once the relevant sterility of the incoming wastewater is determined. As a result, the reactor's optimal location will be determined following a first round of disinfection. To get rid of the remaining microbial community after the treatment, it will be preferable to perform another round of disinfection. The hydraulic retention time (HRT) for the treatment will have an impact on the biodegradation process in addition to the location. The HRT must be carefully chosen for best results because it will be recommended to reduce numerous pharmaceuticals in the same reactor.

Very few microorganisms have the ability to use NSAIDs as their only source of carbon. The polluted water needs to be examined for the existence of the ideal supplementary source of carbon because the maximal microbial breakdown of NSAIDs occurs in co-metabolic circumstances employing a supplementary carbon source. It is not viable to add the specified source externally if it is not available in sufficient concentration; hence, an alternate additional carbon source that is abundantly present in the wastewater must be investigated for the microorganisms.

When a particular strain or microbial consortium scales up laboratory studies of degrading to full application scale, a number of disadvantages can be observed. Among these include treatment conditions that are problematic for use in large-scale applications and the uncertain maintenance of pure growth circumstances or even specified consortia due to cyclical conditions in WWTPs. When taking into account the quantity of microorganisms required to break down all of the key medications present in the hospital or even typical WWTPs, it is vital for the survival of microorganisms to maintain optimal circumstances (like pH and temperature), which might vary for each microbe.

Additionally, it is difficult to predict the degradation metabolites for the entire WWTPs, since the biodegradation pathways of one microbe cannot be readily extended to the degradation abilities of microbial consortia. Besides that, pure cultures or microbial consortiums have been used in lab studies to degrade a variety of pharmaceutical compounds; the similar microorganism introduced into operative WWTPs are less efficient or completely useless due to the extremely heterogeneous mixtures of toxic elements and varied microbial consortium that may interfere with the degradation. The enzyme systems in bacteria that are in charge of degradation may also completely fail as a consequence of this. This can be fixed by investigating the synergistic impact of several drugs while figuring out the microorganisms' potential for biodegradation in real-time WWTPs.

Studies on the degradation of NSAIDs in the presence of metabolites are severely lacking, despite the fact that parent compound and metabolites are present in real-time WWTPs. In these circumstances, feedback inhibition cannot be disregarded. Furthermore, these metabolites may be more hazardous in the reactor, which could have negative effects on the surroundings and marine ecosystems. Even though considerable research has been done on the toxicity of metabolites in lab settings, further study is required to determine and keep track of the toxic effects in the reactor during treatment. Lastly, another limitation is that bioaccumulation occurs in the presence of live cells; reuse is limited; therefore, expensive costs are required for wider-scale use requires greater costs; however, procedures involving bioaccumulation typically outperform other processes.

2.7 CONCLUSIONS

NSAIDs were considered to be a future environmental risk due to their widespread use in daily life around the world and the associated consequences on ecosystems. The following is an overview of the key findings from the analysis of recent research:

- Individual NSAID concentrations are typically far below 1 μg/L in surface water bodies. Consequently, the main cause of the waterways' contamination was the discharge of these low concentration NSAIDs effluents.
- The best wastewater treatment procedures should be designed by correlating the physical and chemical properties of the pollutant with its environmental fate. These properties include solubility, molecular weight, water coefficient, sludge dispersion coefficient, and its reactivity with free radicals. The development of effective and affordable technologies has also received a lot of attention, especially as they relate to wastewater treatment.
- Different NSAIDs were found to be removed more effectively by biological processes.
- IBP and DCF are more hazardous than NAP and KPF when it comes to their toxicity at environmental relevant concentrations. The NSAID metabolites' toxicity was less than that of the parent compounds'.
- NSAIDs can be decomposed by a number of pure and mixed microbial strains in both laboratory and natural environments.
- Different supplemental carbon sources and certain microbial consortia can be used to trigger the distinct variable degradation pathways of enzymes that lead to various degradation mechanisms.
- The vulnerability of microbially-enriched bioremediation approaches to additional implicit and explicit factors and fluctuations in the parameters makes field-scale implementation difficult. It would be essential to conduct more research to fully develop the bioremediation method and commission it in order to ensure that technology gets transferred from the lab to the field.

BIBLIOGRAPHY

Aissaoui S, Ouled-Haddar H, Sifour M, Beggah C, Benhamada F (2017). Biological removal of the mixed pharmaceuticals: diclofenac, ibuprofen, and sulfamethoxazole using a bacterial consortium. *Iran. J. Biotechnol.* 15.

Alfaro RA, Davis DD (2022). Diclofenac. In *StatPearls [Internet]*. Treasure Island (FL): StatPearls Publishing; January. Available from: www.ncbi.nlm.nih.gov/books/NBK557879/.

Almeida B, Oehmen A, MaRques R, Brito D, Carvalho G, Barreto Crespo MT (2013). Modelling the biodegradation of non-steroidal anti-inflammatory drugs (NSAIDs) by activated sludge and a pure culture. *Bioresour. Technol.* 133, 31–37.

Aracagök YD, Göker H, Cihangir NZ, Naturforsch (2017). Biodegradation of micropollutant naproxen with a selected fungal strain and identification of metabolites. *Z. Naturforsch.* 72, 173–179.

Balciunas, E. M., Kappelmeyer, U., Harms, H., & Heipieper, H. J. (2020). Increasing ibuprofen degradation in constructed wetlands by bioaugmentation with gravel containing biofilms of an ibuprofen-degrading Sphingobium yanoikuyae. *Engineering in Life Sciences*, 20(5-6), 160–167.

Banerjee S, Kamila B, Barman S, Joshi SR, Mandal T, Halder G (2019). Interlining Cr(VI) remediation mechanism by a novel bacterium Pseudomonas brenneri isolated from coalmine wastewater. *J. Environ. Manage.* 233, 271–282.

Bankole PO, Adekunle AA, Jeon BH, Govindwar SP (2020). Novel cobiomass degradation of NSAIDs by two wood rot fungi, *Ganodermaapplanatum* and *Laetiporussulphureus*: ligninolytic enzymes induction, isotherm and kinetic studies. *Ecotoxicol. Environ. Saf.* 203, 110997.

Barkin RL (2015). Topical nonsteroidal anti-inflammatory drugs: the importance of drug, delivery, and therapeutic outcome. *J. Ther.* 22(5), 388–407.

Bessa, V. S., Moreira, I. S., Tiritan, M. E., & Castro, P. M. L. (2017). Enrichment of bacterial strains for the biodegradation of diclofenac and carbamazepine from activated sludge. *International Biodeterioration & Biodegradation*, 120, 135–142.

Bindu S, Mazumder S, Bandyopadhyay U (2020). Non-steroidal anti-inflammatory drugs (NSAIDs) and organ damage: a current perspective. *Biochem. Pharmacol.* doi: 10.1016/j.bcp.2020.114147.

Bragança I, Danko AS, Pacheco J, Frascari D, Delerue-Matos C, Domingues VF (2016). Cometabolic degradation of anti-inflammatory and analgesic pharmaceuticals by a pentane enrichment culture. *Water Air Soil Pollut.* 227, 227. https://doi.org/10.1007/s11270-016-2933-9.

Chaiamnuay S, Allison JJ, Curtis JR (2006). Risks versus benefits of cyclooxygenase-2-selective nonsteroidal antiinflammatory drugs. *J. Health Syst. Pharm.* 63(19), 1837–1851.

Cruz-Ornelas R, Sanchez-Vazquez JE, Amaya-DeLgado L, Guillen-Navarro K, Calixto-Romo A (2019). Biodegradation of NSAIDs and their effect on the activity of ligninolytic enzymes from *Pleurotusdjamor. Biotech.* 9, 373. https://doi.org/10.1007/s13205-019-1904-4.

Dawas-Massalha A, Gur-Reznik S, Lerman S, Sabbah I, Dosoretz CG (2014). Co-metabolic oxidation of pharmaceutical compounds by a nitrifying bacterial enrichment. *Bioresour. Technol.* 167, 336–342.

Ding T, Lin K, Yang B, Yang M, Li J, Li W, Gan J (2017). Biodegradation of naproxen by freshwater algae *Cymbella* sp. and *Scenedesmus quadricauda* and the comparative toxicity. *Bioresour. Technol.* 238, 164–173. https://doi.org/10.1016/j. biortech.2017.04.018.

Domaradzka D, GuziK U, Hupert-Kocurek K, Wojcieszy´nska D (2015). Cometabolic degradation of naproxen by *Planococcus* sp. strain S5. *Water Air Soil Pollut.* 226, 297.

Fortunato MS, Abril NPF, Martinefski M, Trípodi V, Papalia M, Rádice M, Gutkind G, GAllego A, Korol SE (2016). Aerobic degradation of ibuprofen in batch and continuous reactors by an indigenous bacterial community. *Environ. Technol.* 37.

Fritsche W, Hofrichter M (2008). *Aerobic Degradation by Microorganisms*, Second Edition. Weinheim, Germany: Wiley-VCH Verlag GmbH.

Girijan, S., & Kumar, M. (2019). Immobilized biomass systems: an approach for trace organics removal from wastewater and environmental remediation. *Current Opinion in Environmental Science & Health*, 12, 18–29.

Hata T, Kawai S, Okamura H, Nishida T (2010). Removal of diclofenac and mefenamic acid by the white rot fungus *Phanerochaete sordida* YK-624 and identification of their metabolites after fungal transformation. *Biodegradation* 21, 681–689.

Ismail MM, Essam TM, Ragab YM, Mourad FE (2016). Biodegradation of ketoprofen using a microalgal–bacterial consortium. *Biotechnol. Lett.* 38, 1493–1502.

Ivshina IB, Tyumina EA, Kuzmina MV, Vikhareva EV (2019). Features of diclofenac biodegradation by *Rhodococcusruber* IEGM 346. *Sci. Rep.* 9, 9159.

Kao CM, Chen SC, Liu JK, Wang YS (2001). Application of microbial enumeration technique to evaluate the occurrence of natural bioremediation. *Water Residual.* 35(8), 1951–1960.

Kim JY (2001). Measurement of Ketoprofen in horse urine using gas chromatography-mass spectrometry. *J. Vet. Pharmacol. Therap.* 24, 315–319.

Langenhoff, A., Inderfurth, N., Veuskens, T., Schraa, G., Blokland, M., Kujawa-Roeleveld, K., & Rijnaarts, H. (2013). Microbial removal of the pharmaceutical compounds ibuprofen and diclofenac from wastewater. *BioMed research international*, 2013.

Li X, AlveS de Toledo R, Wang S, Shim H (2015). Removal of carbamazepine and naproxen by immobilized *Phanerochaetechrysosporium* under non-sterile condition. *N. Biotech.* 32, 282–289.

López-Serna R, Posadas E, GarcíaEncina PA, Mu˜noz R (2019). Removal of contaminants of emerging concern from urban wastewater in novel algal-bacterial photobioreactors. *Sci. Total Environ.* 662, 32–40.

Lu Z, Sun W, Li C, Ao X, Yang C, Li S (2019). Bioremoval of non-steroidal anti-inflammatory drugs by *Pseudoxanthomonas sp.* DIN-3 isolated from biological activated carbon process. *Water Res.* 161, 459–472. https://doi.org/10.1016/j.watres.2019.05.065

Marchlewicz A, Domaradzka D, Guzik U (2016). *Bacillus thuringiensis* B1(2015b) is a gram-positive bacteria able to degrade naproxen and ibuprofen. *Water Air Soil Pollut.* 227, 197. https://doi.org/10.1007/s11270-016-2893-0

Marco-Urrea, E., Aranda, E., Caminal, G., & Guillén, F. (2009). Induction of hydroxyl radical production in Trametes versicolor to degrade recalcitrant chlorinated hydrocarbons. *Bioresource technology*, 100(23), 5757–5762.

Marco-Urrea E, P´erez-Trujillo M, Bl´anquez P, Vicent T, Caminal G (2010). Biodegradation of the analgesic naproxen by Trametesversicolor and identification of intermediates using HPLC-DAD-MS and NMR. *Bioresour. Technol.* 101, 2159–2166.

Mathias FT, Fockink DH, Disner GR, Prodocimo V, Ribas JLC, Ramos LP, Cestari MM, Silva de Assis HC (2018). Effects of low concentrations of ibuprofen on freshwater fish Rhamdiaquelen. *Environ. Toxicol. Pharmacol.* 59, 105–113.

Moreira IS, BeSsa VS, Murgolo S, PiCcirillo C, Mascolo G, Castro PML (2018). Biodegradation of diclofenac by the bacterial strain Labrysportucalensis F11. *Ecotoxicol. Environ. Saf.* 152, 104–113. https://doi.org/10.1016/j.ecoenv.2018.01.040

Palli L, Castellet-Rovira F, Perez-Trujillo M, Caniani D, Sarra-Adroguer M, Gori R (2017). Preliminary evaluation of *Pleurotusostreatus* for the removal of selected pharmaceuticals from hospital wastewater. *Biotechnol. Prog*. 33, 1529–1537.

Picquet M (2013). Organometallics as catalysts in the fine chemical industry. *Platinum Metals. Rev*. 75, 272e280.

Qureshi O, Dua A (2022). COX inhibitors. In *StatPearls [Internet]*. Treasure Island, FL: StatPearls Publishing; January. Available from: www.ncbi.nlm.nih.gov/books/NBK549795/.

Ramezani M, Rad FA, Ghahari S, Ghahari S, Ramezani M (2021). Nano-bioremediation application for environment contamination by microorganism. In *Microbial Rejuvenation of Polluted Environment, Microorganisms for Sustainability*, 26. Springer, 349–378. doi: 10.1007/978-981-15-7455-9_14.

Rasheed T, Bilal M, Nabeel F, Adeel M, Iqbal HMN (2019). Environmentally-related contaminants of high concern: potential sources and analytical modalities for detection, quantification, and treatment. *Environ. Int*. 122, 52–66.

Rastogi A, Tiwari MK, Ghangrekar MM (2021). A review on environmental occurrence, toxicity and microbial degradation of non-steroidal anti-inflammatory drugs (NSAIDs). *J. Environ. Manage*. 300. https://doi.org/10.1016/j.jenvman.2021.113694

Samanta S, Datta D, Halder G (2020). Biodegradation efficacy of soil inherent novel sp. Bacillus tropicus (MK318648) onto low density polyethylene matrix. *J. Polym. Res*. 27, 324. https://doi.org/10.1007/s10965-020-02296-x.

Sandri A (2014). Diclofenac: update on tolerableness and spinal anti-inflammatory action. *Minerva Med*. 105(4), 313–318.

Santaeufemia S, Torres E, Abalde J (2018). Biosorption of ibuprofen from aqueous solution using living and dead biomass of the microalga *Phaeodactylumtricornutum. J. Appl. Phycol*. 30, 471–482.

Seifollahi Z, Rahbar-Kelishami A (2017). Diclofenac extraction from aqueous solution by an emulsion liquid membrane: parameter study and optimization using the response surface methodology. *J. Mol. Liq*. 231, 1–10.

Sharma K, Kaushik G, Thotakura N, Raza K, Sharma N, Nimesh S (2019). Fate of ibuprofen under optimized batch biodegradation experiments using *Micrococcus yunnanensis* isolated from pharmaceutical sludge. *Int. J. Environ. Sci. Technol*. 1–14.

Show S, ChakraBorty P, Karmakar B, Halder G (2021). Sorptive and microbial riddance of micro-pollutant ibuprofen from contaminated water: a state of the art review. *Sci. Total Environ*. 786. https://doi.org/10.1016/j.scitotenv.2021.147327.

Sibeko PA, Naicker D, Mdluli PS, Madikizela LM (2019). Naproxen, ibuprofen, and diclofenac residues in river water, sediments and *Eichhorniacrassipes* of Mbokodweni river in South Africa: an initial screening. *Environ. Forensics*. 20, 129–138. https://doi.org/10.1080/15275922.2019.1597780.

Stylianou K, Hapeshi E, Vasquez M, Fatta-Kassinos D, Vyrides I (2018). Diclofenac biodegradation by newly isolated *Klebsiella* sp. KSC: microbial intermediates and ecotoxicological assessment. *J. Environ. Chem. Eng*. 6, 3242–3248. https://doi.org/10.1016/j.jece.2018.04.052.

Sutherland, D. L., & Ralph, P. J. (2019). Microalgal bioremediation of emerging contaminants-Opportunities and challenges. *Water research*, 164, 114921.

Vandraas KF, Spigset O, Mahic M, Slordal L (2010). Non-steroidal anti-inflammatory drugs: use and co-treatment with potentially interacting medications in the elderly. *Eur. J. Clin. Pharmacol*. 66, 823–829.

Vieno N, Sillanpää M (2014). Fate of diclofenac in municipal wastewater treatment plant—a review. *Environ. Int*. 69, 28–39.

Wojcieszyńska D, GuzIk U, Greń I, Perkosz M, Hupert-Kocurek K (2011). Inductionof aromatic ring e cleavage dioxygenases in Stenotrophomonas maltophiliastrain KB2 in cometabolic systems. *World J. Microbiol. Biotechnol*. 27, 805e811.

Wojcieszyńska, D., Domaradzka, D., Hupert-Kocurek, K., & Guzik, U. (2014). Bacterial degradation of naproxen–undisclosed pollutant in the environment. *Journal of environmental management*, 145, 157–161.

Woolridge A, Hoboy S (2019). Medical waste. In *Waste*. Academic Press, 517–530.

Zur J, Pi´nski A, Wojcieszy´nska D, Smułek W, Guzik U (2020). Diclofenac degradation—enzymes, genetic background and cellular alterations triggered in diclofenac-metabolizing strain *Pseudomonas moorei* KB4. *Int. J. Mol. Sci*. 21, 6786.

3 Biodegradation of Catechol in Sequential Batch Reactor

Vipul Sharma, Ashish Yadav, Jai Prakash Kushwaha, Vimal Chandra Srivastava and Indra Deo Mall

3.1 INTRODUCTION

Phenol, a benzene derivative, is widely used to manufacture antiseptics, disinfectants, cosmetics, hair dye, resins, synthetic fibre, etc. It exhibits very high solubility in water. Physically, it has no colour; however, it turns pink due to oxidation if exposed to the air. Also, it is crystalline, hygroscopic, very toxic, and bears low biodegradability (Swamy, 1997). In this context, phenol derivatives such as resorcinol and catechol are also very toxic and recalcitrant in nature (Chen et al., 2022; Aghapour et al., 2013; Moussavi et al., 2015). From the toxicity comparison viewpoint, catechol has been demonstrated to have higher toxicity than resorcinol and phenol (Othmer, 1995; Sun et al., 2005). If exposed, catechol produces irritation on the skin and eyes. Acute intoxication may occur if taken orally, producing nausea or vomiting, diarrhoea, and central nervous system depression. In some cases, decreased oxygen-carrying capacity of haemoglobin – i.e., methemoglobinemia and acute liver injury – can be detected after a few days of acute intoxication of catechol (Othmer, 1995).

Resorcinol and catechol are used as raw materials/generated by products in industries like textile, pulp and paper, petroleum refinery, rubber, dyes and pigments, petrochemical, pharmaceutical, etc. (Lobo et al., 2016; Aghapour et al., 2015, 2013; Sun et al., 2005; Othmer, 1995). Therefore, the wastewater generated from these industries contains phenol and its derivatives, like resorcinol and catechol. These phenolic compounds have also been found in the wastewater generated from coal conversion to coke/natural gas (from very low to 2,000 mg/L), wood tar, and coal tar processing (Moussavi et al., 2014; Kumaran and Paruchuri, 1997). However, the concentration of the wastewater generated from coal carbonization at low temperatures has been reported to be very high (5,300 mg/L) (Subramanyam and Mishra, 2008). If discharged into the environment without treatment, this wastewater can harm aquatic life and pollute the drinking water resources like surface water/groundwater.

Biological (aerobic and anaerobic) techniques have been used for wastewater treatment for the past few decades. In this regard, activated-sludge-process, moving-bed-biofilm-reactor, wetlands, trickling-filters, etc., techniques have been widely explored in the last few decades as viable alternatives for the treatment of wastewater (Singh et al., 2022; Rashid et al., 2021). A sequential batch reactor (SBR), which is nothing but the aerobic activated sludge treatment system, occupies a considerably smaller area for its installation. In SBR, the various sequences – filling of wastewater, treating, settling, and clarification (Figure 3.1) – that are followed in conventional activated sludge systems are performed in a single reactor (Figure 3.2) (EPA, 1999). SBRs are very successfully used to treat both municipal and industrial wastewater. The major advantage of the SBR is the enhanced nitrogen/ammonia/phosphorus removal. This can be achieved by following the suitable wastewater filling method to the reactor already containing biomass. The filling methods may be static (the wastewater is not mixed with biomass), mixed (wastewater is mixed with biomass), and aerated (aeration during the static/mixed filling). Therefore, the biomass action on the pollutants can occur under anaerobic, anoxic, or aerobic conditions, which is responsible for enhanced nutrient/organics removal.

 DOI: 10.1201/9781003391487-3

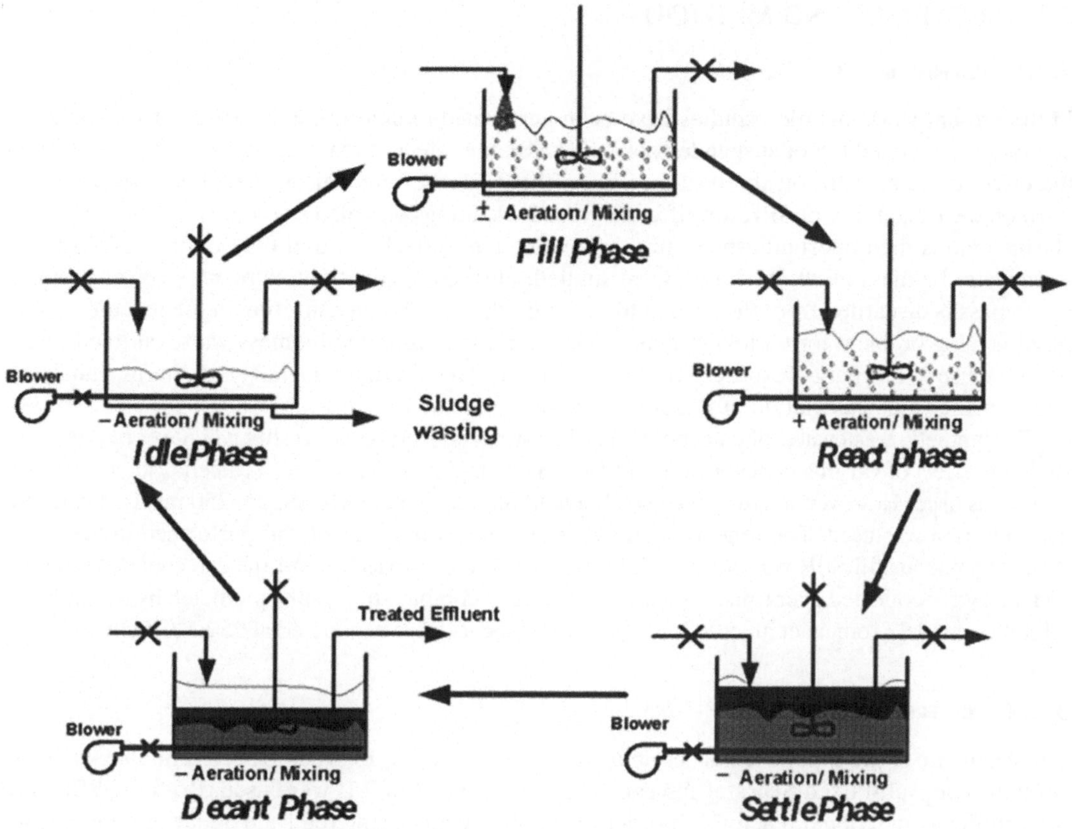

FIGURE 3.1 Sequential batch reactor (SBR) design principle.

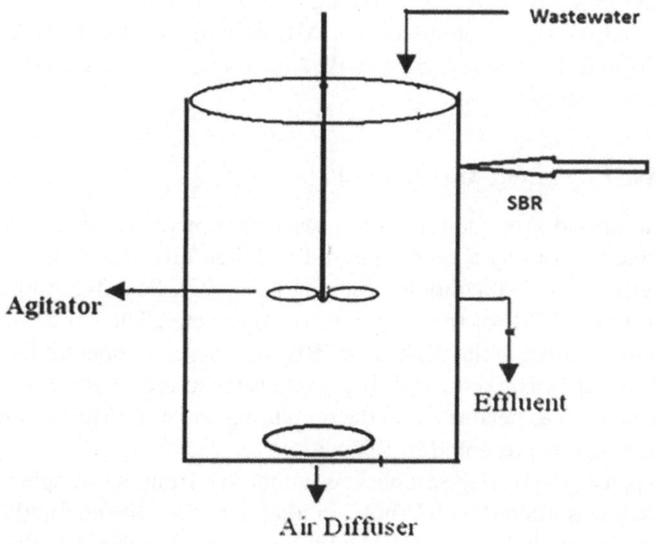

FIGURE 3.2 Schematic of experimental set-up of SBR.

3.2 MATERIALS AND METHODS

3.2.1 General

In the present work, detailed studies have been performed to determine the effect of varying concentrations of mixed liquor suspended solid (MLSS), the effect of reacting to settle phase ratio, and the effect of aerated fill on the performance of SBR for the removal of phenolic waste catechol. Further, the treatability of SBR for different catechol loadings was also studied. After the treatment, clarification is the important aspect. In this context, the settleability and filterability of the treated wastewater-biomass mixture (slurry) were studied. Further, during the treatment, a certain volume of biomass is discarded from the system to maintain the MLSS concentration in the reactor. Its disposal is a big problem for treatment plants. Therefore, the discarded biomass was examined for its potential use as an energy resource. To fulfil this requirement, proximate analysis, thermal analysis, and calorific value calculation of the biomass were performed.

The synthetic wastewater of catechol in distilled water was prepared freshly whenever required. To study the effect of varying concentrations of MLSS and the aerated fill time, a catechol solution of 100 mg/L was used. However, to study the catechol loading rate at fixed MLSS, 25–250 mg/L of catechol concentration was used. The kinetics study of the treatment of catechol was performed in SBR, and the instantaneous fill SBR was employed. Experiments were carried out for the catechol degradation, and data were collected over time. This data was used to compare the results predicted by the Haldane inhibition model (Tomeia et al., 2003). All the experiments were conducted at $25 \pm 1^{o}C$ temperature.

3.2.2 Experimental Set-Up

The schematic of the experimental set-up is shown in Figure 3.2. In order to accommodate the influent flow, two well-mixed SBRs of the same working volume of 2 L were used. Both the SBRs are performing as individual reactors. In order to supply the air, a sintered sand aquarium air diffuser fitted with an air pump was used and fixed at the bottom. To introduce the wastewater to the reactor, a peristaltic pump was used. The biomass wasting was performed from the tapping provided at the bottom of the reactor.

Dissolved oxygen (DO) concentration in the reactor is an important parameter in activated sludge treatment systems. In this study, a DO level of 2 mg/L was maintained in the SBR by regulating the air flow rate with the help of an air rotameter. The MLSS in the reactor was maintained from 3,000 to 5,000 mg/L with 5–10 d sludge age (Tsang et al., 2007). The sludge wasting was performed every three cycles of the SBR operation.

3.2.3 Experimental Procedure and Analysis

The seed biomass was provided by the Haridwar sewage treatment plant in India. The collected seed biomass was first screened for any foreign material and then mixed with the synthetic wastewater for acclimatization after its 1–2 d aeration (Tsang et al., 2007). Post 3–5 d of acclimatization, the biomass was utilized in the SBR for the treatment of wastewater. The MLSS was varied by taking the appropriate biomass volume in the SBR. The SBR was chosen to operate for 300-min total cycle time out of which 150 min was fixed for filling wastewater to the SBR. Other than the fill phase, react, settle, and idle phase was performed in the remaining 150 min (Metcalf and Eddy, 2006). The operating parameters range is presented in Table 3.1.

After the settling phase, the treated sample was withdrawn from the sample withdrawal port, and in subsequent phases, it was filtered with Whatman filter paper no. 42 and finally tested for the residual catechol concentration with the help of a UV/vis spectrophotometer. For this purpose, a calibration curve was prepared with known catechol concentration solutions at 275 nm. Further, residual catechol concentration in the samples was determined using this calibration curve. The analysis of various parameters was carried out using different equipment/methods, as listed in Table 3.2.

TABLE 3.1
SBR Operating Conditions (Eddy and Metcalf, 2006)

Parameters	Values
pH	7.5–9
Fill phase (h)	2.5
React phase (h)	1–2
Settle phase (h)	0.5–1.5
Hydraulic retention time (h)	15–20
Solid retention time (d)	3–5

TABLE 3.2
Analytical Techniques and Instrumentation.

Parameter	Analytical Technique/method	Make/method
Morphology	Scanning electron microscope	LEO 435 VP, England
EDAX	Energy dispersive X-ray spectrometer (FE-SEM)	QUANTA 200 FEG
Proximate analysis	IS 1350	As described in Part I.
Heating value	Bomb calorimeter, IS 1350	Toshniwal, Bombay
Oxidation characteristics	TG/DTG/DTA analyzer	Perkin Elmer Pyris Diamond
SEM micrograph	Scanning electron microscope	LEO 435 VP, England

3.2.4 Physicochemical and Thermal Characterization

The surface morphology and the elemental composition of the biomass before and after its use in SBR were examined using a scanning electron microscope (SEM) (LEO 435 VP, England) and energy dispersive X-ray (EDX) analysis. For SEM and EDX analysis (using QUANTA, Model 200 FEG, USA), the samples were gold coated using Sputter Coater (Edwards S150) to increase the surface conductivity, and then SEM micrographs and EDX spectra were recorded.

In view of the potential use of discarded biomass as an energy resource, the calorific value was estimated with the help of a standard adiabatic bomb calorimeter equipped with a digital firing unit (Toshniwal, Bombay). Further, its proximate analysis was performed as per Indian Standards (IS 1350) and moisture content, volatile matter, ash, and fixed carbon were calculated. Volatile matter, ash, and fixed carbon were calculated on an "air-dried" basis.

Moreover, the thermal analysis of the dried wasted biomass was also studied by carrying out its thermal analysis using a thermal analyzer (Perkin-Elmer Pyris Diamond), and thermogravimetric analysis (TGA), differential thermogravimetric (DTG), and derivative thermal (DTA) analyses were performed. The thermal analyzer was operated in the air atmosphere with a 0.2 L/min air flow rate and a temperature range from ambient to 1000°C, maintaining a ramp rate of 10°C/min.

3.3 RESULTS AND DISCUSSION

3.3.1 Effect of MLSS Concentration

The MLSS concentration in the SBR represents the concentration of biomass or microorganisms present in the reactor. The effect of MLSS concentration on the biodegradation kinetics of the SBR was assessed for different MLSS concentrations of 2130, 3250, 4200, and 5030 mg/L at a catechol

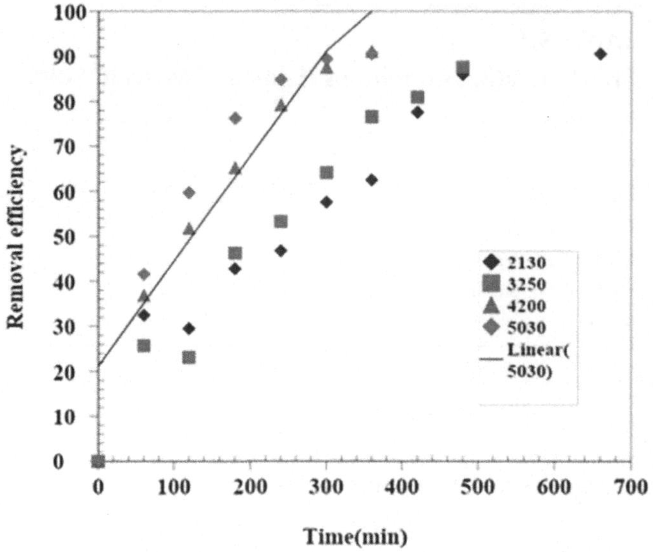

FIGURE 3.3 Variation of catechol-removal efficiency with time for different MLSS concentrations (C_0 = 100 mg/L).

concentration of 100 mg/L. The catechol removal efficiencies with time for varying MLSS concentrations during the treatment in the SBR are shown in Figure 3.3.

It is evident that the treatment efficiency of the catechol-containing wastewater in SBR is increasing with an increase in MLSS concentration. The increased MLSS increases the biomass in the reactor, and hence, removal efficiency is increased. It is also apparent that, as the MLSS concentration increases, the removal rate increases in the SBR. The removal time taken for 2,130 mg/L of MLSS is considerably larger than that taken for the MLSS 5,030 mg/L. It simply shows that, as we increase the MLSS concentration, the removal time decreases for constant initial loading.

3.3.2 Effect of Initial Catechol Loading

The effect of catechol loadings of 25, 50, 100, and 250 mg/L in the SBR on the performance of SBR was studied. The MLSS concentration of 4,200 mg/L was maintained in the SBR. The removal efficiencies during the treatment in the SBR with varying initial concentrations are shown in Figure 3.4.

The results show that, as the catechol loadings in feed to SBR are increased, percentage removal for the catechol is decreased. A higher time is required to remove a particular amount of catechol. It can also be concluded that the catechol-removal rate is also decreasing with increased loadings of catechol. The removal time is more for a high catechol concentration of 250 mg/L than that in the case of a low catechol concentration of 25 mg/L (Aghapour et al., 2013; Subramanyam and Mishra, 2007). The decrease in removal efficiency with an increase in the initial catechol loading was attributed to the inhibition of catechol in the microbial metabolism process. The higher initial concentration led to a greater inhibitory effect, thereby leading to lower removal efficiency. A similar trend of removal efficiency with initial pollutant loading has been amply reported elsewhere (Aghapour et al., 2013; Sarfaraz et al., 2004).

3.3.3 Comparison of Kinetic Model Results with Experimental Kinetic Data

The kinetic experiments were carried out at different MLSS concentrations and for various initial catechol loading in SBR. Figure 3.5 shows the comparison of calculated and experimental results of

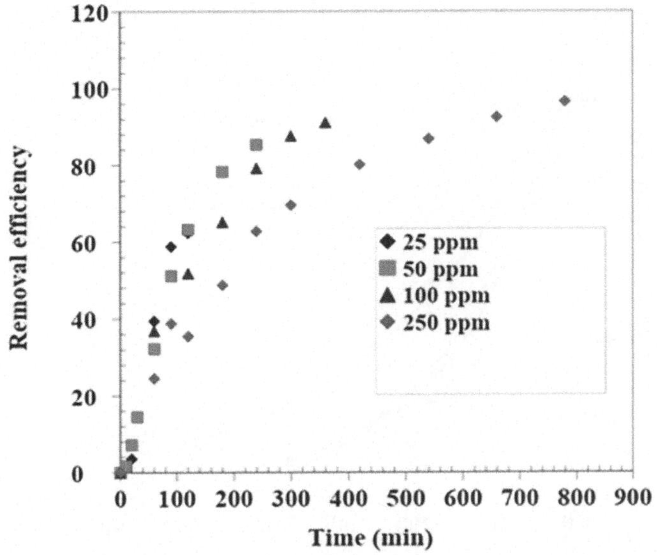

FIGURE 3.4 Variation of catechol-removal efficiency with time for different initial catechol loading (MLSS = 4,200 mg/L).

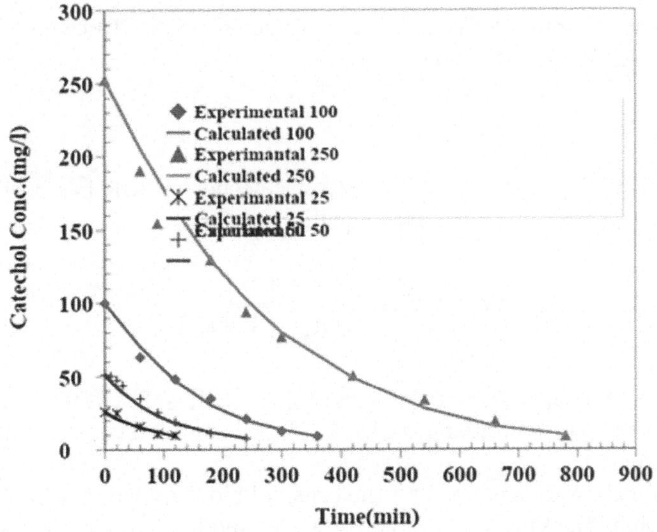

FIGURE 3.5 Comparison of calculated and experimental results at initial differential loading (MLSS = 4,200 mg/L).

a decrease in catechol concentration at constant MLSS of 4,200 mg/L for the different initial loading of 25, 50, 100, and 250 mg/L, respectively. The predicted values of the catechol concentration by the Haldane inhibition model (Tomei et al., 2003) are quite comparable to experimental data with an error value of ±5%.

Figure 3.6 shows the comparison of calculated and experimental results for a decrease in substrate concentration at different MLSS, i.e., 2130, 3250, 4200, and 5030 mg/L. For this study, the initial catechol concentration was fixed at 100 mg/L. The results show that the Haldane inhibition best fitted the experimental data for the biodegradation of catechol. Therefore, the model stands

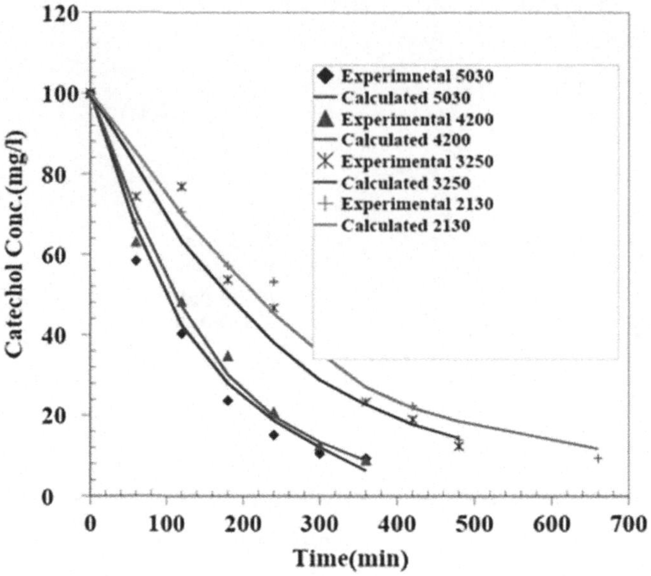

FIGURE 3.6 Comparison of calculated and experimental results of decrease in catechol concentration at different MLSS (C_o = 100 mg/L).

valid for the large range of initial concentrations of catechol (25, 50, 100, and 250) and MLSS concentrations (2130, 3250, 4200, and 5030 mg/L).

3.3.4 Biodegradation Kinetics

In the present study, the biodegradation kinetics of catechol in SBR has been investigated. The kinetic data were represented by the Haldane equation.

$$r_s = Q_{\max} X \frac{S}{S + K_s + \left(S^2 / K_i\right)} \tag{3.1}$$

Here, Q_{max} is the maximum specific removal rate ($M_{Substrate}/M_{VSS}$ t), X is the biomass concentration (mg/L), S is the substrate concentration (mg/L), K_s is the saturation constant (mg/L), and K_i is the inhibition constant (mg/L).

The experimental data were collected for the catechol biodegradation at various MLSS and initial loadings of catechol. The kinetics parameters $Q_{m,}$ $K_{s,}$ and K_i were then calculated for the best-fit curves obtained for all the experiments. Table 3.3 shows the values of kinetic constants obtained for the respective MLSS and initial catechol concentration. Along similar lines, such type of Haldane inhibition model along with the evaluation of kinetic parameters for the degradation of recalcitrant phenolic pollutants employing biological treatment techniques has been extensively discussed and reported in the literature (Khan et al., 2011a, 2011b; Pramparo et al., 2012).

It is evident from the table that the average value of the maximum specific growth rate (Q_m) value varies from 7.99×10^{-4} h^{-1} to 18.04×10^{-4} h^{-1}. The average value can be considered as $12.89 \times 10–^{4}$ h^{-1} for all the experiments. The values obtained from K_s show the constant value of 667 mg/L for all the experiments, except for two cases with initial catechol concentration of 50 mg/L and MLSS concentrations of 2,130 and 3,250 mg/L, respectively. In these two cases, the value of K_s comes out to be 301 mg/L. Similarly, the values of K_i were also evaluated for different experiments and a

TABLE 3.3
Values of Kinetic Parameters at Various MLSS and Initial Catechol Concentrations

Initial conc. (mg/L)	MLSS (mg/L)	Q_m (h^{-1}) ×10–³	K_s (mg/L)	K_i (kg/m^3)
25	4200	1.569	667.5	730
50	2130	1.805	301.1	733
50	3250	0.799	301.1	733
50	4200	1.479	667.5	730
100	2130	1.370	667.5	730
100	3250	1.003	667.5	730
100	4200	1.375	667.5	730
100	5030	1.319	667.5	730
250	4200	0.882	667.5	730

FIGURE 3.7 SEM micrographs of catechol-treated biomass.

constant value of 730 kg/m^3 was obtained except for the same two cases (initial catechol concentration of 50 mg/L and MLSS concentrations of 2,130 and 3,250 mg/L) having different values of K_s. The summarized kinetic parameters calculated with the Haldane inhibition model for the catechol degradation in SBR are $Q_m = 1.29 \times 10^{-3}\ h^{-1}$; $K_s = 667$ mg/L; $K_i = 730\ kg/m^3$.

3.3.5 Sludge Analysis

3.3.5.1 Morphological and Elemental Analysis

The SEM micrographs of the wasted biomass are shown in Figure 3.7. The images present the rough surface of the dried biomass. EDX analysis and spectra showed carbon and oxygen with nutrients as the main elements in the biomass (Figure 3.8). Table 3.4 summarizes the percentage of elements present in the biomass. Biomass contains 50.42% carbon and 19.45% oxygen, respectively. A very minute quantity of other elements like S, Mg, Al, K, Si, Fe, and Ca were also found in the wasted biomass.

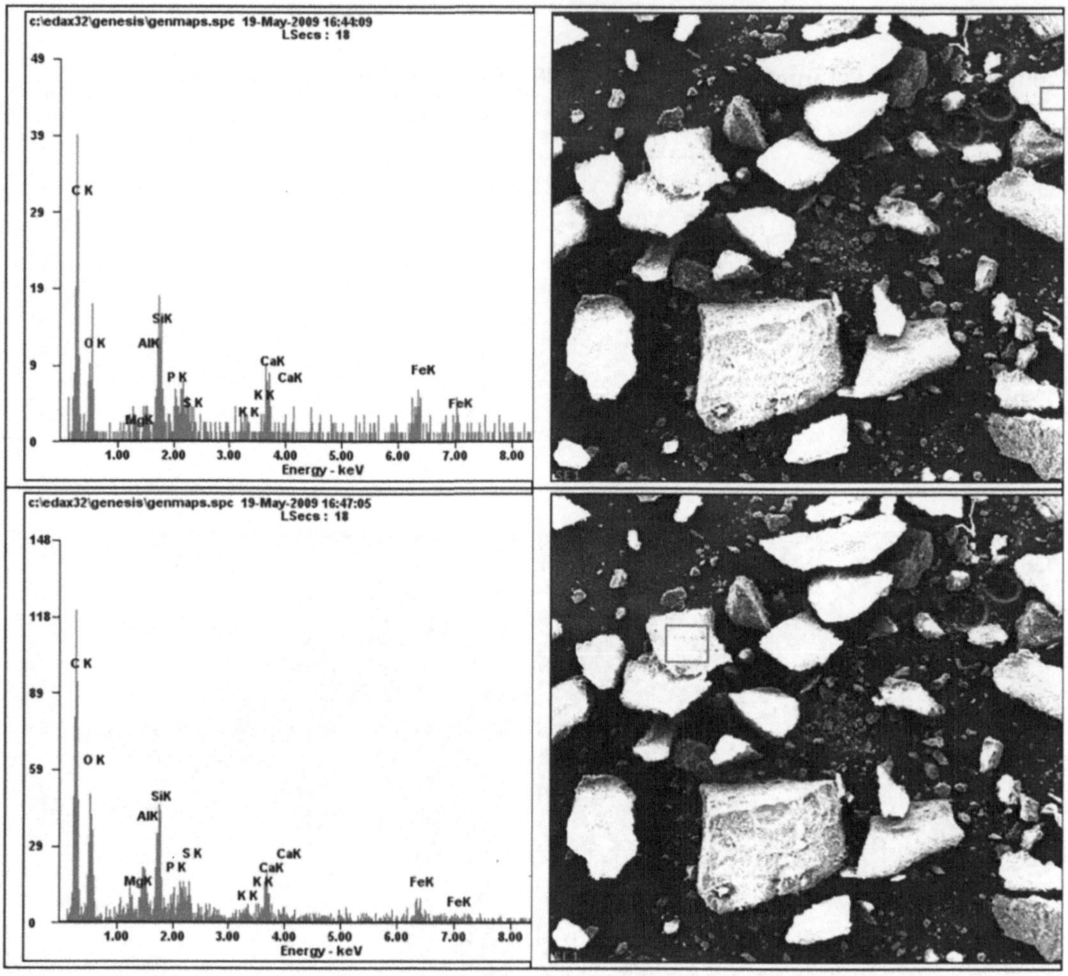

FIGURE 3.8 EDX of catechol-treated activated sludge.

TABLE 3.4
EDX Properties of Catechol-Treated Sludge

Element	C	O	Mg	Al	Si	P	S	K	Ca	Fe
Wt%	50.42	19.45	1.09	2.59	6.18	2.68	1.82	1.21	4.66	9.89

3.3.5.2 Thermal Analysis

The thermal degradation curves, such as TGA, DTA, and DTG of catechol-treated biomass, are presented in Figure 3.9. The %TG curve showed that the wasted biomass is degraded thermally in three zones: ambient-215°C; 215–540°C; and 540–1000°C. Among these three zones, the highest weight loss of 64.1% occurred in the second zone – i.e., 215–540°C. In this zone, carbon and other elements are oxidized resulting in high weight loss. In the first zone (ambient-215°C), the moisture and the light volatile compounds are removed, and 8.7% of weight loss occurred. However, in the third degradation zone (540–1000°C), the sample biomass degraded very slowly, and only 2.6% weight loss was observed. The residue left over at 1000°C is ash, about 24.6% of the initial biomass

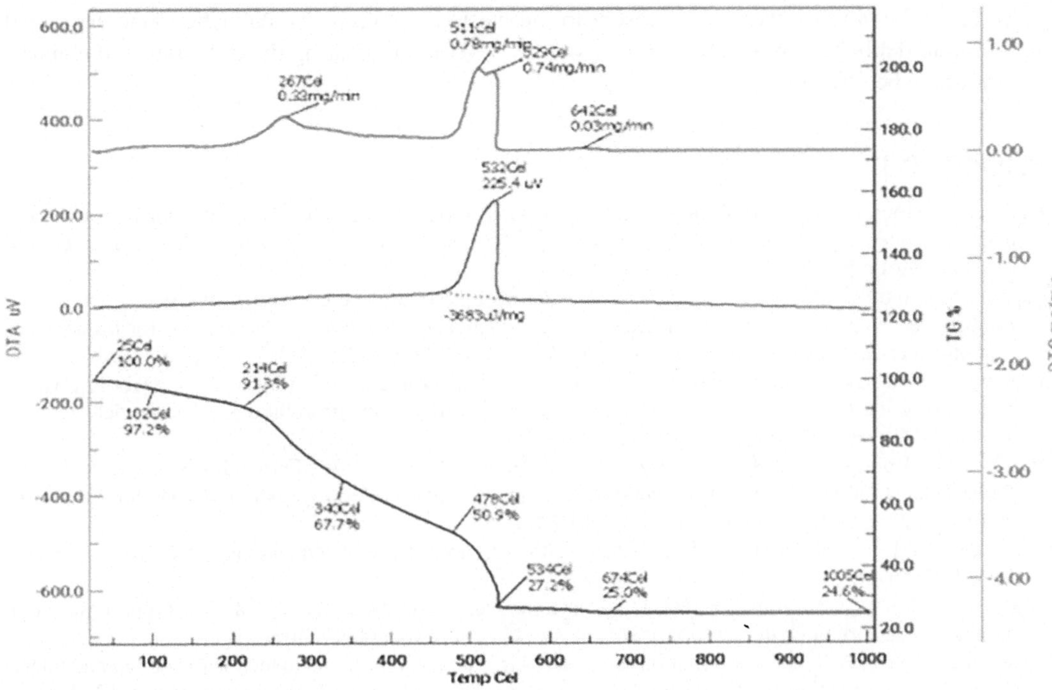

FIGURE 3.9 DTA-DTA-TG plot of catechol-treated activated sludge (heating rate = 10 K/min under air atmosphere).

sample weight. The highest rate of weight loss of 0.78 m/min was observed at 511°C. None of the endothermic peaks was revealed by the DTA, which confirms that there is no crystalline or other phase change occurring in the process (Ng et al., 2002). However, a sharp and strong exothermic peak at 532°C was observed owing to the oxidation of the biomass sample, indicating that the biomass can be utilized as an energy resource. This was confirmed by estimating its calorific value, which was found to be 13.5 MJ/kg. It indicates that the generated biomass can be used as fuel to increase other fuel's calorific value.

3.4 CONCLUSIONS

Catechol is used as raw materials/generated by products in industries like textile, pulp and paper, petroleum refinery, etc., and thus, is present in wastewater generated from these industries. If discharged into the environment without treatment, this wastewater can harm aquatic life and pollute the drinking water resources like surface water/groundwater. Techniques such as activated sludge process, moving bed biofilm reactor, wetlands, trickling filters, etc., have been widely explored in the last few decades as viable alternatives for the treatment of wastewater.

The Haldane substrate inhibition model for the react phase has been validated with the experimental data. The results have indicated that the removal efficiency for catechol is more than 88% for the continuous operation of the SBR system. The results also indicated that the removal efficiency for catechol using SBR increases with increased MLSS concentration. The removal time taken for 2,130 mg/L of MLSS is considerably larger than that of 5,030 mg/L of MLSS. It simply shows that increasing the MLSS concentration decreases the removal time for constant initial loading. Kinetic parameters calculated with the Haldane inhibition model for the catechol degradation in SBR were found to be $Q_m = 0.00129\ h^{-1}$; $K_s = 667.53$ mg/L; $K_i = 730.412\ kg/m^3$. Wasted biomass contained 50.42% carbon and 19.45% oxygen, respectively. A very minute quantity of other elements like S,

Mg, Al, K, Si, Fe, and Ca were also present in the wasted biomass. The calorific value estimated using a standard adiabatic bomb calorimeter was 13.5 MJ/kg, indicating the potential of the generated biomass to be utilized as a fuel.

BIBLIOGRAPHY

Aghapour, A., Moussavi, G., and Yaghmaeian, K. 2013a. Biological degradation of catechol in wastewater using a sequencing continuous-inflow reactor (SCR). *Journal of Environmental Health Science and Engineering* 11:3.

Aghapour, A., Moussavi, G., and Yaghmaeian, K. 2013b. Investigating the performance of a novel cyclic rotating-bed biological reactor compared with a sequencing continuous-inflow reactor for biodegradation of catechol in wastewater. *Bioresource Technology* 138:369–372.

Aghapour, A., Moussavi, G., and Yaghmaeian, K. 2015. Degradation and COD removal of catechol in wastewater using the catalytic ozonation process combined with the cyclic rotating-bed biological reactor. *Journal of Environmental Management* 157:262–266.

Arana, J., Dona Rodrıguez, J. M., Gonzalez Dıaz, O., Herrera Melian, J. A., Fernandez Rodrıguez, C., and Perez Pena, J. 2006. The effect of acetic acid on the photocatalytic degradation of catechol and resorcinol. *Applied Catalysis A: General* 299:274–284.

Arora, M., Barth, E. F., and Umbers, M. B. 1985. Technology evaluation of sequencing batch reactor. *Journal WPCF* 57:867–875.

Bayram, E., Hoda, N., and Ayranci, E. 2009. Adsorption/electrosorption of catechol and resorcinol onto high area activated carbon cloth. *Journal of Hazardous Materials* 168:1459–1466.

Chan, C.-H., and Lim, P.-E. 2007. Evaluation of sequencing batch reactor performance with aerated and unaerated fill periods in treating phenol-containing wastewater. *Bioresource Technology* 98:1333–1338.

Chen, X., Hu, X., Lu, Q., Yang, Y., Linghu, S., and Zhang, S. 2022. Study on the differences in sludge toxicity and microbial community structure caused by catechol, resorcinol and hydroquinone with metagenomic analysis. *Journal of Environmental Management* 302:114027.

Coulson, J. M., and Richardson, J. F. 1979. *Chemical and biochemical reactors and process control.* J. F. Richardson and D. G. Peacock Eds. Pergamon Press Ltd. Volume III.

Dennis, R. W., and Irvine, R. L. 1981. A stoichiometric model of bacterial growth. *Water Research* 15:1363–1373.

EPA. 1999. *Wastewater fact sheet.* SBR.

Fedorak, P. M., Roberts, D. J., and Hrudey, S. E. 1984. The effects of phenol and some alkyl phenolics on batch anaerobic methanogenesis. *Water Research* 18:361–367.

Hays, M. D., Fine, P. M., Geron, C. D., Kleeman, M. J., and Gullett, B. K. 2005. Open burning of agricultural biomass: physical and chemical properties of particle-phase emissions. *Atmosphere Environment* 39:6747–6764.

Huang, J., Huang, K., and Yan, C. 2009. Application of an easily water-compatible hypercrosslinked polymeric adsorbent for efficient removal of catechol and resorcinol in aqueous solution. *Journal of Hazardous Materials* 167:69–74.

Irvine, R. L., and Bush, A. W. 1979. Sequencing batch reactor—an overview. *Journal Water Pollution Control* 51:235.

Irvine, R. L., Miller, G., and Singh, A. 1979. Sequencing batch treatment of wastewater in rural areas. *Journal WPFC* 51:244–254.

IWPCC. 2005. Sequencing batch reactor design and operational considerations. *New England Interstate Water Pollution Control Commission (IWPPC).*

Khan, M. Z., Mondal, P. K., and Sabir, S. 2011a. Bioremediation of 2-chlorophenol containing wastewater by aerobic granules-kinetics and toxicity. *Journal of Hazardous Materials* 190:222–228.

Khan, M. Z., Mondal, P. K., Sabir, S., and Tare, V. 2011b. Degradation pathway, toxicity and kinetics of 2,4,6-trichlorophenol with different co-substrate by aerobic granules in SBR. *Bioresource Technology* 102:7016–7021.

Kulikowska, D., Klimiuk, E., and Drzewicki, A. 2007. BOD_5 and COD removal and sludge production in SBR working with or without anoxic phase. *Bioresource Technology* 98:1426–1432.

Kumaran, P., and Paruchuri, Y. L. 1997. Kinetics of phenol biotransformation. *Water Research* 31:11–12.

Latkar, M., Swaminathan, K., and Chakrabarti, T. 2003. Kinetics of anaerobic biodegradation of resorcinol catechol and hydroquinone in upflow fixed film–fixed bed reactors. *Bioresource Technology* 88:69–74.

Liao, Q., Sun, J., and Gao, L. 2008. The adsorption of resorcinol from water using multi-walled carbon nanotubes. *Colloids and Surfaces A: Physicochemical Engineering Aspects* 312:160–165.

Lobo, C., Bertola, N., and Contreras, E. 2016. Inhibition kinetics during the oxidation of binary mixtures of phenol with catechol, resorcinol and hydroquinone by phenol acclimated activated sludge. *Brazilian Journal of Chemical Engineering* 33:59–71.

MaCabe, W. L., Smith, J., and Harriot, C. P. 2001. *Unit operations of chemical engineering*, 6th ed. New York: McGraw-Hill.

Mahesh, S., Prasad, B., Mall, I. D., and Misha, I. M. 2006. Electrochemical degradation of pulp and paper mill wastewater. Part 2. Characterization and analysis of sludge. *Industrial & Engineering Chemistry Research* 45:5766–5774.

Metcalf and Eddy. 2006. *Wastewater engineering*, 4th ed. Tata McGraw-Hill.

Moussavi, G., Aghapour, A., and Yaghmaeian, K. 2014. The degradation and mineralization of catechol using ozonation catalyzed with MgO/GAC composite in a fluidized bed reactor. *Chemical Engineering Journal* 249:302–310.

Moussavi, G., Jafari, S., and Yaghmaeian, K. 2015. Enhanced biological denitrification in the cyclic rotating bed reactor with catechol as carbon source. *Bioresource Technology* 189:266–272.

Neufeld, R. D., Mack, J. D., and Strakey, J. P. 1980. Anaerobic phenol biokinetics. *Journal Water Pollution Control Federation* 52:2367–2377.

Ng, J. C. Y., Cheung, W. H., and McKay, G. 2002. Equilibrium studies of the sorption of Cu(II) ions chitosan. *Journal of Colloid and Interface Science* 255:64–74.

Orcupold, K., Masirin, A., and Tenno, T. 2001. Estimation of biodegradation parameters of phenolic compounds on activated sludge by respirometry. *Chemosphere* 44:1273–1280.

Othmer, K. 1995. *Encyclopedia of chemical technology,* 4th ed. New York: Wiley-Interscience Publishers, 996.

Pardeshi, S. K., and Patil, A. B. 2009. Solar photocatalytic degradation of resorcinol A model endocrine disrupter in water using zinc oxide. *Journal of Hazardous Materials* 163:403–409.

Phutdhawong, W., Chowwanapoonpohn, S., and Buddhasukh, D. 2000. Electro coagulation and subsequent recovery of phenolic compounds. *Analytical Science* 16:1083–1084.

Pramparo, L., Suarez-Ojeda, M. E., Perez, J., and Carrere, J. 2012. Kinetics of aerobic biodegradation of dihydroxybenzenes by a p-nitrophenol-degrading activated sludge. *Bioresource Technology* 110:57–62.

Puri, B. R., and Walker, P. L. 1966. *Chemistry and physics of carbon*. New York: Marcel Dekker, 51–120.

Qin, L., and Liu, Y. 2004. Effect of settling time on aerobic granulation in sequencing batch reactor. *Biochemical Engineering Journal* 21:47–52.

Rashid, R., Shafiq, I., Akhter, P., Iqbal, M. J., and Hussain, M. 2021. A state-of-the-art review on wastewater treatment techniques: the effectiveness of adsorption method. *Environmental Science and Pollution Research* 28:9050–9066.

Rene, E. R., Kim, S. J., and Park, H. 2007. Suck effect of COD/N ratio and salinity on the performance of sequencing batch reactors. *Bioresource Technology* 99:839–846.

Richardson, J. F., and Zaki, W. N. 1954. Sedimentation and fluidization. Part I. *Transactions of the Institution of Chemical Engineers* 32:35–53.

Rozich, A. F., Gaudy, A. F., and D'Adamo, P. C. 1985. Selection of growth model for activated sludges treating phenol. *Water Research* 19:481–490.

Sarfaraz, S., Thomas, S., Tewari, U. K., and Iyengar, L. 2004. Anoxic treatment of phenolic wastewater in sequencing batch reactor. *Water Research* 38:965–971.

Sax, I. N. 1968. *Dangerous properties of industrial material*, 2nd ed. NY: Van Nostrand Reinhold Co.

Silva, M. R., Coelho, M. A. Z., and Araujo, O. Q. F. 2002. Minimization of phenol and ammonical nitrogen in refinery wastewater employing biological treatment. *Engenharia Térmica, Edição Especial* 33.

Singh, A., Srivastava, A., Saidulu, D., and Gupta, A. K. 2022. Advancements of sequencing batch reactor for industrial wastewater treatment: major focus on modifications, critical operational parameters, and future perspectives. *Journal of Environmental Management* 317:115305.

Stanchev, V., Stoilova, I., and Krastanov, A. 2008. Biodegradation dynamics of high catechol concentrations by Aspergillus awamori. *Journal of Hazardous Materials* 154:396–402.

Subramanyam, R., and Mishra, I. M. 2007. Biodegradation of catechol (2-hydroxy phenol) bearing wastewater in an UASB reactor. *Chemosphere* 69:816–824.

Subramanyam, R., and Mishra, I. M. 2008. Co-degradation of resorcinol and catechol in an UASB reactor. *Bioresource Technology* 99:4147–4157.

Suidan, M. T., Cross, N. I., Pfeffer, J. T., and Wang, Y. 1988. Anaerobic biodegradation of phenol. *Water Research* 114:1359.

Sun, Y., Chen, J., Aimin, L., Fuqiang, L., and Zhang, Q. 2005. Adsorption of resorcinol and catechol from aqueous solution by aminated hypercrosslinked polymers. *Reactive & Functional Polymers* 64:63–73.

Swamy, M. M. 1998. *Studies on the treatment of phenolic wastewater using adsorption and immobilized whole cells*. Ph.D. Thesis, University of Roorkee, India.

Swamy, M. M., Mall, I. D., and Prasad, B. 1997. Resorcinol removal aqueous solution by bagasse fly ash and activated carbon. *Journal of Environmental Engineering Institution of Engineers(I)* 2:49.

Tomei, C. M., Annesini, C. M., Luberti, R., Cento, G., and Senia, A. 2003. Kinetics of 4-nitrophenol biodegradation in a sequencing batch reactor. *Water Research* 37:3803–3814.

Tsang, Y. F., Hua, F. L., Chua, H., Sin, S. N., and Wang, Y. J. 2007. Optimization of biological treatment of paper mill effluent in a sequencing batch reactor. *Biochemical Engineering Journal* 34:193–199.

Uygur, A., and Kargi, F. 2003. Biological nutrient removal in sequencing batch reactor with different number of steps. *Clean Technologies and Environmental Policy* 6:61–65.

Van Duursen, M. B. M., Sanderson, J. T., de Jong, P. C., Kraaij, M., and van den Berg, M. 2004. Phytochemicals inhibit catechol-o-methyltransferase activity in cytosolic fractions from healthy human mammary tissues: implications for catechol estrogeninduced DNA damage. *Toxicological Science* 81:316–324.

Vinitnantharat, S., Cha, W. S., Ishibashi, Y., and Ha, S. R. 2001. Stability of biological activated carbon-sequencing batch reactor (BAC-SBR) to phenol shock loading. *Thammasat International Journal of Science and Technology* 6:27–32.

Wang, Z., Liu, L., Yao, J., and Cai, W. 2006. Effects of extracellular polymeric substances on aerobic granulation in sequencing batch reactors. *Chemosphere* 63:1728–1735.

Zhou, S., Zhang, H., and Yong, S. 2006. Combined treatment of landfill leachate with fecal supernatant in SBR. *Journal of Zhejiang University Science B* 7:397–403.

4 Biodegradation and Decolorization of Textile Wastewater by Sequential Aerobic/Microaerophilic, Aerobic and Microaerophilic Processes

C. Ponnarasi, R. Gandhimathi, S. T. Ramesh, Vishesh Mishra and Ayush Panwar

4.1 INTRODUCTION

Application of dyes to various industries like textile, leather, paper, etc. is well known. But the textile industry accounts for two-thirds of the total dyestuff market produced (Fang et al., 2004; Asad et al., 2007). Due to the difference in fibres, dyes, processes and finishing products used, the wastewater generated from the textile industries are of diverse and great chemical complexities (Philippe et al., 1998). The wastewater containing dyes causes serious environmental and health problems. Several methods are used to treat textile effluents. These includes physical methods such as adsorption (Nidheesh et al. [2011, 2012a, 2013b]; Ramesh et al., 2014), membrane filtration, ion exchange (Constantin et al., 2013) and chemical processes like coagulation (Beltrán-Heredia et al., 2011), electrocoagulation (Anantha Singh and Ramesh, 2014), peroxicoagulation (Nidheesh and Gandhimathi, 2014), photocatalysis (Lodha et al., 2011), catalytic wet air oxidation (Hua et al., 2013), electro Fenton process (Nidheesh et al. [2014a, 2014b]), ozonation (Gao et al., 2012), ultrasonic degradation and electrochemical oxidation. These methods are effective, but they are expensive and involve the formation of a concentrated sludge in most of the processes, which creates a secondary disposal problem (Maier et al., 2004).

In recent years, biological processes, including aerobic and anaerobic bacteria and fungi, for dye degradation and wastewater reutilization have been developed. Algae, yeast, fungi and bacteria are a few examples of the different microbial species that have been used to cleanse dye wastewater. Historically, the favoured microbial agents for dye wastewater remediation have been algae, yeast and fungi. However, bacterial organisms have recently drawn more notice because of the longer development cycles and poor ability of algae, yeast and fungi to remove dyes from materials. Microbial decolourization and degradation have appeared as an environmentally friendly and cost-competitive alternative to the physical and chemical decomposition processes (Verma and Madamwar, 2003; Pandey et al., 2007; Dhanve et al., 2008). Studies have been conducted to assess the potential of dye degradation by various bacterial strains (*Pseudomonas fluorescens, Pseudomonas putida, Bacillus spp.* ETL (2012) in batch system (Anwer and Merkhan 2013; Kannan et al., 2013; Shah et al., 2013). To date, only a few bacterial laccases have been studied. Compared to fungal laccases, bacterial laccases are much more thermotolerant and pH stable. Laccases with these properties have been isolated from certain species of bacteria (*Pseudomonas,*

DOI: 10.1201/9781003391487-4

Bacillus) and are highly useful in textile effluent dye biodegradation (Lu et al., 2012; Wang et al., 2012; Zhang et al., 2012).

Bacterial strains can decolorize azo dyes under aerobic and anaerobic conditions. Based on the available literature, it can be concluded that the microbial decolorization of azo dyes is more effective under anaerobic conditions. On the other hand, these conditions lead to aromatic amine formation, and these are mutagenic and toxic to humans (Chung and Stevens, 1993; Pinheiro et al., 2004), requiring a subsequent oxidative (aerobic) stage for their degradation. In this context, the combined anaerobic/aerobic biological treatments of textile dye effluents using microbial consortia are common in the literature (Van Der Zee and Villaverde, 2005).

However, few commercial or industrial applications have been developed using a single, adaptable microorganism in a sequential anaerobic/aerobic treatment (Isik and Sponza, 2004; Supaka et al., 2004). Even though literatures suggest the use of combined anaerobic/aerobic biological treatment, till date, only limited sequential treatment studies using single adaptable microorganisms are available (Sandhya et al., 2004; Elisangela et al., 2009; Jonstrup et al., 2011). Mixed culture often does not confirm the symbiotic relation between microbial species. So, this became the motivation behind developing a novel biological decolorization process for more effective degradation of azo dyes using a single microorganism that is efficient under both anaerobic/microaerophilic and aerobic conditions.

4.2 DEGRADATION OR DECOLORIZATION STUDY OF THE DYE

Reactive Black 5 (RB5) dye was used for this study. The structure of dye is shown in Figure 4.1.

The characteristics of Reactive Black-5 are given as follows in Table 4.1.

The decolorization/degradation of Reactive Black-5 is carried out in a sequential aerobic/microaerophilic process, aerobic process and microaerophilic process using *Pseudomonas sp.* and *Bacillus sp.* The growth kinetics of the microbes in the sequential aerobic/microaerophilic process and microaerophilic process degradation were analyzed. To monitor the aromatic amine concentration, chemical oxygen demand (COD), pH, total organic carbon (TOC) reduction during biodegradation process and to analyze the degradation, products using UV-visible spectrophotometer and FTIR were carried out.

Pseudomonas sp. and *Bacillus sp.* can be isolated from textile wastewater contaminated site using serial dilution. The slant culture of *Pseudomonas sp.* and *Bacillus sp.* obtained were subcultured for the further use according to standard methods APHA (1998). A subculture is then done in the nutrient broth with the composition as: peptone (10 g/L), NaCl (5 g/L), yeast extract (2 g/L), beef extract (3 g/L), distilled water (1 L). Before inoculation with the microbes, the nutrient media was sterilized in an autoclave at 121°C for 15 minutes. After sterilization, 100 mL of this nutrient

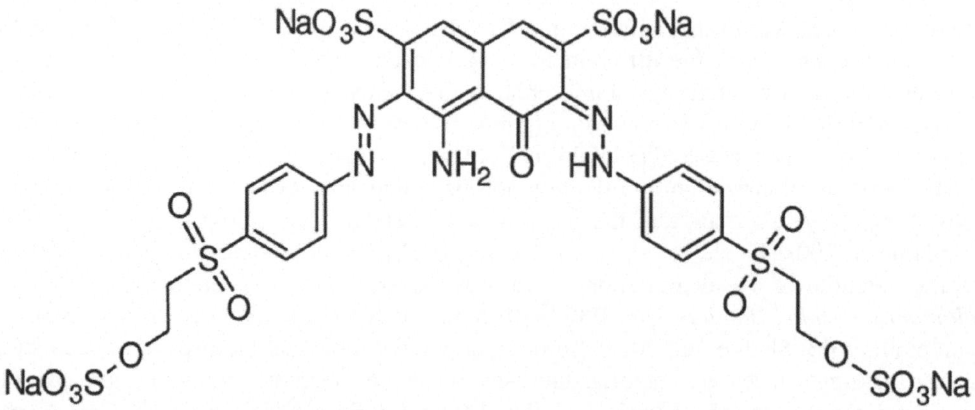

FIGURE 4.1 Structure of Reactive Black-5.

TABLE 4.1
Characteristics of Reactive Black-5

Characteristics	Value/Formula/Group
Colour Index	20505
Molecular wt.	991.8
Molecular formula	$C_{26}H_{21}N_5Na_4O_{19}S_6$
λ_{max}	597 nm
Natural pH	5.8
RB Class	Diazo
Reactive group	Sulfatoethylsufone

broth was taken in two different 250-mL Erlenmeyer flasks (sterilized prior to use). One of the flasks was inoculated with *Pseudomonas sp.* by transferring a loop full of culture from the slant culture, and the other was inoculated with *Bacillus sp.* After inoculation, the broth was incubated at 30°C for 24 h in an orbital shaker. The inoculated broth turns turbid, which indicates the growth of organisms. These microbes can be used for further acclimatization with dyes.

Subculture organisms were used for acclimatization. To avoid shock loading of dye, acclimatization of microbes was done prior to the study. Acclimatization was done by increasing the dye concentration for every 24 hrs. The concentration of Reactive Black-5 was increased from 10 mg/L to 100 mg/L in 5 days. The organism grown in nutrient broth containing 100 mg/L dye concentration was used for biodegradation studies.

4.2.1 Preparation of Synthetic Textile Effluent

Synthetic textile effluent prepared for decolorization/degradation studies has the following composition, like basal mineral media, along with the presence of various nutrients to supplement the growth of microorganisms with carbon source and micronutrients. The mineral salts medium (MM) at pH 7 used in all the batch experiments contains the following minerals: Dipotassium hydrogen phosphate (K_2HPO_4) – 1.6 g/L; Potassium dihydrogen phosphate (KH_2PO_4) – 0.2 g/L; Diammonium sulphate ($(NH_4)_2SO_4$) – 1.0 g/L; Magnesium sulphate heptahydrate ($MgSO_4.7H_2O$) – 0.2 g/L; Ferrous sulphate heptahydrate ($FeSO_4.7H_2O$) – 0.01 g/L; Sodium chloride (NaCl) – 0.1 g/L; Calcium chloride ($CaCl_2.2H_2O$) – 0.02 g/L. Thus, the mineral media was prepared and its was supplemented with 100 mg/L of dye, 10g/L of peptone and 2 g/L of yeast extract to prepare nutrient-rich mineral media (Elisangela et al., 2009). The nutrient-rich mineral media (MMR) was autoclaved at 121°C for 15 min to prevent contamination by other microbes, which are not the interest of this study.

4.2.2 Decolorization/Degradation Batch Study

The prepared synthetic textile effluent was inoculated with the *Pseudomonas sp.* and *Bacillus sp.*, which were acclimatized prior to inoculation with 100 mg/L of RB5 so that the microbes would not die due to a sudden shock load of the dye in the nutrient media. Three different sets of 250 mL Erlenmeyer flask were taken, and each flask was provided with 150 mL of synthetic textile effluent. One from each set was inoculated with *Pseudomonas sp.*, and the other was inoculated with *Bacillus sp.* After inoculation with microbes, the flasks were incubated in three different conditions, like aerobic (AI), microaerophilic (MI), sequential aerobic/microaerophilic (SAMI) conditions at 30^{0}C in an orbital shaker, BOD incubator and orbital shaker respectively. In the SAMI conditions, the microbes were incubated in aerobic state for 12 h, and next, 6 h microaerophilic, and the sequence was repeated until the maximum decolorization was observed. 10 mL of sample was withdrawn from these flasks every 12 h for aerobic and 4 h for MI, SAMI conditions and analyzed for dye decolorization using UV-vis spectrophotometer, dye degradation using Fourier transform infrared spectroscopy (FT-IR), total organic carbon (TOC) reduction and for aromatic amine production by Fourier Transform Infrared spectroscopy (FT-IR). Since the dye acts as the carbon substrate for the growth of microbes, the substrate utilization rate and growth kinetics were calculated by measuring mixed liquor suspended solids (MLSS).

To evaluate the mechanism of colour removal, the sample was withdrawn after maximum colour removal was observed. It was centrifuged for 20 minutes at 8,000 rpm. The supernatant was discarded and the pellet was suspended in distilled water, which was further ultrasonicated in the sonicator for 1 h. After sonication, the sample was centrifuged again, and the supernatant was analyzed for the presence of dye using a UV-Vis spectrophotometer. Sonication helps in cleaving the cells of biomass so that if dye got bio-adsorbed in the biomass, it will reappear in the solution. If the dye is absent in the solution after ultrasonication, then the mechanism of colour removal is due to biodegradation of dye by the microorganisms.

4.3 MICROBIAL ACTION ON THE DYE

Decolorization studies were conducted initially on synthetic textile effluent containing 100 ppm of RB5, and the results are shown in Figure 4.2. As shown in Figure 4.2, the percentage decolourization was higher in SAMI compared to other two processes. But compared with the AI process, decolourization was higher in MI process. Though *Bacillus sp.* has shown only 39% efficiency after 120 h decolourization in aerobic conditions, it has efficiency of 96.8% within 72 h under SAMI, which is highest among all incubation conditions. In MI condition and AI condition, *Pseudomonas sp.* was found to be more effective in decolourizing the azo dye. But *Pseudomonas sp.* under aerobic incubation was found to be less efficient than the *Pseudomonas sp.* under microaerophilic incubation because it took about 120 h for 81% decolourization in aerobic incubation, whereas it took only 72 h for 95.2% decolourization in MI. Thus, among all incubation conditions, *Bacillus sp.* under SAMI was more effective in azo dye decolourization.

4.3.1 Effect of Initial Dye Concentration

To know the effect of initial dye concentration on decolourization/degradation studies were conducted for four different concentrations ranging from 50 to 250 ppm under SAMI and MI because AI was found to be less efficient compared to the other two processes in the study conducted for 100 ppm dye concentration. The decolourization results are shown in the Figure 4.3. As shown in Figure 4.3, the percentage decolourization was higher in SAMI compared to MI process.

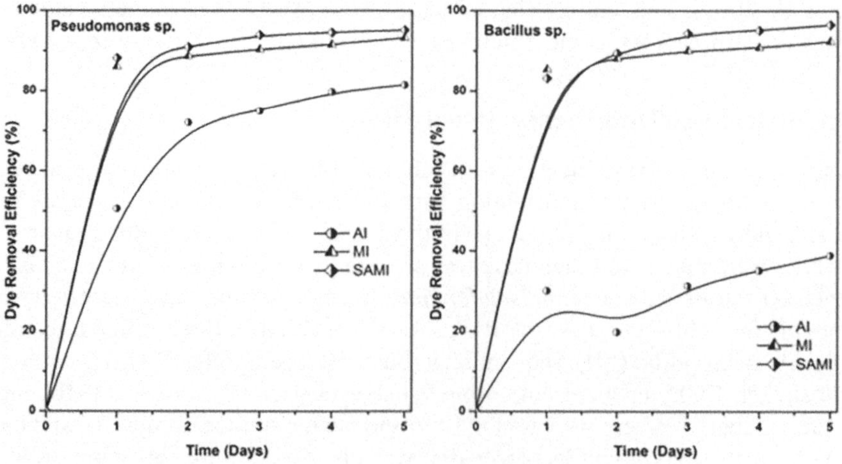

FIGURE 4.2 Decolorization kinetics of bacteria under various process conditions.

Under SAMI, *Bacillus sp.* was found to be more effective compared to *Pseudomonas sp.* Almost the same trend was followed for all the concentrations under various incubation conditions, but with decreasing percentage of colour removal in other incubation conditions compared to *Bacillus sp.* under SAMI. The following results were interpreted from SAMI inoculated with *Bacillus sp.* As the concentration of dye increases from 50 to 250 ppm, the rate of colour removal decreases under all incubation conditions, irrespective of the microbes it is inoculated with. For 50 ppm concentration above 95.6%, colour removal was observed in 48 h, whereas it took 72 h for 96.8% colour removal using 100 ppm dye concentration and the same 72 h for 95.1% colour removal using 150 ppm dye concentration, but it took 120 h for 96.1% colour removal while using 250 ppm dye concentration. Thus, SAMI inoculated with *Bacillus sp.* was found to be more effective in dye decolourization.

The dye degradation products produced during biodegradation of Reactive Black-5 under MI and SAMI conditions were studied by following the change in the UV–vis spectra (from 250 to 1000 nm) using a UV–vis spectrophotometer. As shown in Figure 4.4, untreated RB5

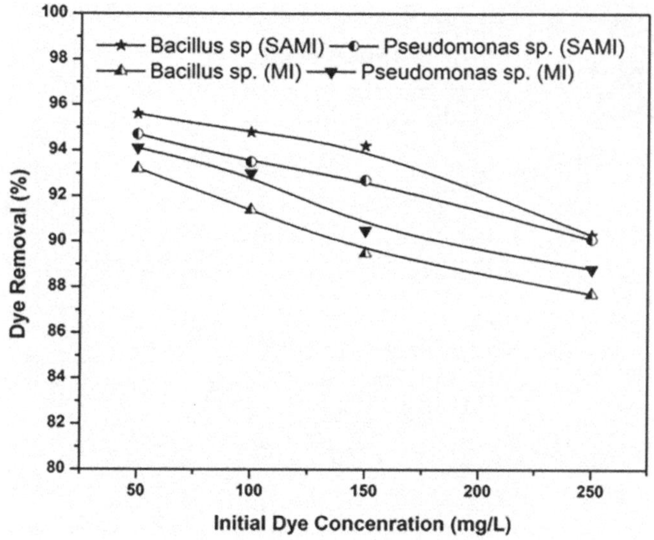

FIGURE 4.3 Effect of initial dye concentration on dye decolourization.

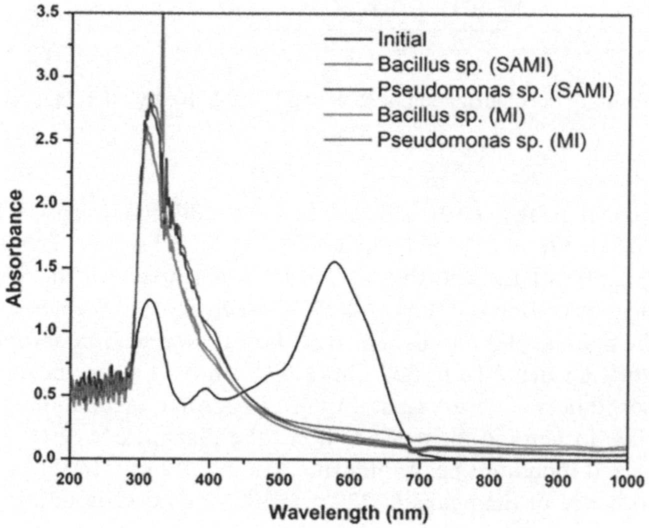

FIGURE 4.4 UV-vis spectra of Reactive Black -5 before and after treatment.

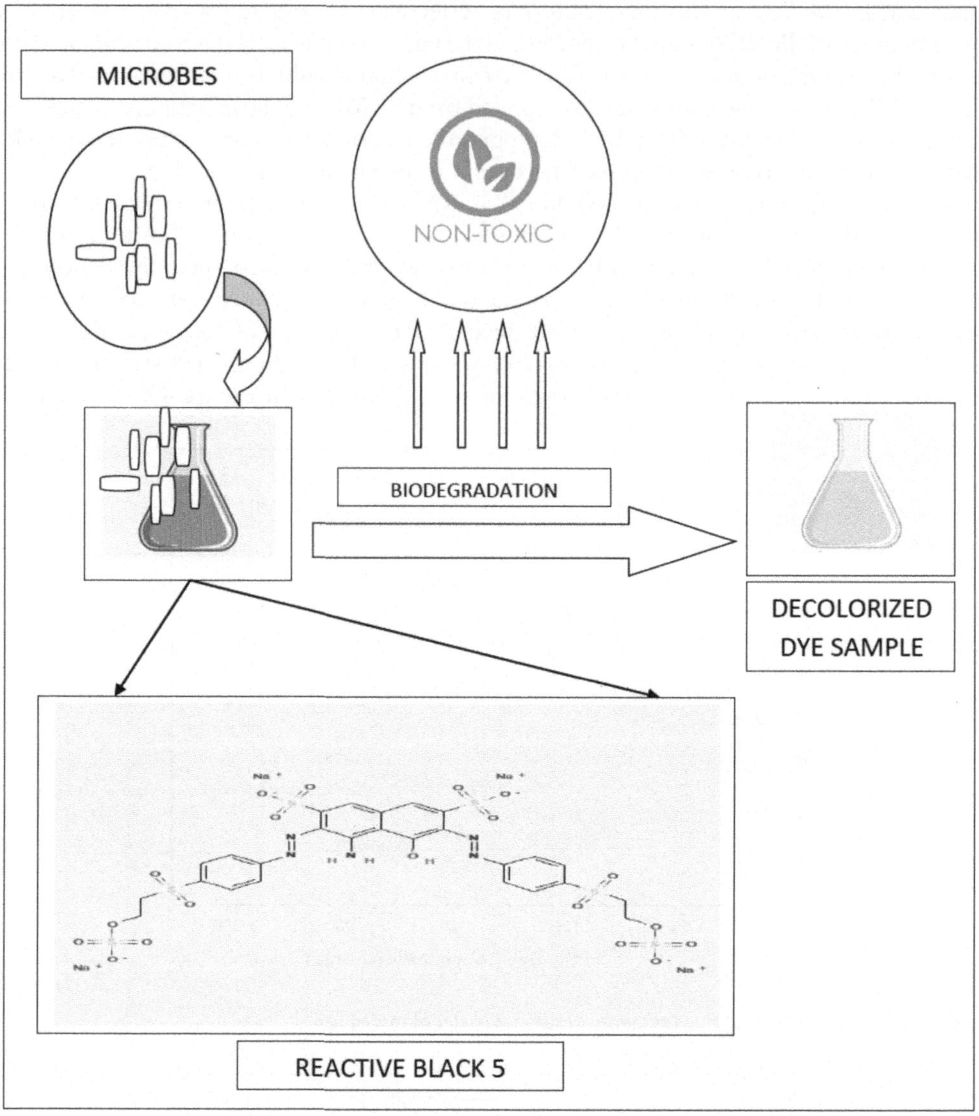

FIGURE 4.5 Mechanism for decolourization of Reactive Black-5 dye by microbial action.

presented two absorbance peaks at 301 and 597 nm. Two additional peaks with low absorbance were observed at 320 and 390 nm. Wide band absorption was observed near 220 nm. For treated dyes, after biodegradation of the azo dye, the absorbance peaks in the visible region disappeared, indicating decolorization of RB5. In the UV spectra, the absorbance of the peaks at 320 nm increased and the peak at 390 nm disappeared. For dyes treated in aerobic incubation using *Bacillus sp.*, peak shift occurred from 597 nm to 573 nm, and a new peak was formed at 307 nm. Wide band absorption was observed at 250 nm in aerobic incubation. The effluent treated with *Pseudomonas sp.* in aerobic incubation shows the disappearance of the peak at 597 nm in the visible region and increased peak intensity around 300 nm. In the SAMI and MI stage, the decrease in absorbance of the peaks at 220 and 390 nm, corresponding to the benzene and

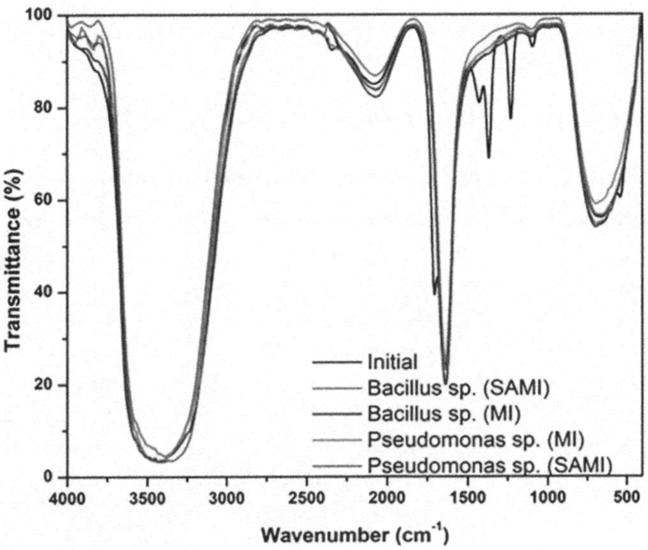

FIGURE 4.6 FT-IR spectra for untreated and treated dye.

naphthalene rings respectively (Yang, 1987), and the formation of a new peak at 260 nm suggested that the reductive destruction of the azo conjugated structure disclosed the narrow multi-peaks of aromatic rings in the spectra.

4.3.2 Mechanism

Decolourization of the azo dye can be due to two different mechanisms. One is due to the adsorption of dye in the biomass, and the other is degradation of dye by bacteria. To find the exact mechanism of dye decolourization, the samples were centrifuged to collect the biomass. The suspension of this extracted biomass in distilled water should be subjected to ultrasonication for 1 h. After ultrasonication, the sample should be centrifuged again, and the supernatant was analyzed in UV-vis spectrophotometer for the presence of azo dye. The obtained results shows the absence of a peak at 597 nm for all the samples under various incubation conditions. If decolourization has happened due to adsorption of dye on the biomass, the peak corresponding to the azo dye would have reappeared because ultrasonication helps in the cleavage of the cell; thus, adsorbed dye will come out, resulting in the peak reappearance. Thus the decolourization is due to the degradation of dye by the bacteria (El Bouraie and El Din, 2016; Zubbair et al., 2018; Tan et al., 2021). This mechanism is represented in Figure 4.5.

As shown in the Figure 4.6, the FT-IR spectra obtained for the untreated dye sample showed several peaks in the region where N–H and O–H stretching is normally observed (3,300–3,500 cm^{-1}). But since no multiple sharp peaks are present in this region, which signifies the presence of –NH-, hence, we can conclude that the peak in 3,300–3,500 cm^{-1} is for –OH- only. After microaerophilic and sequential aerobic/microaerophilic treatments, a significant reduction in absorption was observed in this region. Other bands located in the range from (1,400–1,090 cm^{-1}) disappeared in the MI and SAMI after reductive treatment. In untreated dye, the compounds responsible for bands in the region (1,400–1,090 cm^{-1}) are aromatic carbon, alkyl compounds, aromatic ethers, aliphatic amines, which all degraded after treatment under different incubation conditions

(Elisangela et al., 2009). The characteristic FT-IR spectrum peaks for RB-5 are given as follows in Table 4.2:

TABLE 4.2
FT-IR Peaks with Corresponding Functional Groups

Peaks Corresponding	Functional Groups	References
3300–3500 cm^{-1}	-OH stretch	Lee et al., 2009
2968–2888 cm^{-1}	$-CH_2$	Lee et al., 2009
2161 cm^{-1}	C-S and S-O stretch	Bouraie et al., 2016
1626–1748 cm^{-1}	C=C stretch	Lee et al., 2009
1387–1596 cm^{-1}	N-H	Bouraie et al., 2016
1078–1256 cm^{-1}	O-C	Bouraie et al., 2016
1011–1078 cm^{-1}	Alkenes	Lee et al., 2009
386–644 cm^{-1}	Secondary Amines	Bouraie et al., 2016

Moreover, in the microaerophilic stage and sequential aerobic/microaerophilic stage, two new bands were observed in the carbonyl region at around 697cm^{-1} and 1,100 cm^{-1} attributed to aromatic compounds and ketones respectively. The band located in 1,704 cm^{-1}, corresponding to the carboxyl group, disappeared after treatment. Band shift occurred from 2,080 cm^{-1} to 2,069 cm^{-1} due to transformation of isocyanides into alkyenes group after biodegradation. The increase in the transmittance of 3,974 cm^{-1} region is due to the formation of secondary amines.

The reduction in chemical oxygen demand was monitored for all the concentrations, and obtained results show that the maximum COD removal was shown by *Pseudomonas* sp. compared to *Bacillus* sp. But the *Pseudomonas sp.* grown under MI condition shows the maximum COD removal compared to the same species inoculated in SAMI. The same trend of COD removal was followed for all the concentrations. For 50 ppm, effluent COD removal efficiency varies from 40% to 64% under various incubation conditions, but this percentage got reduced as the concentration of the synthetic textile effluent increased. It was only 38.6% to 49.5% for 100 ppm textile effluent, 36.2% to 45% for 150 ppm synthetic textile effluent and 35.5% to 39.7% for 250 ppm synthetic effluent. Not only was the efficiency of COD removal reduced, but the time taken for average COD removal of 35% in all incubation conditions increased from 48 h for 50 ppm to 120 h for 250 ppm. Among all the incubation conditions, *Bacillus sp.* incubated under microaerophilic conditions has a lower COD removal rate. Thus, the efficiency of the overall process is inversely proportional to the concentration of the azo dye present in the effluent.

The changes in pH during the decolourization/degradation process were monitored throughout the process for various incubation conditions are shown in the Table 4.3.

TABLE 4.3
Changes in Solution pH under Various Experimental Conditions

	Dye concentration in ppm			
Incubation Conditions	**50**	**100**	**150**	**200**
Sequential Aerobic/Microaerophilic *Bacillus sp.*	8.24	8.43	8.53	8.82
Sequential Aerobic/Microaerophilic *Pseudomonas sp.*	8.27	8.51	8.69	8.74
Microaerophilic *Bacillus sp.*	8.12	8.36	8.47	8.61
Microaerophilic *Pseudomonas sp.*	8.40	8.61	8.44	8.67

TABLE 4.4
TOC Removal under Various Incubation Conditions

S. no.	Incubation Conditions	% Removal
1.	Sequential aerobic/microaerophilic *Bacillus sp.*	91.10
2.	Sequential aerobic/microaerophilic *Pseudomonas sp.*	83.4
3.	Microaerophilic *Bacillus sp.*	66.04
4.	Microaerophilic *Pseudomonas sp.*	79.7
5.	Microaerophilic/Aerobic *Bacillus sp.*	89.01
6.	Microaerophilic/Aerobic *Pseudomonas sp.*	81.96

Initial pH of synthetic textile effluent was 6.93; as the process started, pH of the sample withdrawn with respect to time increased constantly. This steady increase in pH is because of an increase in alkalinity, which is due to the prevalence of microaerophilic conditions.

Table 4.4, shows the percentage reduction in total organic carbon (TOC) under various incubation conditions. Total organic carbon can be measured after obtaining complete decolourization. In MI, the samples were aerated after 7 days to oxidize the aromatic amines produced, so total organic carbon was measured before and after the aeration process. TOC reduction was found to be higher in SAMI compared to the MI. Though before aeration, TOC reduction was only 66% in microaerophilic incubation using *Bacillus sp.*, it increased to about 89% after aeration of the sample. But TOC reduction was not promoted by using *Pseudomonas sp.* under aerobic conditions as in the *Bacillus sp.* incubated under aerobic conditions, since *Pseudomonas sp.* is a facultative anaerobe (Shilpa and Shikha, 2015; Srivastava et al., 2022).

TOC reduction using *Bacillus sp.* and *Pseudomonas sp.* was about 66% using *Bacillus sp.* and 79% using *Pseudomonas sp.*, whereas TOC reduction using *Staphylococcus arlettae* was only 15% in microaerophilic conditions, and it reached to the maximum of only 70% after aeration (Elisangela et al., 2009). Though 100% decolourization was obtained in 24 h, TOC reduction was reported to be very less even though incubated for 7 days. Though decolourization will occur in microaerophilic conditions, aeration is necessary for stabilizing the total organic carbon and to oxidize the aromatic amines produced during the microaerophilic incubation into simpler non-toxic compounds.

4.3.3 Kinetic Modeling for Substrate Utilization

Monod- and Contois-type equations and kinetics for different reaction orders results suggest second-order kinetics with regards to dye concentration.

$$r_{su} = -kS^2 \tag{4.1}$$

By integrating the differential equation

$$(1/S) = (1/S_o) + Kt \tag{4.2}$$

Where,

K - Coefficient of substrate utilization in L/mg day
r_{su} - Substrate utilization rate in mg/L day
S - Final substrate concentration in mg/L
S_o - Initial substrate concentration in mg/L
t - Time in days

By plotting the graph 1/S vs. t, the K values (substrate utilization rate constant values) can be found. The kinetic coefficients for aerobic, sequential aerobic/microaerophilic and microaerophilic incubation using *Bacillus sp.* and *Pseudomonas sp.* were obtained from the plot between 1/S and t and are given in Table 4.5. The maximum kinetic coefficient indicates the maximum dye degradation, since the kinetic coefficient is directly proportional to the substrate utilization, whereas, in these biodegradation reactions, dye stuff acts as the substrate for the microbes. From Table 4.5, the kinetic coefficient for substrate utilization for all concentrations was found to be higher for SAMI using *Bacillus sp.*, and thus, the rate of dye degradation will be higher for it compared with other

TABLE 4.5
Kinetic Coefficient for Substrate Utilization

	Sequential Aerobic/microaerophilic Incubation		Microaerophilic Incubation	
Dye Concentration	K in L/mg days (*Bacillus sp.*)	K in L/mg days (*Pseudomonas sp.*)	K in L/mg days (*Bacillus sp.*)	K in L/mg days (*Pseudomonas sp.*)
50ppm	0.0089	0.0076	0.0057	0.0061
100ppm	0.0033	0.0025	0.0022	0.0024
150ppm	0.0019	0.0017	0.0015	0.0016
250ppm	0.0007	0.0006	0.0005	0.0006

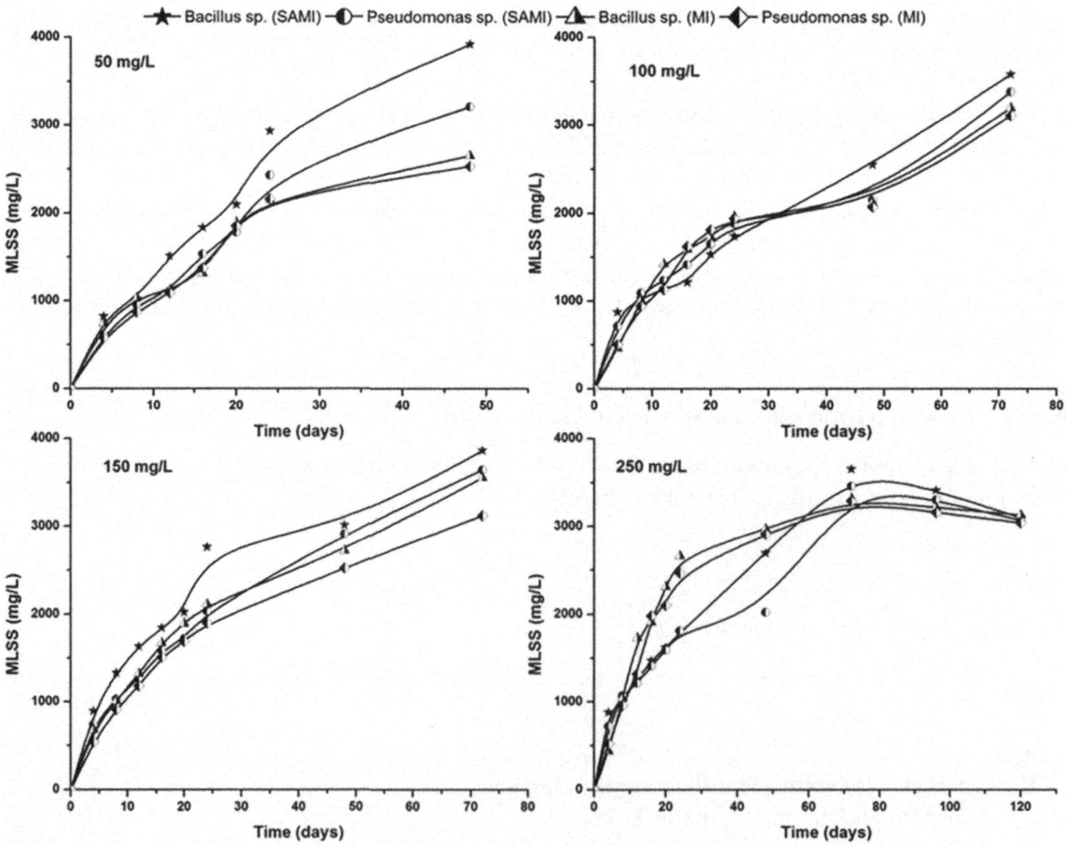

FIGURE 4.7 Growth curve for various initial dye concentrations.

incubation conditions. The kinetic coefficient was found to be much less for *Bacillus sp.* under MI, in which the decolourization was also found to be less. *Pseudomonas sp.* was more efficient in decolourizing the dye under MI. Kinetic coefficient for substrate utilization steadily decreased as the concentration of the effluent increased, which results in overall efficiency reduction with increase in the concentration. So, it can be concluded that *Bacillus sp.* under SAMI is more efficient in the azo dye decolourization/degradation process.

Figure 4.7 shows the growth for *Bacillus sp.* and *Pseudomonas* sp. under various incubation conditions. The growth was found to be higher for *Bacillus sp.* in the sequential aerobic/microaerophilic stage in which maximum decolourization was obtained. This is because the growth is directly proportional to the dye degradation. At dye concentration of 250 mg/L, it can be seen from Figure 4.7 that, after almost 90 days, there is a stationary phase followed by a death phase of the microbes for all of the experimental set-ups observed. This may be owed to the increased concentration of the dye, which inhibits the growth due to the high toxicity. For all other given dye concentrations, the toxicity levels are within the tolerable levels for the microbe community, and hence, the stationary phase was not observed in the given period of study.

4.4 CONCLUSIONS

Three distinct incubation conditions, including sequential aerobic/microaerophilic incubation, microaerophilic incubation and aerobic incubation, were used for the decolourization/biodegradation studies. Studying the original dye concentration had an impact. The following findings were reached after experiments were conducted to examine the decolourization, TOC reduction, COD reduction, mechanism of dye decolorization, substrate utilization kinetics and degradation products produced during the reaction. The most colour was removed from Reactive Black-5 using *Bacillus sp.* under sequential aerobic/microaerophilic incubation out of all incubation settings. The effect of initial dye concentration demonstrates that, as dye concentration rises, dye degradation efficiency diminishes. *Bacillus sp.* exhibits the best kinetic coefficient for substrate utilization under sequential aerobic/microaerophilic conditions for all concentrations.

This suggests the cause of the greater colour removal relative to all other incubation conditions. As dye concentration rises, the kinetic coefficient for substrate usage falls, resulting in a slower rate of dye deterioration. The peak at 597 nm in the visible area of UV-vis spectra disappears, indicating dye decolourization. In comparison to *Bacillus sp.*, *Pseudomonas sp.* is more efficient at decolourizing under microaerophilic culture. Incubation under successive aerobic and microaerophilic conditions reduced TOC by about 90%. After aeration, the TOC reduction in microaerophilic incubation rose by 10%, suggesting that aeration is required for the stabilization of TOC. Under microaerophilic incubation inoculated with *Pseudomonas sp.*, COD reduction was observed to be higher. By ultrasonication of the degraded sample, the mechanism of decolourization was found to be degradation of the azo dye by the bacteria. FT-IR and UV-vis spectra shows the production of aromatic amines, but the peaks corresponding to aromatic amines were absent in aerobic incubation.

BIBLIOGRAPHY

Ali, N. F. and R. S. R. E. Mohamedy. 2012. Microbial decolorization of textile waste water. *Journal of Saudi Chemical Society*, **16**, 117–123.

Andleeb, Saadia, Naima Atiq, Muhammad Ishtiaq Ali, Raja Razi-Ul-Hussnain, Maryam Shafique, Bashir Ahmad, PirBux Ghumro, Masroor Hussain, Abdul Hameed and Safia Ahmad. 2010. Biological treatment of textile effluent in stirred tank bioreactor. *International Journal of Agriculture & Biology*, **12**(2), 256–260.

Anwer, Sewgil Saaduldeen and Sawan Merkhan. 2013. Removal of different dyes by *Pseudomonas fluorescens. Journal of Life Sciences*, **7**, 51–56.

Asad, S., M. A. Amoozegar, A. A. Pourbabaee, M. N. Sarbolouki and S. M. M. Dastgheib. 2007. Decolorization of textile azo dyes by newly isolated halophilic and halo tolerant bacteria. *Bioresource Technology*, **98**, 2082–2088.

Blumel, S., H. J. Knackmuss and A. Stolz. 2001. Molecular cloning and characterization of the gene coding for the aerobic azoreductase from *Xenoplylusazovorans* KF46F. *Applied and Environmental Microbiology*, **68**, 3948–3955.

Borchert, M. and Judy A. Libra. 2001. Decolorization of reactive dyes by the white Rot Fungus Trametes versicolor in sequencing batch reactors. *Biotechnology and Bioengineering*, **75**(3), 313–321.

Bumpas, J. A. 2004. *Biodegradation of azo dyes by fungi*. Marcel Dekker, New York, 457–480.

Çelik, Öztürk A. and M. I. Abdullah. 2012. Biodegradation of reactive red 195 azo dye by the bacterium Rhodopseudomonaspalustris 51ATA. *African Journal of Microbiology Research*, **6**(1), 120–126.

Chang, J. S. and Y. C. Lin. 2000. Fed-batch bioreactor strategies for microbial decolorization of azo dye using a *Pseudomonas luteola* strain. *Biotechnology Progress*, **16**, 979–985.

Christie, R. M. 2001. *Colour chemistry*. Royal Society of Chemistry, Cambridge.

Chung, K. T. and S. E. J. Stevens. 1993. Degradation of azo dyes by environmental and helminthes. *Environmental Toxicology and Chemistry*, **12**, 2121–2132.

Dhanve, R. S., U. U. Shedbalkar and J. P. Jadhav. 2008. Biodegradation of diazo reactive dye Navy blue HE2R (reactive blue 172) by an isolated Exiguobacterium sp. RD3. *Biotechnology and Bioprocess Engineering*, **13**, 53–60.

Do, T., J. Shen, G. Cawood and R. Jeckins. 2002. Biotreatment of textile effluent using Pseudomonas spp. Immobilized on polymer supports. In: *Advances in bio treatment for textile processing*. Hardin, I. R., Akin, D. E. and Wilson, J. S. (Eds). University of Georgia Press.

Dos Santos, A. B., J. F. Cervantes and B. J. Van Lier. 2007. Review paper on current technologies for decolourisation of textile wastewaters: perspectives for anaerobic biotechnology. *Bioresource Technology*, **98**, 2369–2385.

El Bouraie, M. and W. S. El Din. 2016. Biodegradation of reactive black 5 by Aeromonas hydrophila strain isolated from dye-contaminated textile wastewater. *Sustainable Environment Research*, **26**(5), 209–216.

Elisangela, F., Z. Andrea, D. G. Fabio, R. M. Cristiano, D. L. Regina and C. P. Artur. 2009. Biodegradation of textile azo dyes by a facultative *Staphylococcus arlettae* strain VN-11 using a sequential microaerophilic/aerobic process. *International Bio deterioration & Biodegradation*, **63**, 280–288.

Fang, H., H. Wenrong and L. Yuezhong. 2004. Biodegradation mechanisms and kinetics of azo dye 4BS by a microbial consortium. *Chemosphere*, **57**, 293–301.

Forgacs, E., T. Cserhati and G. Oros. 2004. Removal of synthetic dyes from wastewaters: a review. *Environmental International*, **30**, 953–971.

Forootanfar, H., A. Moezzi, M. Aghaie-Khozani, Y. Mahmoudjanlou, A. Ameri, F. Niknejad and M. A. Faramarzi. 2012. Synthetic dye decolorization by three sources of fungal laccase. *Iranian Journal of Environmental Health Sciences & Engineering*, **9**, 27.

Gogate, P. R. and A. B. Pandit. 2004a. A review of imperative technologies for wastewater treatment II: hybrid methods. *Advance in Environmental Research*, **8**, 553–597.

Gogate, P. R. and A. B. Pandit. 2004b. A review of imperative technologies for wastewater treatment I: oxidation technologies at ambient conditions. *Advance in Environmental Research*, **8**, 501–551.

Harazono, K. and K. Nakamura. 2005. Decolorization of mixtures of different reactive textile dyes by the white-rot basidiomycete, *Phanerochaetesordida* and inhibitory effect of polyvinyl alcohol. *Chemosphere*, **59**, 63–68.

Hsueh, C. C. and B. Y. Chen. 2008. Exploring effects of chemical structure on azo dye decolorization characteristics by Pseudomonas luteola. *Journal of Hazardous Materials*, **154**, 703–710.

Isik, M. and D. T. Sponza. 2003. Aromatic amine degradation in an UASB/CSTR sequential system treating congo red dye. *Journal of Environmental Science and Health Part A*, **38**, 2301–2315.

Isik, M. and D. T. Sponza. 2004. Monitoring of toxicity and intermediates of CI Direct black 38 azo dye through decolorization in an anaerobic/aerobic sequential reactor system. *Journal of Hazardous Materials*, **114,** 29–39.

Jonstrup, M., N. Kumar, M. Murto and B. Mattiasson. 2011. Sequential anaerobic–aerobic treatment of azo dyes: decolourisation and amine degradability. *Desalination*, **280,** 339–346.

Kannan, S., K. Dhandayuthapani and M. Sultana. 2013. Decolorization and degradation of azo dye—Remazol Black B by newly isolated Pseudomonas putida. *International Journal of Current Microbiology and Applied Sciences*, **2,** 108–116.

Kapdan, I. K. and F. Kargi. 2002. Biological decolorization of textile dyestuff containing wastewater by Coriolus versicolor in a rotating biological contactor. *Enzyme Microbial Technology*, **30**, 195–199.

Kapdan, I. K., F. Kargi, G. Mcmullan and R. Marchant. 2000. Comparison of white rot fungi cultures for decolorization of textile dyestuffs. *Bioprocess Engineering*, **22**, 347–351.

Khalid, A., M. Arshad and D. E. Crowley. 2008. Accelerated decolorization of structurally different azo dyes by newly isolated bacterial strains. *Applied Microbiology and Biotechnology*, **78**, 361–369.

Khan, J. A. 2011. Azo dye biodegradation by azoreductase from *Bacillus megaterium. Advanced Biotech*, **10**, 23.

Kiran, Shumaila, Shaukat Ali, M. Asgher and Farooq Anwar. 2012. Comparative study on decolorization of reactive dye 222 by white rot fungi Pleurotusostreatus IBL-02 and Phanerochaetechrysosporium IBL-03. *African Journal of Microbiology Research*, **6**(15), 3639–3650.

Kudlich, M., A. Keck, J. Klein and A. Stolz. 1997. Localization of the enzyme system involved in anaerobic reduction of azo dyes by Sphingomonas sp. strain BN6 and effect of artificial redox mediators on the rate of azo dye reduction. *Applied and Environmental Microbiology*, **63**, 3691–3694.

Lee, Young-Hee, Kang Bo-Kyung, Kim Han-Do, Yoo Hye-Jin, Kim Jung-Soo, Huh Jae-Ho, Jung Young-Jin and Lee Dong-Jin. 2009. Effect of hot pressing/melt mixing on the properties of thermoplastic polyurethane. *Macromolecular Research*, **17**, 616–622.

Lin, Y. H. and J. Y. Leu. 2008. Kinetics of reactive azo-dye decolorization by Pseudomonas luteola in a biological activated carbon process. *Biochemical Engineering Journal*, **39**, 457–467.

Lodato, A., F. Alfieri, G. Olivieri, A. D. Donato, A. Marzocchella and P. Salatino. 2007. Azo-dye conversion by means of *Pseudomonas* sp. OX1. *Enzyme and Microbial Technology*, **41**, 646–652.

Lu, Lei, Min Zhao, Tian-Nv Wang, Li-Yan Zhao, Mei-Hui Du, Tai-Lun Li and De-Bin Li. 2012. Characterization and dye decolorization ability of an alkaline resistant and organic solvents tolerant laccase from Bacillus licheniformis LS04. *Bioresource Technology*, **115**, 35–40.

Madhavi, S., S. Revankar and S. Lele. 2007. Synthetic dye decolorization by white rot Ganoderma sp. WR-1. *Bioresource Technology*, **98,** 775–780.

Maier, J., A. Kandelbauer, A. Erlacher, A. C. Paulo and M. G. Gubitz. 2004. A new alkali thermo stable azoreductase from *Bacillus* sp. *Applied and Environmental Microbiology*, **70**, 837–844.

Mane, U. V., P. N. Gurav, A. M. Deshmukh and S. P. Govindwar. 2008. Degradation of textile dye reactive navy—blue Rx (reactive blue -59) by an isolated Actinomycete *Streptomyces krainskii*SUK—5. *Malaysian Journal of Microbiology*, **4**, 1–5.

McMullan, G., C. Meehan, A. Conneely, N. Kirby, T. Robinson, P. Nigam, I. M. Banat, R. Marchant and W. F. Smyth. 2001. Microbial decolourisation and degradation of textile dyes. *Applied Microbiology and Biotechnology*, **56**, 81–87.

Mielgo, I., M. T. Moreira, G. Feijoo and J. M. Lema. 2001. A packed-bed fungal bioreactor for continuous decolourisation of azo-dyes (orange II). *Journal of Biotechnology*, **89**, 99–106.

Mohorcic, M., S. Teodorovic, V. Golob and J. Fiedrich. 2006. Fungal and enzymatic decolorization of artificial textile dye baths. *Chemosphere*, **63**, 1709–1717.

Moosvi, S., H. Keharia and D. Madamwar. 2005. Decolourization of textile dye reactive violet 5 by a newly isolated bacterial consortium RVM 11.1. *World Journal of Microbiology and Biotechnology*, **21**, 667–672.

Olukanni, O. D., A. AOsuntoki and G. O. Gbenle. 2006. Textile effluent biodegradation potentials of textile effluent-adapted and non-adapted bacteria. *African Journal of Biotechnology*, **5**, 1980–1984.

Ozdemir, G., B. Pazarbasi, A. Kocyigit, E. E. Omeroglu, I. Yasa and I. Karaboz. 2008. Decolorization of acid black 210 by Vibrio harveyi TEMS1, a newly isolated bioluminescent bacterium from Izmir Bay, Turkey. *World Journal of Microbiology and Biotechnology*, **24**, 1375–1381.

Pandey, A., P. Singh and L. Iyengar. 2007. Bacterial decolorization and degradation ofazo dyes. *International Biodeterioration and Biodegradation*, **59**, 73–84.

Pazarlioglu, N. K., R. O. Urek and F. Ergun. 2005. Biodecolourization of direct blue 15 by immobilized *Phanerochaetechrysosporium. Process Biochemistry*, **40**, 1923–1929.

Philippe, C. V., Roberto and B. Willy. 1998. Treatment and reuse of wastewater from the textile wet processing industry a review of technology. *Biotechnology*, **22**, 289–302.

Pinheiro, H. M., E. Touraud and O. Thomas. 2004. Aromatic amines from azo dye reduction: status review with emphasis on direct UV spectrophotometric detection in textile industry wastewaters. *Dyes and Pigments*, **61**, 121–139.

Ponraj, M., K. Gokila and V. Zambare. 2011. Bacterial decolorization of textile dye-orange 3R. *International Journal of Advanced Biotechnology and Research*, **2**, 168–177.

Praveen, S., S. Lakhvinder and M. Jyoti. 2010. COD reduction and colour removal of simulated textile mill wastewater by mixed bacterial consortium. *Rasayan*, **3**, 731–735.

Puvaneswari, N., J. Muthukrishnan and P. Gunasekaran. 2006. Toxicity assessment and microbial degradation of azo dyes. *Indian Journal of Experimental Biology*, **44**, 618–626.

Rafii, F., W. Franklin and C. E. Cerniglia. 1990. Azoreductase activity of anaerobic bacteria isolated from human intestinal microflora. *Applied and Environmental Microbiology*, **56**, 2146–2151.

Robinson, T., B. Chandran and P. Nigam. 2001. Studies on the production of enzymes by white-rot fungi for the decolorization of textile dyes. *Enzyme and Microbial Technology*, **29**, 575–579.

Sandhya, S., K. Padmavathy, Y. V. Subrahmanya and N. S. Kaul. 2004. Microaerophilic–aerobic sequential batch reactor for treatment of azo dyes containing simulated wastewater. *Process Biochemistry*, **40**, 885–890.

Shah, M. P., K. A. Patel, S. S. Nair and A. M. Darji. 2013. Microbial degradation of Textile dye (Remazol Black B) by Bacillus spp. ETL-2012. *Journal of Bioremediation and Biodegradation*, **4**, 180.

Shilpa, S. and R. Shikha. 2015. Biodegradation of dye reactive black—5 by a novel bacterial endophyte. *International Research Journal of Environment Sciences*, **4**(4), 44–53.

Siddiqui, Muhammad Faisal, Saadia Andleeb, Naeem Ali, PirBux Ghumro and Safia Ahmed. 2010. Biotreatment of anthraquinone dye Drimarene Blue K_2RL. *African Journal of Environmental Science and Technology*, **4**(1), 45–50.

Slokar, Y. M. and M. A. Marechal. 1998. Methods of decolorization of textile wastewater. *Dyes and Pigments*, **37**, 335–356.

Srivastava, A., L. K. Dangi, S. Kumar and R. Rani. 2022. Microbial decolorization of reactive black 5 dye by Bacillus albus DD1 isolated from textile water effluent: kinetic, thermodynamics & decolorization mechanism. *Sustainable Environment Research*, **26**(5), 209–216.

Supaka, Nuttapun, Kanchana Juntongjin, Somsak Damronglerd, Marie-Line Delia and Pierre Strehaiano. 2004. Microbial decolorization of reactive azo dyes in asequential anaerobic–aerobic system. *Chemical Engineering Journal*, **99**, 169–176.

Suzuki, Y., T. Yoda, A. Ruhul and W. Sugiura. 2001. Molecular cloning and characterization of the gene coding for azoreductase from Bacillus sp. OY1-2 isolated from soil. *Journal of Biological Chemistry*, **276**, 9059–9065.

Tan, S. M., S. A. Ong, L. N. Ho, Y. S. Wong, C. Z. A. Abidin, W. E. Thung and T. P. Teoh. 2021. Biotreatment of sulfonated dyestuffs with energy recovery in microbial fuel cell: influencing parameters, kinetics, degradation pathways, mechanisms, and phytotoxicity assessment. *Journal of Environmental Chemical Engineering*, **9**(4), 105525.

Tekere, M., I. Ncube, J. S. Read and R. Zvauya. 2001. Biodegradation of the organochlorine pesticide, lindane by a subtropical white rot fungus in batch and packed bed bioreactor system. *Environment Technol*ogy, **23**, 199–206.

Tetsch, L., J. Bend, M. Janben and U. Holker. 2005. Evidence for functional laccases in the acidophilic ascomycete Hortaeaacidophila and isolation of laccase-specific gene fragments. *FEMS Microbiology Letter*, **245**, 161–168.

Toh, Y. C., J. J. L. Yen, J. P. Obbard and Y. P. Ting. 2003. Decolourisation of azo dyes by white-rot fungi (WRF) isolated in Singapore. *Enzyme and Microbial Technology*, **33**, 569–575.

Van Der Zee, F. P. and S. Villaverde. 2005. Combined anaerobic–aerobic treatment of azodyes—a short review of bioreactor studies. *Water Research*, **39**, 1425–1440.

Verma, P. and D. Madamwar. 2003. Decolourization of synthetic dyes by a newly isolatedstrain of Serratia marcescens. *World Journal of Microbiology and Biotechnology*, **19**, 615–618.

Vijaykumar, M. H., P. A. Vaishampayan, S. Y. Shouche and T. B. Karegoudar. 2007. Decolourization of naphthalene-containing sulfonated azo dyes by *Kerstersia* sp. strain VKY1. *Enzyme and Microbial Technology*, **40,** 204–211.

Wang, Tian-Nv, Lei Lu, Guo-Fu Li, Jun Li, Teng-Fei Xu and Min Zhao. 2011. Decolorization of the azo dye reactive black 5 using laccase mediator system. *African Journal of Biotechnology*, **10**(75), 17186–17191.

Wang, Wei, Zhen Zhang, Hong Ni, Xiaomeng Yang, Qianqian Li and Lin Li. 2012. Decolorization of industrial synthetic dyes using engineered Pseudomonas putida cells with surface-immobilized bacterial laccase. *Microbial Cell Factories*, **11**, 75–81.

Wesenberg, D., I. Kyriakides and S. N. Agathos. 2003. White-rot fungi and their enzymes for the treatment of industrial dye effluents. *Biotechnology Advances*, **22**, 161–187.

Whiteley, C. G. 2007. Bioremediation of textile dyes. *Industrial Bioprocessing*, **29**, 7–15.

Xu, H., T. M. Heinze, S. Chen, C. E. Cerniglia and H. Chen. 2007. Anaerobic metabolism of 1-amino-2-naphthol-based azo dyes (Sudan dyes) by human intestinal microflora. *Applied Environment Microbiology*, **73**, 7759–7762.

Yang, J. (Ed.). 1987. *Analysis of dye*. Chemical Industry Press, Beijing, 156–163.

Zhang, C., Hanwen Diao, Fengxia Lu, Xiaomei Bie, Yufeng Wang and Zhaoxin Lu. 2012. Degradation of triphenylmethane dyes using a temperature and pH stable spore laccase from a novel strain of Bacillus vallismortis. *Bioresource Technology*, **126**, 80–86.

Zubbair, N. A., A. T. Ajao, E. O. Adeyemo and D. O. Adeniyi. 2018. Biotransformation and detoxification of reactive black dye by Ganoderma tsugae. *African Journal of Environmental Science and Technology*, **12**(5), 158–171.

5 Bio-Adsorbent-Based Integrated System

Silvia Escudero-Curiel, Marta Pazos and María Ángeles Sanromán

5.1 INTRODUCTION

Adsorption-based techniques for the removal of toxic and hazardous chemicals in aqueous media are an excellent approach to be considered, thanks to certain characteristics of their process that make them quite interesting from a functional and industrial scale application point of view. Some of the outstanding characteristics offered by these adsorption-based techniques include simplicity of design and operation, easy maintenance, affordability and reduction of chemical or biological sludge, among others (Rajapaksha et al., 2019). In addition, the adsorption process is highly effective and efficient, which has been more than proven, even at very low concentrations (Gavrilescu, 2004). Additionally, it is non-destructive, it is consistent and, in the case of metal removal, it facilitates the recovery of metals that are often economically exploited (Rial & Ferreira, 2022; Torres, 2020). Due to the nature of bio-adsorbents, modifying and tailoring them to target specific pollutants opens up a new approach for adsorption-based techniques to meet the changing needs of the industry (Baskar et al., 2022).

Nowadays, there exist a myriad of adsorbents from diverse sources that have been successfully applied to remove toxic and hazardous chemicals from water (Hayoun et al., 2022). Adsorbents can be classified according to their origin or provenance. Thus, the broadest classification of adsorbent materials that could be applied is based on their nature as organic, inorganic and synthetic (Rajapaksha et al., 2019). Bio-adsorbents fall into the organic classification and will be the focus of this chapter.

Bio-adsorbent category includes dead/inactive biological material ranging from microorganisms, fungi, algae, and forestry, agricultural and agro-industrial wastes (Adewuyi, 2020; Fomina & Gadd, 2014; Vijayaraghavan & Yun, 2008). These bio-adsorbents are often widely available, renewable and often locally sourced, which helps to be in line with the economics of environmental remediation. Furthermore, using bio-adsorbents contributes to an increasingly necessary circular economy of resources and reduces the carbon footprint in most cases.

Bio-adsorbents, as a branch of biotechnology, have been the focus of numerous studies based on the removal and/or recovery of organic and inorganic substances from aqueous media (Ahmad et al., 2014; Akhtar et al., 2016). The achieved uptakes can be ranged from 1–2 mg/g to 2,700 mg/g depending on the nature of the pollutant and the bio-adsorbent used (Torres, 2020).

Among the key influencing factors for adsorption by bio-adsorbents are those porosity-related and their structural characteristics, both intrinsically linked to the type of material. The specific surface area, size and morphology of the pore contribute to the presence of more or fewer specific adsorption sites for chemical and/or physical adsorption of contaminants. Regarding structural characteristics, chemical and mechanical stability are essential requirements (Baskar et al., 2022). Besides, it is crucial the type of functional groups (Figure 5.1) and their density on the surface of the bio-adsorbent, especially carboxyl, hydroxyl, sulfate, phosphate, and amine groups (He & Chen, 2014). An inherently related parameter to the content of these functional groups that is decisive in bio-adsorption is the point of zero charge (PZC) of the bio-adsorbent surface. PZC is defined as the pH at which the net charge of the total adsorbent surface becomes zero. On these terms, values of

 DOI: 10.1201/9781003391487-5

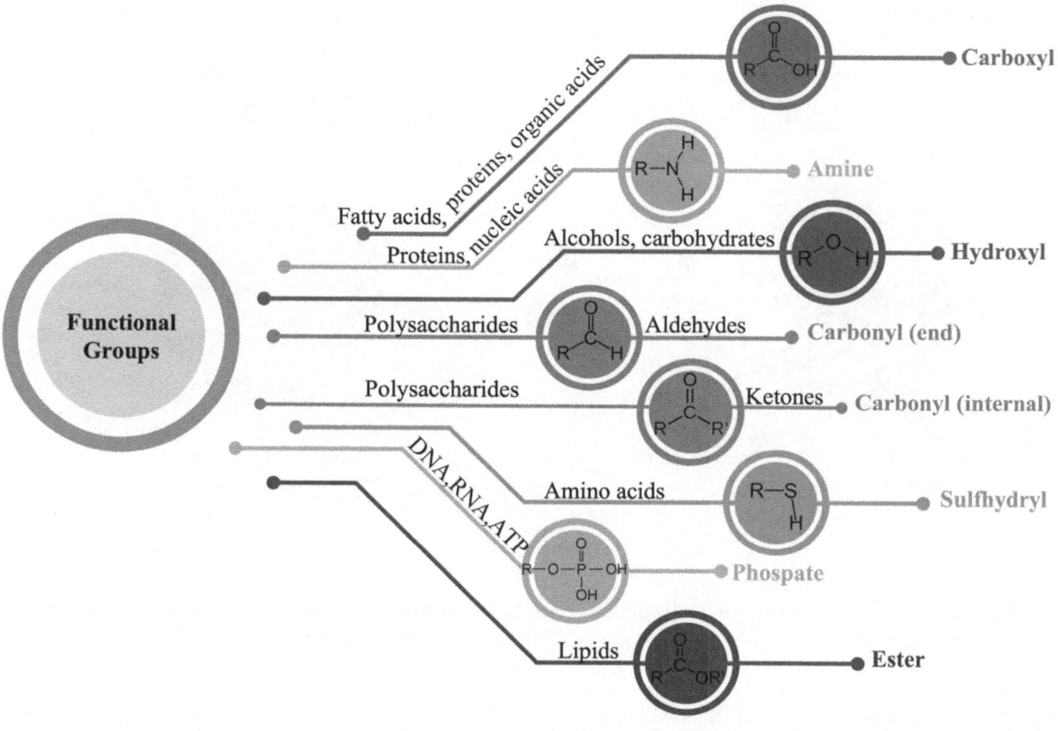

FIGURE 5.1 Characteristic functional groups present in bio-adsorbents surface for their application to wastewater treatment.

solution pH lower than the PZC indicate a higher contribution of positive charges and a favorable adsorption of negatively charged pollutants. Conversely, at solution pH values higher than PZC, there is mainly a contribution of anions on the adsorbent surface charge, which is favorable for adsorbing positively charged pollutants (Torres, 2020). Hence, controlling the solution pH allows adjusting the adsorbent's net surface charge. This point provides a potential for selective adsorption of the charged pollutants. Accordingly, some experimental parameters affect the adsorption yield, as is the case of the pH of the solution, temperature, contact time, adsorbent load, pollutant concentration and simultaneous presence of different contaminants (Akhtar et al., 2016). Therefore, the fact that the adsorption capacity of the bio-adsorbent is highly dependent on its prehistory and experimental conditions means that it is important to take these factors into account when comparing adsorption results (Fomina & Gadd, 2014).

Bio-adsorption refers to a physicochemical process of mass transfer that is metabolically independent (Adewuyi, 2020). Insight into the mechanisms involved in the adsorption process is valuable to improve the adsorption capacity by changing the experimental conditions or by using bio-adsorbents specifically tailored for each occasion. In this way, bio-adsorption involves different mechanisms such as pore filling, electrostatic interaction, hydrogen bonding, reduction, ion exchange, surface complexation and π-π interaction (see Figure 5.2 left) (Fdez-Sanromán et al., 2020). First, pore filling and electrostatic interactions between the compounds and bio-adsorbent surface would be crucial for the adsorption process to be initiated. Second, the rest of the mechanism could occur individually or simultaneously. The main or mixture of them governing the process might be different, depending on both the contaminant and the bio-adsorbent (de Freitas et al., 2019). Once attractive forces take place, the most common mechanism of adsorption

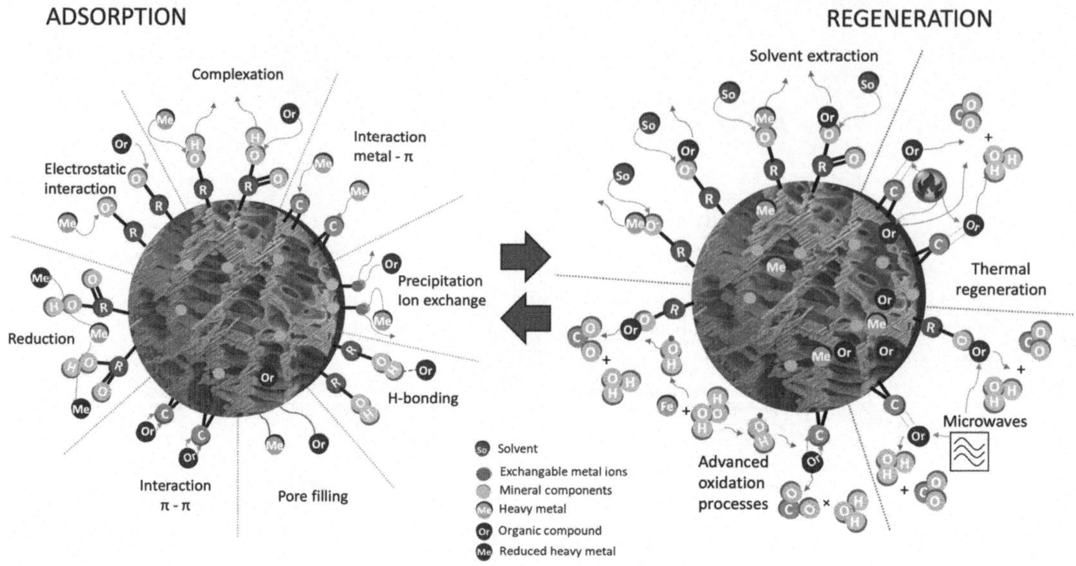

FIGURE 5.2 The main adsorption (left) and regeneration (right) mechanisms follow when bio-adsorbents are used. Reproduced from Fdez-Sanromán et al. (2020), with permission of MDPI.

is hydrogen bonding. Subsequently, several compounds may be reduced when attached to the binding groups by surrounding electron-donor groups, thus occasionally reducing their toxicity and mobility (microprecipitation can occur) (Bahafid et al., 2017). Ion exchange happens when adsorption comes along with the release of the counterion from the bio-adsorbent matrix, mainly polysaccharides, with which metal ions are exchanged (Crist et al., 1994). Combining metal cations with molecules or anions that contain free electron pairs is referred to as coordination or complex formation. The metal cation binds to the central atom and surrounding anions or molecules with which it forms a coordination compound. Thereby, complexation or coordination results from electrostatic interaction between a metallic ion-chelating agent and a polymer present in the bio-adsorbent with free electron pairs such as polysaccharides, biosurfactants, proteins and nucleic acids (Gahlout et al., 2021). When hazardous chemicals have a neutral charge, as is the case of many organic pollutants, π-π interaction between aromatic rings by electron transfer is the dominant adsorption mechanism in some carbonaceous bio-adsorbents (Cheng et al., 2021).

At the end of an adsorption process, spent bio-adsorbent is obtained that turns itself into a solid waste that must be treated. Sustainable management of spent adsorbents, therefore, must be examined in the current context of resource recovery and circular economy (Baskar et al., 2022). Recovery and regeneration of the bio-adsorbents must be addressed when applying adsorption-based techniques in the long run and should be driven toward their reuse (Rajapaksha et al., 2019). Thus, integrated solutions for wastewater decontamination need to be offered by the scientific community. Regeneration techniques can be divided into destructive and non-destructive from the point of view of the contaminating compound that can be desorbed or degraded (Dai et al., 2019). Solvent extraction, thermal regeneration, microwaves and advanced oxidation processes may be the most typical regeneration techniques applied for bio-adsorbents (Figure 5.2 right) (Fdez-Sanromán et al., 2020).

In this chapter, bio-adsorbents, engineered bio-adsorbents and their mechanism of action will be reviewed. Furthermore, the regeneration techniques and their limitations are presented and discussed.

5.2 BACTERIA, FUNGI AND ALGAE

An exceptionally large ratio of surface-area-to-volume and a remarkable richness in functional groups on the surface make microorganisms, including microalgae, yeast and bacteria, outstanding bio-adsorbents for several compounds (Afrooz et al., 2022). Moreover, the cell wall composition of filamentous fungi and algae makes them great candidates for bio-adsorption (Vieira & Volesky, 2000). Although living biomass can be used, dead biomass offers advantageous characteristics when it comes to its industrial applicability. Some of the more remarkable advantages of using non-living/inactive microorganisms are detailed in Figure 5.3 (Torres, 2020).

5.2.1 Bacteria

Bacteria dead biomass has been widely used as bio-adsorbent for removing heavy metals from water (Torres, 2020). The solution pH has turned out to be the critical factor for adsorption in most studies, as it marks the state of the metal and the predominancy of positive and negative functional

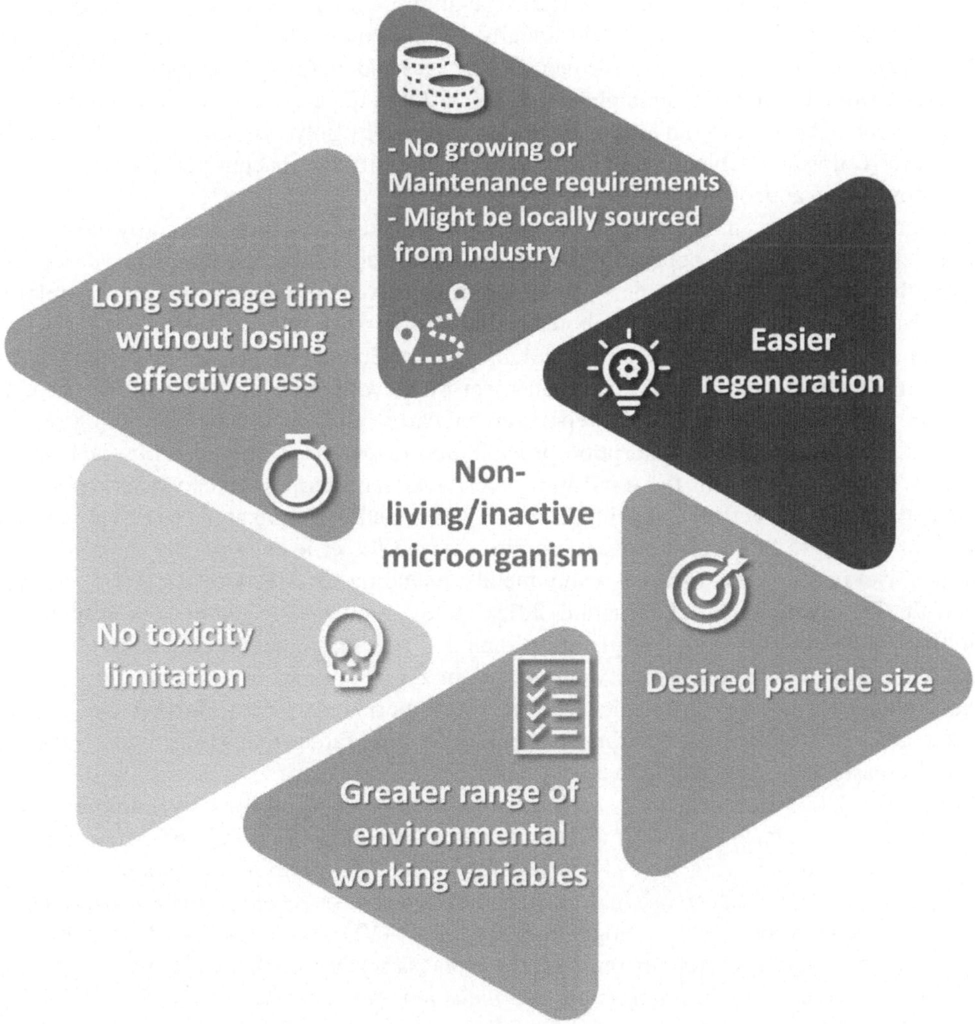

FIGURE 5.3 Advantages of using non-living/inactive microorganisms in an adsorption process.

groups on the bio-adsorbent surface depending on PZC (Aryal & Liakopoulou-Kyriakides, 2015). For metal cations, if the pH of the solution is too low, protons' competition for functional groups causes the adsorption capacity of the metals to be reduced. On the other hand, if the pH is too high, insoluble precipitates may be formed. Conversely, in the case of metal anions species, the bio-adsorption efficiency is enhanced at low pH values. The reason for this is that there is an overall rise of positive functional groups on the surface of the bio-adsorbent. On the contrary, when negatively charged groups increase due to higher pH, repulsive forces negatively affect the adsorption of metal anions (de Freitas et al., 2019). Therefore, electrostatic interactions and hydrogen bonding seem to govern the chemisorption process for metals in bacteria.

Depending on the characteristics and composition of the cell membrane, two main groups of bacteria are distinguished: gram-positive and gram-negative. In Figure 5.4 is disclosed the cell walls of the two types of bacteria. Gram-positive bacteria cell wall (Figure 5.4 a) is mainly composed of a peptidoglycan layer linked by amino acid groups and polyols, known as teichoic acids, sandwiched within the lattice. Teichoic and teichuronic acids along with anionic groups from peptidoglycans confer an overall negative charge to the membrane in gram-positive bacteria, and they are the responsible of metal-binding capability of this group (Aryal & Liakopoulou-Kyriakides, 2015). In contrast, the cell wall of gram-negative bacteria (Figure 5.4 b) lacks so many peptidoglycans, and there is no presence of teichoic acids. Additionally, it has an outer layer composed of phospholipids and lipopolysaccharides. In this case, the negative charges and, in turn, the metal-binding capability come mainly from the latter (Vijayaraghavan & Yun, 2008). Adsorption of heavy metals is linked to hydroxyl, sulfate, phosphoryl and amino groups contributed by polysaccharides, proteins and lipids present in the cell walls. Table 5.1 provides a summary of different bacteria dead biomass used for metal removal from aqueous media.

Gram-positive bacteria have been broadly used to better adsorb metals, especially genus *Bacillus* due to its elevated content of peptidoglycans and teichoic acid in the cell wall (Gavrilescu, 2004). However, there exist species gram-positive and gram-negative that may produce extracellular polysaccharides and small peptides, siderophores, that contribute to capturing metals, as is the case of the genera *Pseudomonas, Actinomyces*, and *Azotobacter* (Elnahas et al., 2021; Gavrilescu, 2004). Lyophilized *Pseudomonas stutzeri* (gram-negative) was investigated as bio-adsorbent for Cd (II), Cu (II) and Pb (II) by Oh et al. (2009), reporting that carboxylic, hydroxyl and amino groups were the groups involved in the bio-adsorption process, and below pH 2, the adsorption capacity was drastically reduced because of the repulsion forces. The maximum uptakes values reported were 43.5 and 36.2 mg/g, respectively, at pH 5 (Cd (II), Cu (II)) and 142 mg/g at pH 6 (Pb (II)) just before precipitates started to be formed. Several studies showed the efficiency of *Streptomyces rimosus* (gram-positive) in removing different heavy metals, including Cr (III), Cd (II), Ag (I), Ni (II), Pb (II) and Zn (II), among others (Sahmoune, 2018). Acidic groups were found to be responsible for metal binding, and the adsorption at pH lower than 2–3 was significantly reduced in all cases. For instance, amino and carboxyl groups were crucial in Zn (II) bio-adsorption, having its optimum uptake, 30 mg/g, at pH 7.5 and precipitating after 8 (Mameri et al., 1999). The differences found in cations uptakes may be attributed to the ionic radius. Smaller radius tends to be hydrolyzed, which leads to decreased bio-adsorption (Oh et al., 2009).

Zhou et al. (2007) used dead cells of *Bacillus licheniformis* (gram-positive) biomass, isolated from contaminated soils, to remove Cr (VI) from aqueous media. *B. licheniformis* showed a remarkable performance in removing Cr (VI), achieving, at pH 2.5, a maximum uptake of 60.5 mg/g. Increasing the pH caused a decrease in the adsorption capacity, since, at low pH, the overall positive charge of the bio-adsorbent surface might enhance the Cr (VI) oxoanions bio-adsorption. Besides, at low pH, hydronium ions layer surrounding the biomass surface could help attract Cr (VI) to the active sites for chemisorption. *Bacillus amyloliquefaciens* was studied for U (VI) removal by Liu et al. (2019), finding that at an optimum pH of 6 an uptake of 179.5 mg/g was achieved. Oxygen-containing groups, especially carboxyl and hydroxyl, played the main role in the bio-adsorption process forming UO_2^{2+} complexes.

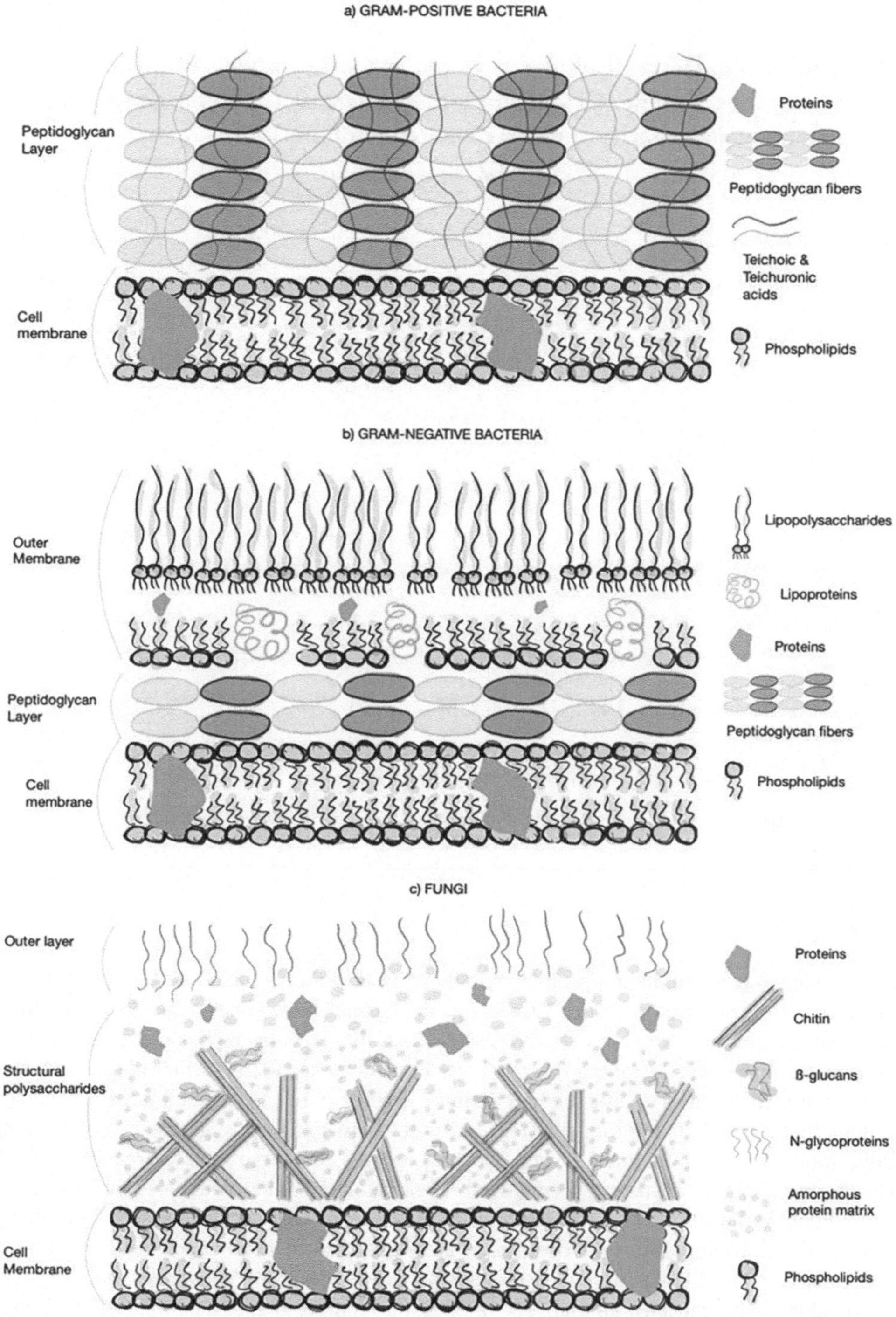

FIGURE 5.4 Typical cell wall structure schemes of gram-positive (a) and gram-negative (b) bacteria and fungi (c).

TABLE 5.1
Several Examples of Bacteria Dead Biomass Used as Bio-Adsorbents of Different Hazardous Chemicals (t= time, T= temperature, *X*= biomass, C_0= initial concentration, q_{max}= maximum uptake capacity and RT= room temperature)

		Operational conditions						
Pollutant	Bio-adsorbent	pH	t (min)	T (°C)	*X* (g/L)	C_0 (mg/L)	q_{max} (mg/g)	Reference
Cd (II)	*Bacillus cereus*	5	30	28	1	50	31.95	Huang et al., 2013
Cu (II)	*Bacillus thioparans*	6	420	30	0.5	40	42.3	Rodríguez-Tirado et al., 2012
Pb (II)		4	240				210.1	
Cr (VI)	*Ochrobactrum anthropi*	2	120	30	1	-	86.2	Ozdemir et al., 2003
Cd (II)		8					37.3	
Cu (II)		3					32.6	
Pb (II)	*Bacillus xiamenensis*	6	240	35	1	100	207.4	Mohapatra et al., 2019
Pb (II)	*Pseudomonas sp.*	5	2400	RT	1	500	182.2	Vélez et al., 2021
Pb (II)	*Pseudomonas stutzeri*	6	30	30	1	300	43.5	Oh et al., 2009
Cd (II)		5					36.2	
Cu (II)		5					142	
Zn (II)	*Streptomyces rimosus*	7.5	240	RT	3	100	30	Mameri et al., 1999
Cr (VI)	*Bacillus licheniformi*	2.5	140	50	1	300	60.5	Zhou et al., 2007
U (VI)	*Bacillus amyloliquefaciens*	2.5	180	30	0.5	-	179.5	Liu et al., 2019

5.2.2 Fungi

Fungi are divided into two main groups: filamentous and yeast. Filamentous fungi and yeasts are mostly saprophytes, being found free in nature, especially in organic matter in decomposition. Filamentous fungi are multicellular and are characterized by branched filaments or hyphae that develop and intertwine to form the mycelium. There is a vegetative mycelium attached to the surface of the substrate (soil, plants, food) and an aerial or reproductive mycelium, where spores are formed (asexual and sexual). Filamentous fungi include macrofungal species or mushrooms. Yeasts are globular, oval or elongated unicellular fungi, 2–6 μm in diameter, which means they possess a large surface-area ratio for adsorption. They form smooth, moist-textured, creamy or membranous colonies (without aerial hyphae) similar to bacteria. Fungi cell walls (Figure 5.4 c) are composed predominantly of structural polysaccharides as β-glucans, chitin, chitosan and/or cellulose along with N-glycoproteins embedded in an amorphous protein matrix that can form filaments on the external surface. Moreover, polyphosphates and inorganic ions are also found in the fungal cell wall (Sağ, 2001). Thus, this composition of the fungi cell walls provides abundant functional groups that could help in bio-adsorption such as carboxyl, hydroxyl, phosphate and amino groups, among others. Amino groups are predominantly provided by the polysaccharide chitin, since the protein content of the fungal biomass is not particularly elevated (Tsezos & Volesky, 1982). Fungal biomass can be obtained from nature or easily cultivated. Besides, it can be produced as a residue from several industries, making these fungi an environmentally friendly bio-adsorbent. Fungi dead biomass has been studied by many authors for the removal of heavy metals, dyes and micropollutants

(Legorreta-Castañeda et al., 2020), and some examples are listed in Table 5.2. Iqbal & Saeed (2007) demonstrated the efficiency of a white-rot mushroom, *Phanerochaete chrysosporium*, removing the reactive dye remazol brilliant blue R with a maximum capacity of 102 mg/g at pH 2. At this pH, below its PZC, the surface of the fungal biomass is protonated, acquiring an overall positive charge, thus attracting the anionic reactive dye to the active sites. Bio-adsorption capacity of *Rhizopus arrhizus* was studied by Fourest et al. (1994). The maximum capacity was observed at pH 7 for all studied metals Ni (II), Zn (II), Cd (II), Ag (I) and Pb (II) being 22.18, 42.82, 43.80, 76.58 and 211.34 mg/g, respectively. The adsorption mechanism that played the most important role in the adsorption of metals in fungi is the formation of coordinates with amino groups and the simultaneous formation of hydroxides within the microcrystalline structure provided by the chitin cell wall (Tsezos & Volesky, 1982). In addition, it has been proposed that, at pH between 6 and 7, an increment in hydroxide concentration could lead to the appearance of new binding sites in the hyphae walls, such as phosphate monoester

TABLE 5.2
Several Examples of Fungi Dead Biomass Used as Bio-Adsorbents for Different Hazardous Chemicals (t= time, T= temperature, X= biomass, C_0= initial concentration, q_{max}= maximum uptake capacity)

		Operational conditions						
Pollutant	**Bio-adsorbent**	**pH**	**t (min)**	**T (°C)**	***X* (g/L)**	**C_0 (mg/L)**	**q_{max} (mg/g)**	**Reference**
Cd (II)	*Pleurotus platypus*	6	60	28	1	10	34.96	Vimala & Das, 2009
Pb (II)	*Agaricus bisporus*	5	360	28	0.1	10	33.78	Vimala & Das, 2009
Cr (VI) Pb (II) Ni (II)	*Saccharomyces cerevisiae*	1 5 5	1440	25	1	200	44.4 144 23	Özer & Özer, 2003
U (IV) Th (IV)	*Rhizopus arrhizus*	4	-	5 40	-	30	132 143	Tsezos & Volesky, 1981
Cu (II)	*Gliomastix murorum*		720	25	2		12.08	Li et al., 2009
Benzylpenicillin	*Rhizopus arrhizus*	6	1440	35	1	1000	459	Aksu & Tunç, 2005
Orlamar red BG Orlamar blue G Orlamar red GTL	*Fomitopsis carnea*	4	2880	-	0.15	1000	503.1 545.2 634.9	Mittal & Gupta, 1996
Remazol brilliant blue R	*Phanerochaete chrysosporium*	2	40	30	1	500	102	Iqbal & Saeed, 2007
Ni (II) Zn (II) Cd (II) Ag (I) Pb (II)	*Rhizopus arrhizus*	7 7 7 7 6	120	-	3	-	22.18 42.82 43.80 76.58 211.34	Fourest et al., 1994
Bleomycin Vincristine	*Fomes fomentarius*	-	240	22.5	10	10	35 10	Jureczko & Przystaś, 2021

groups from sugars or phospholipids, whose dissociation constant occurs between pH 6.1 and 6.75. In contrast, at low pH, proton competition for binding sites showed an inhibitory response for metal bio-adsorption. The uptakes for Ag (I) and Pb (II) are remarkably high. In the case of Ag (I), it is explained by the fact that this monovalent cation has a low charge density (charge/ionic radius = 0.8 Å), which translates into a low hydration energy in solution, and, in turn, good stability of the silver complexes formed on the fungal surface. In the case of Pb (II), the stability of complexes of Pb (II)-bio-adsorbent decreases at pH 6 and above, thus increasing the adsorption of new species formed after their disappearance. Several yeast species have been used for Cr (VI) removal from aqueous solutions (Bahafid et al., 2017) such as *Candida tropicalis*, *Pichia anomala* and *Saccharomyces cerevisiae* achieving uptake values of 44.4 mg/g for the latter (Özer & Özer, 2003). In yeast, the first step among the adsorption mechanisms for Cr (VI) is reduction, and it has described Glutathione and cysteine as the most prevailing non-enzymatic chromate reductants involved in the process. In this way, Cr (VI) is reduced to Cr (III) either by direct contact with electron-donating groups on the yeast cell wall or is first adsorbed by positively charged groups and reduced by adjacent electron-donating groups. Afterwards, Cr (III) might be adsorbed by surrounding negatively charged groups. Five white-rot fungi were tested as bio-adsorbent for two cytostatic drugs (drugs widely used in chemotherapy), bleomycin and vincristine, achieving 35 and 10 mg/g, respectively, by, at the same time, *Fomes fomentarius* (Jureczko & Przystaś, 2021). On this occasion, π-π interactions seem to be the mechanism involved in the adsorption process, as the organic molecules do not have an overall positive or negative charge.

5.2.3 Algae

Algae are a promising candidate material to be used as a bio-adsorbent due to several factors such as their abundance in the oceans, their ubiquity, their high production and growth rate and their high content of components that donate active functional groups for adsorption. The broadest classification of algae without falling into error is microalgae and macroalgae. Microalgae are unicellular organisms that include some groups such as Chlorophyta (green algae), Rhodophyta (red algae), Bacillariophyceae (diatoms) and Dinophyceae (dinoflagellates), which are naturally part of phytoplankton, while pluricellular macroalgae or seaweeds fall into the groups Chlorophyta, Rhodophyta and Phaeophyta (brown algae). Macroalgae Chlorophyta, Rhodophyta and Phaeophyta have been studied as bio-adsorbents for many hazardous chemicals such as heavy metals, pharmaceuticals, dyes and phenols, among others, although some microalgae have also been tested (Coimbra et al., 2018; He & Chen, 2014; Kiran Marella et al., 2020; Lin et al., 2020; Silva et al., 2019). When it comes to bio-adsorption, the fundamental differences between microalgae, macroalgae and within the different groups is present in their cell wall composition (Kumar et al., 2016). The general composition of the algae cell wall provides a myriad of active functional groups, including carboxyl, sulfhydryl, sulphates, phosphates, imidazoles and amino associated with polysaccharides, alginic acid and proteins for binding various pollutants (Aksu & Tezer, 2005). Figure 5.5 illustrates the principal active components of the cell wall of each macroalgae group involved in the bio-adsorption process.

The elevated alginate content of brown algae is particularly conducive to heavy metal chelates due to the unique concave distribution pattern provided by the alternating guluronic and mannuronic acids in the polysaccharide chain (Navarro et al., 2008). Thus, brown algae are the most widely studied group of algae as bio-adsorbents, not only because of their high content of functional groups and binding sites (provided mainly by alginate and sulfonate groups) but also for being the group of algae with a potential harvest estimated at 16 million tons worldwide (He & Chen, 2014). Generally, ion exchange and/or complex formation are the mechanisms underlying the bio-adsorption efficiency of algae. Typically, the entire process occurs in steps. First, the physical attraction is caused by Van der Waals forces. This is followed by the formation of complexes with the active functional groups, and then, mainly with metals, ion exchange with the ions (sodium, potassium, calcium or magnesium) contained in polysaccharides can occur (Crist et al., 1994). For instance, Murphy et al. (2008) investigated the adsorption of Cr (III and VI) by different types of

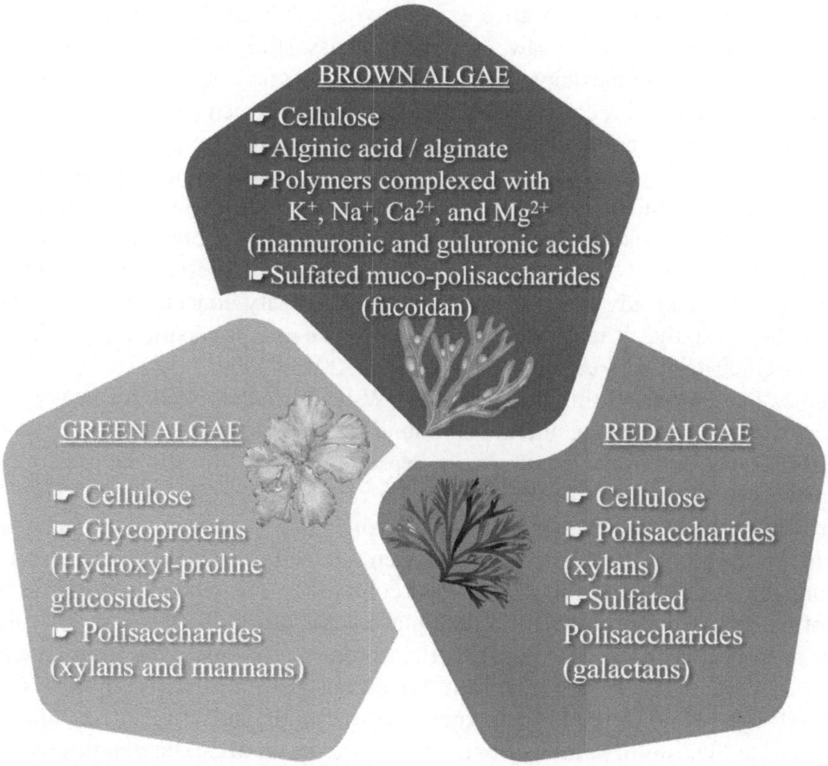

FIGURE 5.5 Main components of the cell wall for green, red and brown macroalgae.

green, red and brown macroalgae. The brown algae *Fucus vesiculosus* showed the best removal performance for Cr (III), whereas, for Cr (VI), it was the red algae *Polysiphonia lanosa*. One more time, it was stated that the pH was the key factor when it comes to adsorption for Cr (III and VI). The maximum uptake for Cr (III), 62.92 mg/g, was achieved at pH 4.5 with carboxyl and sulfonate groups playing the main role. For Cr (VI), pH 2 was the optimum for achieving the maximum uptake, 46.18 mg/g, the point at which amino and carboxyl groups might be protonated, attracting the anionic species of Cr (VI) in the solution. On the other hand, Navarro et al. (2008) studied the adsorption of phenolic compounds by two brown algae, determining that the main mechanism of adsorption was the formation of hydrogen bonds with the hydroxyl groups within the lattice formed by the polysaccharide chains and, again, strongly pH-dependent. As an example of the adsorption of dyes by algae, a study was carried out by Abou Oualid et al. (2020) with green macroalgae of massive growth, *Codium decorticatum*, achieving maximum uptakes of 278.46 mg/g and 191.01 mg/g for crystal violet (cationic) and congo red (anionic), respectively, at pH 6. Hydroxyl, amino and sulfonated groups were intensively involved in the adsorption process. Congo red adsorption was favored below the PZC of the algal surface (6.4), but the presence of the sulfonated groups, thanks to their low pKa (1–2.5), may contribute to the fact that there is no drastic drop in the bio-adsorption capacity of the algae at pH 6 (Murphy et al., 2008).

5.3 AGRICULTURAL, AGRO-INDUSTRIAL AND FORESTRY WASTES (AFWs)

Forestry, agriculture and agro-industrial activities generate tons of amounts of waste per year in both developing and developed countries. AFWs are environmentally friendly and are widely distributed and readily available for local use, thus contributing to a zero-waste circular economy

(Rajapaksha et al., 2019). Some illustrative examples of AFWs are husk, bran, hull and different parts of plants, such as barks, leaves, sawdust, fruit biomass, skin, and stones (Nguyen et al., 2013). AFWs are great candidates to participate in adsorption processes due to their content of functional groups such as amino, phenolic, carboxyl, hydroxyl and ester, which have the ability to form complexes with a wide range of pollutants. All the functional groups are provided by the main components of these AFW, which are cellulose, hemicellulose, lignin, lipids, simple sugars, proteins and water hydrocarbons (Bailey et al., 1999; Saka et al., 2012). In addition, since they are derived from superior organisms whose composition is rich in structural components, they present porosity and a large surface area that benefits the adsorption processes by presenting a greater exposure of functional groups. Another advantage of this increased porosity concerning other bio-adsorbents previously mentioned is that it may present a higher regeneration capability, which is interesting when it comes to industrial applicability (Nguyen et al., 2013).

Many studies have reported the capacity of AFWs as bio-adsorbents for heavy metals, dyes, endocrine-disrupting and pharmaceutical compounds (Bakyayita et al., 2014; Feizi & Jalali, 2015; Kareem, 2016). Some illustrative examples are hereafter presented, as is the case of Martín-Lara et al. (2014), who performed adsorption studies for removing Cr (IV), Cu (II) and Ni (II) from a real electroplating plant wastewater using olive stones in a continuous two fixed-bed columns system. In this study, Cr (IV) was adsorbed and then reduced to Cr (III) by the surrounding olive stone environment at pH 2 in the first fixed bed column. In the second column, at pH 4–5, the adsorption of Cr (III), Cu (II) and Ni (II) took place achieving 5.19, 1.97 and 3.61 mg/g, respectively. Other authors used powdered peanut hulls to remove anionic dyes from an aqueous solution (Gong et al., 2005). These authors referred to an adsorption capacity of 14.90 mg, 13.99 mg and 15.60 mg dye per gram of the bio-adsorbent for amaranth, sunset yellow and fast green FCF, respectively. These maximum values of uptake were achieved at acidic conditions (pH below 2) where the peanut hull was positively charged, and electrostatic interactions took place between the surface and the anionic dyes. Liu et al. (2013) used rice straw to remove clofibric acid and carbamazepine from an aqueous solution. Clofibric acid adsorption was extremely influenced by pH, achieving 126.3 mg/g at pH 2 and plunging to 11.6 at pH 3.5 because of electrostatic interactions, whereas carbamazepine adsorption (40 mg/g) was not affected by pH of the solution but by the dosage of the bio-adsorbent. Notwithstanding, a larger dose of bio-adsorbents could result in a larger dose of dissolved organic carbon in the water increasing color, chemical oxygen demand and biological oxygen demand, leading to depletion of the oxygen content of the water (Liu et al., 2013). Thus, when AFWs are used as bio-adsorbents, this is likely to occur and therefore raises concerns about secondary environmental pollution and should be taken into account (Abegunde et al., 2020).

5.4 ENGINEERED BIO-ADSORBENTS

The excellent characteristics and outstanding performance of bio-adsorbents that have been discussed throughout this chapter also offer the advantage to the ease of modification (Adewuyi, 2020). Engineering the bio-adsorbent surface consists of adapting the bio-adsorbent by tailoring the physical or chemical characteristics of the starting material so as to be suitable for the intended purpose. Thus, tailoring bio-adsorbents offers great simplicity of design and adaptability to meet industrial requirements. The modification technique applied and the nature of the precursor have a direct correlation to surface functionalities (Abegunde et al., 2020). The bio-adsorbent modification is intended to be comprehensive, not only improving the adsorption capacity but also helping to overcome some drawbacks for its practical use such as the lack of selectivity, chemical and thermal stability, low mechanical strength and/or safety of the biomaterial generated by preventing the introduction of unwanted organic compounds into the treated water, as is the case of dissolved organic carbon, mentioned earlier with AFWs (Wan Ngah & Hanafiah, 2008). Mechanisms for engineering bio-adsorbents can be divided in chemical treatments/modifications and thermal treatments (biochars/hydrochar) (Zoroufchi Benis et al., 2020). Some of the mechanisms of modification are related in the next lines.

5.4.1 Chemical Treatments

Acid modification is an oxidative process that implies the application of acids at the bio-adsorbents' surface. Typically, strong acids (HCl, H_2SO_4, HNO_3, H_3PO_4) have been used, but lately, some more environmentally friendly acids, such as acetic, citric or oxalic acid, have gained prominence (Bhatti et al., 2021; Hayoun et al., 2022). Thus, acid treatment increases oxygen-rich active functional groups, such as carboxyl, phenolic hydroxyl, ester and carbonyl, and enhances the polarity and hydrophilicity of the surface (mainly negative) (Figueiredo et al., 1999). As a good example of acid modification, it is worth mentioning the study carried out by Mavioglu et al. (2012). These authors modified *Oreganum onites* L. stalks with two different acids, HNO_3 and H_3PO_4, to improve its adsorption capacity for two basic dyes – basic red 18 and methylene blue – and one acid – acid red 11. The adsorption capacity increased 56%, 63% and 125%, respectively, achieving after the treatment maximum uptakes of 280.73 mg/g, 147.06 mg/g and 112.36 mg/g. The authors suggested that the main mechanism involved in the adsorption of basic dye were surface charge and the acidic groups, while non-electrostatic forces gained influence in the process of acid dye adsorption.

Alkaline modification means treating the bio-adsorbent with a reducing agent. Using alkaline substances like NaOH, KOH, LiOH, Na_2SiO_3, Na_2CO_3 depletes the oxygen-containing functional groups on the bio-adsorbent surface and, hence, increases the alkalinity (Abegunde et al., 2020). An example of an alkalinization reaction on the bio-adsorbent surface is given in Equation 5.1. This process could increase the non-polar surface area, thus improving the adsorption performance of bio-adsorbents for non-polar substances. Additionally, it induces a positive character on the surface, which may boost the interaction between the surface and positive compounds through dipole-dipole interactions, hydrogen bonding and covalent bonding (Rehman et al., 2019). Ofudje et al. (2015) reported a 25.6% increase in Pb (II) adsorption for a coconut shaft treated with KOH over the untreated one. The authors suggested that the alkaline treatment triggered a series of hydrolysis reactions on the surface of the bio-adsorbent, thus converting the inactive surface groups into active ones available for interaction with the heavy metals. In another study, *Streptomyces rimosus* biomass was treated with 1M of NaOH during 1.5 h to increase Zn (II) adsorption from 30 to 80 mg/g due to a Na (I)/Zn (II) exchange, although a 35% of biomass loss was observed after the treatment (Mameri et al., 1999).

$$Surface - OH + NaOH \rightarrow surface - O^{-}Na^{+} + H_2O \tag{5.1}$$

Solvent extraction is usually an easy way to clean the material that avoids pore clogging and usually leads to better adsorption via pore-filling or partitioning increasing specific surface area, pore volume and modifying hydrophobicity or polarity (Schreiter et al., 2020). It might be performed even by applying temperature to the process (Wang et al., 2017). Extraction with organic solvents is found in many studies, as, for example Göksungur et al. (2005), who treated yeast biomass, *Saccharomyces cerevisiae*, with ethanol to improve the adsorption of Cd (II) and Pb (II). They succeeded in shifting from a maximum uptake of about 5 mg/g for both metals to a maximum uptake of 31.75 mg/g for Cd (II) and 60.24 mg/g for Pb (II). This improvement could be explained due to an increase in the access of metal ions to new metal binding sites available on the yeast cell walls.

Metal impregnation is a process of modification of a porous material with a metallic component that involves the uniform distribution of the metallic component on the internal surface. Impregnation can occur either with both components in a solid state or by dissolving the metallic component in a solution (wet impregnation) (Abegunde et al., 2020). These modification techniques are not generally associated with a significant change in surface PZC and are an emerging field of research (Rehman et al., 2019). In a recent study, Wu et al. (2017) investigated the As (III) adsorption capacity of sawdust impregnated with aluminum oxide. They reported that the oxygen functional group present in aluminum oxides was primarily responsible for the adsorption of the pollutant and was due to the complexation mechanism. In addition, using iron to impregnate bio-adsorbent surface could lead to a magnetic adsorbent and would help to overcome some disadvantages, like the difficulty of separating the powdered biochar from the aqueous matrix after treatment (Rajapaksha et al., 2016).

Surface grafting consists of introducing new active functional groups or a greater number of existing ones, which may lead to a modification of the PZC of the bio-adsorbent (Deng & Ting, 2005). Anionic or cationic surfactants and nitrogen-rich compounds such as urea, ammonia or polyethyleneimine are often used for tailoring bio-adsorbents (Bhatti et al., 2021). In the study carried out by Inyinbor et al. (2016), a modified epicarp of *Raphia hookerie* was used to remove rhodamine B from aqueous solution. Aware that the dye reacts mainly with the amino groups present in the textile fibers, the authors grafted amino groups using urea to improve adsorption performance. Maximum uptake of 434.78 mg/g was achieved at pH 3 and for and bio-adsorbent dosage of 2 g/L.

5.4.2 Thermal Treatment

Heating is the simplest and most commonly used engineering technique applicable to bio-adsorbents for adsorption purposes (Rehman et al., 2019). Heating bio-adsorbent causes a thermochemical conversion that provokes the carbonization of the bio-adsorbent and helps to develop a large, specific surface area with a porous structure of molecular dimensions (Alaya et al., 2000). The main three heating methods performed to bio-adsorbents are pyrolysis, hydrothermal carbonization (HTC) and, more recently, microwaves. Pyrolysis is usually applied to biomass having lower water content (< 50%), whereas HTC is a promising treatment for biomass with higher water content (Pala et al., 2014).

Pyrolysis is defined as the thermal decomposition of a starting material in an inert atmosphere at temperatures above 300°C for a certain time. After a pyrolysis process, biochar or carbon is obtained, depending on the conditions (> 80% of carbon). The main factors during production affecting the characteristics of biochar as bio-adsorbent are temperature, residence time and feedstock (Ahmad et al., 2014). Iamsaard et al. (2022) produced biochar at 600°C from pineapple leaves for removing Ni (II), Zn (II) and Cu (II) from aqueous solution. The biochar adsorption capacity was 44.88, 46.00 and 53.14 mg/g for Ni (II), Zn (II) and Cu (II), respectively, at a pH above 2, which was the PZC of the biochar and, thus, promoted the metal cation adsorption and a pH lower than 6 to avoid metal hydroxide precipitation. Cation exchange and surface complexation mechanisms were proposed based on the surface characterization and cation leaching during the assays.

HTC, also sometimes called wet pyrolysis, is applied to avoid the costly pre-drying step, and the obtained solid is termed hydrochar. Hydrochar (55–70% of carbon) is obtained by applying low temperatures (180–250°C) in subcritical water at saturated vapor pressure in a high-pressure reactor for a few minutes to several hours (Missaoui et al., 2017). Several simultaneous reactions take place in the reactor such as hydrolysis, dehydration, decarboxylation, aromatization and recondensation (Benavente et al., 2015). A typical experimental set-up is presented in Figure 5.6. In the production of hydrochar, the factors that determine its subsequent application as bio-adsorbent are temperature, residence time, solid-liquid ratio and feedstock (Pala et al., 2014). For instance, Delgado-Moreno et al. (2021) prepared hydrochar from wet olive mill wastes at 190°C to remove three pharmaceutical compounds. The produced hydrochar at low temperatures developed a great number of oxygen-rich groups, and it achieved great adsorption rates for triclosan (98%), ibuprofen (43%) and diclofenac sodium salt (63%), all comparable to a commercial activated carbon.

5.4.2.1 Pre- and Post-Treatment

Many biochars or hydrochars are further treated to remove organic matter deposited in the pores during the thermal process, develop more pores and activate or include new functional groups.

Washing, leaching or extracting with distilled water is often applied to chars and carbons after the production and activation process for removing the labile/soluble carbon fraction deposited in the pores that cause pore-clogging and also the unreacted activating product (Figure 5.6). This is the case of Escudero-Curiel et al. (2021), who performed a post-treatment and hydrothermal extraction at 121°C for 30 min in an autoclave, achieving a fluoxetine adsorption improvement of 20%. Other authors, as in the case of Guo & Rockstraw (2007), deposited the carbon obtained after modification with H_3PO_4 in a Soxhlet extractor. The carbon was then rinsed with distilled water for 48 h until phosphate ion leaching stopped. On the other hand, other processes can be applied for the activation of carbonaceous materials and, depending on the purposes, might be pre-treatment, post-treatment or at the same time with the carbonization process (Zoroufchi Benis et al., 2020). Activation mechanisms are varied in the literature but are generally the same as the chemical ones listed in the previous section (4.1) and, more recently, microwave activation and physical activation (Rajapaksha et al., 2016). Physical activation consists of using steam or CO_2 as the activation agent, which is environmentally friendly, but this craft demands higher activation temperature and longer activation time, which leads to relatively lower yield and higher energy cost (Xin-hui et al., 2011).

Whereas pyrolysis and HTC can be applied to a broad range of bio-adsorbents, microwaves need to be applied to microwave absorbers (Menéndez et al., 2010). Biomass is generally a poor receptor of microwave energy, so it cannot be heated to the required temperatures of pyrolysis processes, although it might be induced if mixed with dielectric material acting as microwave receptors or absorbers, as is the case of chars (Menéndez et al., 2002). Thus, this method is mostly applied to activate chars in combination with other chemical methods. Microwave irradiation interacts directly with the particles inside the material, producing a transfer of electromagnetic energy into the heat source within the dielectric materials. Therefore, the energy is received through dipole rotations and ionic conduction, allowing shortened treatment time and saving energy. In addition, since the sample is not driven by an external heat source, rapid volumetric and uniform heating is achieved as well as precise and immediate control of start-up and shut-down (Xin-hui et al., 2011). Therefore, the use of microwaves increased drastically the specific surface area and pore volume.

The following are some examples of carbonaceous materials enhanced by the previously described methods. Activated carbon from peanut shells pre-treated with microwaves and H_2SO_4 were used for removing direct black 38 and reactive red 141 from aqueous solutions by Georgin et al. (2016). The specific surface area increased from 2 cm^2/g to 395.8 cm^2/g, the average pore volume had a ten-point increase and the maximum uptake achieved at pH 2.5 was 110.6 and 284.5 mg/g when starting from values close to 15 and 50 mg/g for direct black 38 and reactive red 141, respectively. Iron-modified biochar was produced to remove Cu (II) and sulfadiazine, changing from

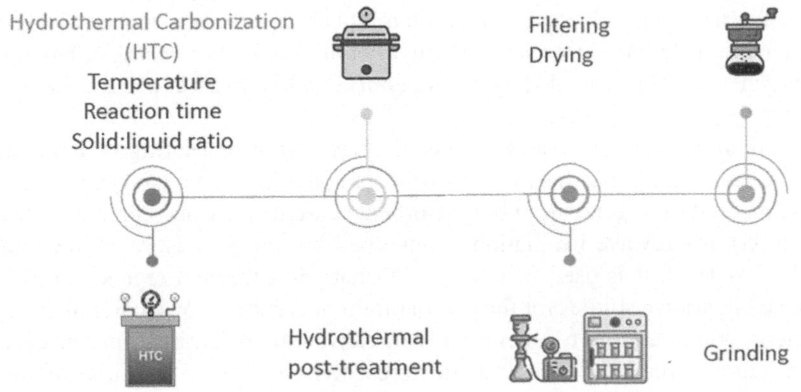

FIGURE 5.6 HTC's general production scheme.

competitive to synergistic behavior in the adsorption process (Yan et al., 2022). The study showed that the treatment not only increased the specific surface area of the biochar but also impregnated the biochar surface with nanoparticles of iron oxide. Iron oxide resulted to act as a physical barrier between the adsorption sites for Cu (II) and sulfadiazine, thus inhibiting competition and improving the total adsorption capacity for both pollutants (46.85 and 45.43 mg/g, respectively). Xu et al. (2022) modified wheat straw biochar with $KOH/KMnO_4$ to adsorb tetracycline. They obtained biochar with a hierarchical structure that presented a specific surface area of 1524.6 cm^2/g and achieved a great adsorption capacity of 589.19 mg/g at a wide range of pH (3–10). The authors indicated that the plausible mechanism of adsorption was ascribed to the pore filling and π-π interaction, followed by hydrogen bonding and metal complexation. In another study, Jiang et al. (2019) performed two different modifications, H_3PO_4 and polyethyleneimine, to a corn-straw hydrochar for Pb (II) adsorption. Both modified hydrochar exhibited excellent adsorption yield for Pb (II) adsorption with maximum uptakes of 353.4 mg/g and 214.0 mg/g, respectively, whereas the adsorption capacity of untreated hydrochar was 32.67 mg/g. The adsorption mechanisms attributed to acid-modified hydrochar were ion exchange and hydrogen bonding, and for polyethyleneimine-modified hydrochar, they were coordination bonding and hydrogen bonding.

5.5 RECOVERY AND REGENERATION

Recovery and sustainable management of spent adsorbents are one of the main problems facing the application of adsorption processes for the removal of pollutants from wastewater streams (Baskar et al., 2022). Thus, only by applying coupled adsorption-regeneration techniques will environmental science be in line with the circular economy and the principles of zero-waste (Figure 5.7). Ideal adsorbents exhibit reusability and recoverability for commercial and industrial applications and potentially minimize the cost associated with adsorbent manufacturing. Notwithstanding, the regeneration may limit the lifetime of the adsorbent itself, reducing its adsorption capacity after several applications (Rial & Ferreira, 2022). The adsorbent and its stability, pollutant and toxicity are factors of great importance when selecting regeneration techniques from the point of view of industrial feasibility (Baskar et al., 2022). Some of the regeneration techniques successfully applied for bio-adsorbents are listed after Figure 5.7.

Solvent Extraction, alone or combined with heating, is a much used and simple technique for pollutant desorption from loaded bio-adsorbents, usually implying no degradation of the pollutant, which is especially interesting for recovering substances economically exploited like metals. It consists in breaking the adsorption equilibrium by changing several solution parameters such as pH and temperature. Some of the extraction solutions include acids (HNO_3, HCl), bases (NaOH) and organic solvents (ethanol, methanol, acetonitrile). Each is selected depending on the nature of the bio-adsorbent and pollutant (Fdez-Sanromán et al., 2020). It is worth mentioning that some authors have pointed out that solvent regeneration might change the physicochemical properties and adsorption capacity of the bio-adsorbents (Reguyal et al., 2017). For example, Huang et al. (2013) used HCl to regenerate Cd (II)-loaded *Bacillus cereus* dead biomass at pH 1, achieving a 91.2% of Cd (II) recovery.

Thermal and microwave regeneration techniques comprise heating or irradiating to break the physical and chemical bonds between adsorbents and adsorbates. Conventional heating is the most extensive method for regenerating bio-adsorbents and the only one applied in treatment plants (Ania et al., 2005). Microwave irradiation is adsorbed by the bio-adsorbent and transferred into heat at a molecular level. It is used as a faster alternative to thermal regeneration (Baskar et al., 2022). In some comparative studies for the regeneration of carbonaceous materials, rapid heating by microwave energy led to a rapid regeneration than conventional heating. Furthermore, microwave facilitated the reuses several times, as it better preserves the porous structure of the regenerated adsorbents compared to conventional thermal treatments (Menéndez et al., 2010). Rocha et al. (2021) performed the regeneration of a recoverable magnetic iron-oxide functionalized activated carbon

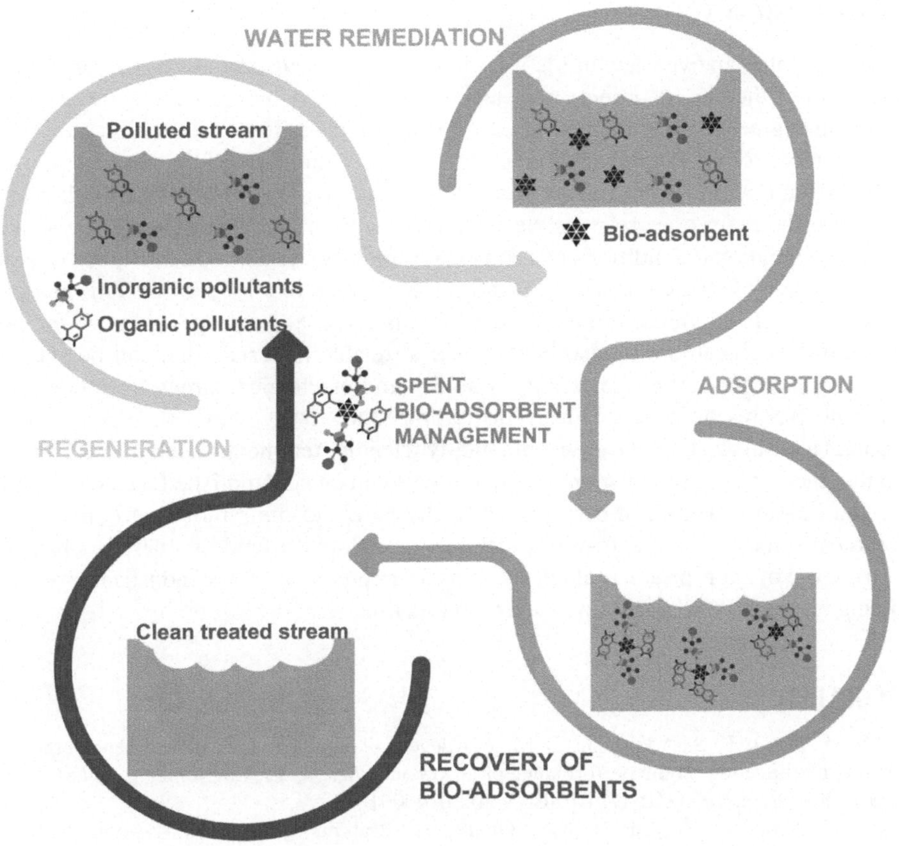

FIGURE 5.7 Diagram of the remediation of polluted streams by sustainable adsorption processes.

loaded with diclofenac and venlafaxine by microwave at 400°C for 20 minutes. They achieved a complete regeneration without affecting the adsorption capacity of the carbon in a second adsorption cycle of the same compounds.

Advanced oxidation processes are applied in an attempt to achieve complete mineralization of contaminants on the adsorbent. Thus, the regeneration of the porous matrix by generating oxidizing radicals (mainly sulfate or hydroxyl radicals) is produced. Various techniques like Fenton, Fenton-like, photo-Fenton, electro-Fenton and ozonation, alone or combined with other treatments, are the typical processes applied to adsorbent regeneration (Gallego-Ramírez et al., 2022; Omorogie et al., 2014). Thus, Acevedo-García et al. (2020) accomplished the electro-Fenton regeneration of biochar, where hydroxyl radicals generated by the electric field action react with the pollutant in the porous matrix and effective regeneration was achieved. Furthermore, the regeneration of biochar can be produced by several cycles, demonstrating the sustainability of the process. In some cases, the bio-adsorbent has even been used as a catalyst for the advanced oxidation process as a result of its nature. Accordingly, Escudero-Curiel et al. (2021) accomplished the effective regeneration of biochar polluted with the drug fluoxetine by the action of sulfate radicals (Fenton-like). They found that the mineral content of the biochar acts as a catalyst enhancing the sulphate generation of biochar in presence of citric acid as a complexing agent. Furthermore, the applicability and possibility of scaling up the system proposed were efficiently evaluated in a continuous treatment using a fixed-bed column reactor.

5.6 CONCLUSIONS

In the present chapter, an overview of the use of biomass as an adsorbent was presented. It was disclosed that most of the bio-adsorbents presented a non-selective nature, and by modifying and tailoring them, an engineered adsorbent with adsorptive properties to target specific pollutants can be obtained. The most common mechanisms of adsorption for unmodified biomass from microorganisms are hydrogen bonding, ion exchange and π-π interaction. On the other hand, the electrostatic interactions are the main ones responsible for the adsorption process by using AFW. It was disclosed that the operational conditions can drastically modify the adsorption capacity, and solution pH has been pointed out to be the critical factor for adsorption in most studies. This factor modifies the predominancy of functional groups on the bio-adsorbent and the speciation of the pollutants. Furthermore, the engineered bio-adsorbent provides tunable materials that can be synthetized ad hoc, enhancing their properties (selectivity, stability and mechanical strength); therefore, this has aroused a powerful research line. However, the procedure for synthesis should be considered in order to obtain a cost-effective and environmentally friendly treatment. In addition, it was also demonstrated that the regeneration of spent bio-adsorbents can be accomplished. Hence, the adsorption process using bio-adsorbents can be a plausible alternative to the removal of contaminants from polluted effluents in the way of zero-waste technologies. Notwithstanding, there is a lack of investigation into the use of appropriate, scalable systems for applications at the industrial level. Therefore, the following research should also be focused on this line.

BIBLIOGRAPHY

Abegunde, S. M., Idowu, K. S., Adejuwon, O. M., & Adeyemi-Adejolu, T. (2020). A review on the influence of chemical modification on the performance of adsorbents. *Resources, Environment and Sustainability*, *1*, 100001. https://doi.org/10.1016/J.RESENV.2020.100001

Abou Oualid, H., Abdellaoui, Y., Laabd, M., el Ouardi, M., Brahmi, Y., Iazza, M., & Abou Oualid, J. (2020). Eco-efficient green seaweed codium decorticatum biosorbent for textile dyes: Characterization, mechanism, recyclability, and rsm optimization. *ACS Omega*, *5*(35), 22192–22207. https://doi.org/10.1021/ACSOMEGA.0C02311/ASSET/IMAGES/MEDIUM/AO0C02311_M013.GIF

Acevedo-García, V., Rosales, E., Puga, A., Pazos, M., & Sanromán, M. A. (2020). Synthesis and use of efficient adsorbents under the principles of circular economy: Waste valorisation and electroadvanced oxidation process regeneration. *Separation and Purification Technology*, *242*, 116796. https://doi.org/10.1016/J.SEPPUR.2020.116796

Adewuyi, A. (2020). Chemically modified biosorbents and their role in the removal of emerging pharmaceuticalwaste in the water system. In *Water (Switzerland)* (Vol. 12, Issue 6). MDPI AG. https://doi.org/10.3390/W12061551

Afrooz, M. R., Moghadas, B. K., & Tamjidi, S. (2022). Performance of functionalized bacterial as bio-adsorbent for intensifying heavy metal uptake from wastewater: A review study. *Journal of Alloys and Compounds*, *893*, 162321. https://doi.org/10.1016/J.JALLCOM.2021.162321

Ahmad, M., Rajapaksha, A. U., Lim, J. E., Zhang, M., Bolan, N., Mohan, D., Vithanage, M., Lee, S. S., & Ok, Y. S. (2014). Biochar as a sorbent for contaminant management in soil and water: A review. *Chemosphere*, *99*, 19–33. https://doi.org/10.1016/j.chemosphere.2013.10.071

Akhtar, J., Amin, N. A. S., & Shahzad, K. (2016). A review on removal of pharmaceuticals from water by adsorption. *Desalination and Water Treatment*, *57*(27), 12842–12860. https://doi.org/10.1080/19443994.2015.1051121

Aksu, Z., & Tezer, S. (2005). Biosorption of reactive dyes on the green alga Chlorella vulgaris. *Process Biochemistry*, *40*(3–4), 1347–1361. https://doi.org/10.1016/J.PROCBIO.2004.06.007

Aksu, Z., & Tunç, Ö. (2005). Application of biosorption for penicillin G removal: Comparison with activated carbon. *Process Biochemistry*, *40*(2), 831–847. https://doi.org/10.1016/J.PROCBIO.2004.02.014

Alaya, M. N., Girgis, B. S., & Mourad, W. E. (2000). Activated carbon from some agricultural wastes under action of one-step steam pyrolysis. *Journal of Porous Materials*, *7*, 509–517.

Ania, C. O., Parra, J. B., Menéndez, J. A., & Pis, J. J. (2005). Effect of microwave and conventional regeneration on the microporous and mesoporous network and on the adsorptive capacity of activated carbons. *Microporous and Mesoporous Materials*, *85*(1–2), 7–15. https://doi.org/10.1016/J.MICROMESO.2005.06.013

Aryal, M., & Liakopoulou-Kyriakides, M. (2015). Bioremoval of heavy metals by bacterial biomass. *Environmental Monitoring and Assessment*, *187*(1), 1–26. https://doi.org/10.1007/S10661-014-4173-Z/TABLES/3

Bahafid, W., Joutey, N. T., Asri, M., Sayel, H., Tirry, N., & el Ghachtouli, N. (2017). Yeast biomass: An alternative for bioremediation of heavy metals. In *Yeast—industrial applications*. InTech. https://doi.org/10.5772/intechopen.70559

Bailey, S. E., Olin, T. J., Bricka, R. M., & Adrian, D. D. (1999). A review of potentially low-cost sorbents for heavy metals. *Water Research*, *33*(11), 2469–2479. https://doi.org/10.1016/S0043-1354(98)00475-8

Bakyayita, G. K., Norrström, A. C., Nalubega, M., & Kulabako, R. N. (2014). Kinetic studies of Cd (II) and Pb (II) ions biosorption from aqueous media using untreated and chemically treated biosorbents. *Water Science and Technology*, *69*(11), 2230–2236. https://doi.org/10.2166/wst.2014.147

Baskar, A. V., Bolan, N., Hoang, S. A., Sooriyakumar, P., Kumar, M., Singh, L., Jasemizad, T., Padhye, L. P., Singh, G., Vinu, A., Sarkar, B., Kirkham, M. B., Rinklebe, J., Wang, S., Wang, H., Balasubramanian, R., & Siddique, K. H. M. (2022). Recovery, regeneration and sustainable management of spent adsorbents from wastewater treatment streams: A review. *Science of the Total Environment*, *822*. https://doi.org/10.1016/J.SCITOTENV.2022.153555

Benavente, V., Calabuig, E., & Fullana, A. (2015). Upgrading of moist agro-industrial wastes by hydrothermal carbonization. *Journal of Analytical and Applied Pyrolysis*, *113*, 89–98. https://doi.org/10.1016/J.JAAP.2014.11.004

Bhatti, H. N., Sadaf, S., Naz, M., Iqbal, M., Safa, Y., Ain, H. U., Nawaz, S., & Nazir, A. (2021). Enhanced adsorption of Foron Black RD 3GRN dye onto sugarcane bagasse biomass and na-alginate composite. *Desalination and Water Treatment*, *216*, 423–435. https://doi.org/10.5004/dwt.2021.26893

Cheng, N., Wang, B., Wu, P., Lee, X., Xing, Y., Chen, M., & Gao, B. (2021). Adsorption of emerging contaminants from water and wastewater by modified biochar: A review. *Environmental Pollution*, *273*, 116448. https://doi.org/10.1016/J.ENVPOL.2021.116448

Coimbra, R. N., Escapa, C., Vázquez, N. C., Noriega-Hevia, G., & Otero, M. (2018). Utilization of non-living microalgae biomass from two different strains for the adsorptive removal of diclofenac from water. *Water*, *10*(10), 1401. https://doi.org/10.3390/W10101401

Crist, R. H., Martin, J., Carr, D., Watson, J. R., Clarke, H. J., & Crist, D. L. R. (1994). Interaction of metals and protons with algae. 4. Ion exchange vs adsorption models and a reassessment of scatchard plots; ion-exchange rates and equilibria compared with calcium alginate. *Environmental Science and Technology*, *28*(11), 1859–1866. https://doi.org/10.1021/es00060a016

Dai, Y., Zhang, N., Xing, C., Cui, Q., & Sun, Q. (2019). The adsorption, regeneration and engineering applications of biochar for removal organic pollutants: A review. *Chemosphere*, *223*, 12–27. https://doi.org/10.1016/J.CHEMOSPHERE.2019.01.161

de Freitas, G. R., da Silva, M. G. C., & Vieira, M. G. A. (2019). Biosorption technology for removal of toxic metals: A review of commercial biosorbents and patents. *Environmental Science and Pollution Research*, *26*(19), 19097–19118. https://doi.org/10.1007/s11356-019-05330-8

Delgado-Moreno, L., Bazhari, S., Gasco, G., Méndez, A., El Azzouzi, M., & Romero, E. (2021). New insights into the efficient removal of emerging contaminants by biochars and hydrochars derived from olive oil wastes. *Science of The Total Environment*, *752*, 141838. https://doi.org/10.1016/j.scitotenv.2020.141838

Deng, S., & Ting, Y. P. (2005). Polyethylenimine-modified fungal biomass as a high-capacity biosorbent for Cr(VI) anions: Sorption capacity and uptake mechanisms. *Environmental Science & Technology*, *39*(21), 8490–8496. https://doi.org/10.1021/es050697u

Elnahas, M. O., Hou, L., Wall, J. D., & Majumder, E. L.-W. (2021). Bioremediation potential of streptomyces sp. MOE6 for toxic metals and oil. *Polysaccharides*, *2*(1), 47–68. https://doi.org/10.3390/POLYSACCHARIDES2010004

Escudero-Curiel, S., Penelas, U., Sanromán, M. Á., & Pazos, M. (2021). An approach towards zero-waste wastewater technology: Fluoxetine adsorption on biochar and removal by the sulfate radical. *Chemosphere*, *268*, 129318. https://doi.org/10.1016/j.chemosphere.2020.129318

Fdez-Sanromán, A., Pazos, M., Rosales, E., & Sanromán, M. A. (2020). Unravelling the environmental application of biochar as low-cost biosorbent: A review. *Applied Sciences*, *10*(21), 7810. https://doi.org/10.3390/APP10217810

Feizi, M., & Jalali, M. (2015). Removal of heavy metals from aqueous solutions using sunflower, potato, canola and walnut shell residues. *Journal of the Taiwan Institute of Chemical Engineers*, *54*, 125–136. https://doi.org/10.1016/J.JTICE.2015.03.027

Figueiredo, J. L., Pereira, M. F. R., Freitas, M. M. A., & Órfão, J. J. M. (1999). Modification of the surface chemistry of activated carbons. *Carbon*, *37*(9), 1379–1389. https://doi.org/10.1016/S0008-6223(98)00333-9

Fomina, M., & Gadd, G. M. (2014). Biosorption: Current perspectives on concept, definition and application. *Bioresource Technology*, *160*, 3–14. https://doi.org/10.1016/J.BIORTECH.2013.12.102

Fourest, E., Canal, C., & Roux, J.-C. (1994). Improvement of heavy metal biosorption by mycelial dead biomasses (Rhizopus arrhizus, Mucor miehei and Penicillium chrysogenum): pH control and cationic activation. *FEMS Microbiology Reviews*, *14*(4), 325–332. https://doi.org/10.1111/j.1574-6976.1994.tb00106.x

Gahlout, M., Prajapati, H., Tandel, N., & Patel, Y. (2021). *Biosorption: An eco-friendly technology for pollutant removal*. In D. G. Panpatte & Y. K. Jhala (eds.), *Microbial Rejuvenation of polluted environment. Microorganisms for sustainability* (Vol. 26). Springer. https://doi.org/10.1007/978-981-15-7455-9_9

Gallego-Ramírez, C., Chica, E., & Rubio-Clemente, A. (2022). Coupling of advanced oxidation technologies and biochar for the removal of dyes in water. *Water*, *14*(16), 2531. https://doi.org/10.3390/W14162531

Gavrilescu, M. (2004). Removal of heavy metals from the environment by biosorption. *Engineering in Life Sciences*, *4*(3), 219–232. https://doi.org/10.1002/ELSC.200420026

Georgin, J., Dotto, G. L., Mazutti, M. A., & Foletto, E. L. (2016). Preparation of activated carbon from peanut shell by conventional pyrolysis and microwave irradiation-pyrolysis to remove organic dyes from aqueous solutions. *Journal of Environmental Chemical Engineering*, *4*(1), 266–275. https://doi.org/10.1016/J.JECE.2015.11.018

Göksungur, Y., Üren, S., & Güvenç, U. (2005). Biosorption of cadmium and lead ions by ethanol treated waste baker's yeast biomass. *Bioresource Technology*, *96*(1), 103–109. https://doi.org/10.1016/J.BIORTECH.2003.04.002

Gong, R., Ding, Y., Li, M., Yang, C., Liu, H., & Sun, Y. (2005). Utilization of powdered peanut hull as biosorbent for removal of anionic dyes from aqueous solution. *Dyes and Pigments*, *64*(3), 187–192. https://doi.org/10.1016/J.DYEPIG.2004.05.005

Guo, Y., & Rockstraw, D. A. (2007). Physicochemical properties of carbons prepared from pecan shell by phosphoric acid activation. *Bioresource Technology*, *98*(8), 1513–1521. https://doi.org/10.1016/J.BIORTECH.2006.06.027

Hayoun, B., Escudero-Curiel, S., Bourouina, M., Bourouina-Bacha, S., Angeles Sanromán, M., & Pazos, M. (2022). Preparation and characterization of high performance hydrochar for efficient adsorption of drugs mixture. *Journal of Molecular Liquids*, *353*, 118797. https://doi.org/10.1016/J.MOLLIQ.2022.118797

He, J., & Chen, J. P. (2014). A comprehensive review on biosorption of heavy metals by algal biomass: Materials, performances, chemistry, and modeling simulation tools. *Bioresource Technology*, *160*, 67–78. https://doi.org/10.1016/J.BIORTECH.2014.01.068

Huang, F., Dang, Z., Guo, C. L., Lu, G. N., Gu, R. R., Liu, H. J., & Zhang, H. (2013). Biosorption of Cd(II) by live and dead cells of Bacillus cereus RC-1 isolated from cadmium-contaminated soil. *Colloids and Surfaces B: Biointerfaces*, *107*, 11–18. https://doi.org/10.1016/J.COLSURFB.2013.01.062

Iamsaard, K., Weng, C. H., Yen, L. T., Tzeng, J. H., Poonpakdee, C., & Lin, Y. T. (2022). Adsorption of metal on pineapple leaf biochar: Key affecting factors, mechanism identification, and regeneration evaluation. *Bioresource Technology*, *344*, 126131. https://doi.org/10.1016/J.BIORTECH.2021.126131

Inyinbor, A. A., Adekola, F. A., & Olatunji, G. A. (2016). Kinetics and isothermal modeling of liquid phase adsorption of rhodamine B onto urea modified Raphia hookerie epicarp. *Applied Water Science*, *7*(6), 3257–3266. https://doi.org/10.1007/S13201-016-0471-7

Iqbal, M., & Saeed, A. (2007). Biosorption of reactive dye by loofa sponge-immobilized fungal biomass of Phanerochaete chrysosporium. *Process Biochemistry*, *42*(7), 1160–1164. https://doi.org/10.1016/J.PROCBIO.2007.05.014

Jiang, Q., Xie, W., Han, S., Wang, Y., & Zhang, Y. (2019). Enhanced adsorption of Pb(II) onto modified hydrochar by polyethyleneimine or H3PO4: An analysis of surface property and interface mechanism. *Colloids and Surfaces A: Physicochemical and Engineering Aspects*, *583*, 123962. https://doi.org/10.1016/J.COLSURFA.2019.123962

Jureczko, M., & Przystaś, W. (2021). Removal of two cytostatic drugs: Bleomycin and vincristine by white-rot fungi—a sorption study. *Journal of Environmental Health Science and Engineering*, *19*(1), 651–662. https://doi.org/10.1007/S40201-021-00635-8/TABLES/5

Kareem, K. A. (2016). Removal and recovery of methylene blue dye from aqueous solution using Avena fatua seed husk. *Ibn Al-Haitham Journal for Pure and Applied Science*, *29*(3), 179–194.

Kiran Marella, T., Saxena, A., & Tiwari, A. (2020). Diatom mediated heavy metal remediation: A review. *Bioresource Technology*, *305*, 123068. https://doi.org/10.1016/J.BIORTECH.2020.123068

Kumar, D., Pandey, L. K., & Gaur, J. P. (2016). Metal sorption by algal biomass: From batch to continuous system. *Algal Research*, *18*, 95–109. https://doi.org/10.1016/J.ALGAL.2016.05.026

Legorreta-Castañeda, A. J., Lucho-Constantino, C. A., Beltrán-Hernández, R. I., Coronel-Olivares, C., & Vázquez-Rodríguez, G. A. (2020). Biosorption of water pollutants by fungal pellets. *Water*, *12*(4), 1155. https://doi.org/10.3390/W12041155

Li, X., Xu, Q., Han, G., Zhu, W., Chen, Z., He, X., & Tian, X. (2009). Equilibrium and kinetic studies of copper(II) removal by three species of dead fungal biomasses. *Journal of Hazardous Materials*, *165*(1–3), 469–474. https://doi.org/10.1016/J.JHAZMAT.2008.10.013

Lin, Z., Li, J., Luan, Y., & Dai, W. (2020). Application of algae for heavy metal adsorption: A 20-year meta-analysis. *Ecotoxicology and Environmental Safety*, *190*, 110089. https://doi.org/10.1016/J.ECOENV.2019.110089

Liu, L., Liu, J., Liu, X., Dai, C., Zhang, Z., Song, W., & Chu, Y. (2019). Kinetic and equilibrium of U(VI) biosorption onto the resistant bacterium Bacillus amyloliquefaciens. *Journal of Environmental Radioactivity*, *203*, 117–124. https://doi.org/10.1016/J.JENVRAD.2019.03.008

Liu, Z., Zhou, X., Chen, X., Dai, C., Zhang, J., & Zhang, Y. (2013). Biosorption of clofibric acid and carbamazepine in aqueous solution by agricultural waste rice straw. *Journal of Environmental Sciences (China)*, *25*(12), 2384–2395. https://doi.org/10.1016/S1001-0742(12)60324-6

Mameri, N., Boudries, N., Addour, L., Belhocine, D., Lounici, H., Grib, H., & Pauss, A. (1999). Batch zinc biosorption by a bacterial nonliving *Streptomyces rimosus* biomass. *Water Research*, *33*(6), 1347–1354. https://doi.org/10.1016/S0043-1354(98)00349-2

Martín-Lara, M. A., Blázquez, G., Trujillo, M. C., Pérez, A., & Calero, M. (2014). New treatment of real electroplating wastewater containing heavy metal ions by adsorption onto olive stone. *Journal of Cleaner Production*, *81*, 120–129. https://doi.org/10.1016/J.JCLEPRO.2014.06.036

Mavioglu Ayan, E., Secim, P., Karakaya, S., & Yanik, J. (2012). Oreganum stalks as a new biosorbent to remove textile dyes from aqueous solutions. *Clean—Soil, Air, Water*, *40*(8), 856–863. https://doi.org/10.1002/CLEN.201100153

Menéndez, J. A., Arenillas, A., Fidalgo, B., Fernández, Y., Zubizarreta, L., Calvo, E. G., & Bermúdez, J. M. (2010). Microwave heating processes involving carbon materials. *Fuel Processing Technology*, *91*(1), 1–8. https://doi.org/10.1016/J.FUPROC.2009.08.021

Menéndez, J. A., Inguanzo, M., & Pis, J. J. (2002). Microwave-induced pyrolysis of sewage sludge. *Water Research*, *36*(13), 3261–3264. https://doi.org/10.1016/S0043-1354(02)00017-9

Missaoui, A., Bostyn, S., Belandria, V., Cagnon, B., Sarh, B., & Gökalp, I. (2017). Hydrothermal carbonization of dried olive pomace: Energy potential and process performances. *Journal of Analytical and Applied Pyrolysis*, *128*, 281–290. https://doi.org/10.1016/j.jaap.2017.09.022

Mittal, A. K., & Gupta, S. K. (1996). Biosorption of cationic dyes by dead macro fungus Fomitopsis carnea: Batch studies. *Water Science and Technology*, *34*(10), 81–87. https://doi.org/10.1016/S0273-1223(96)00700-7

Mohapatra, R. K., Parhi, P. K., Pandey, S., Bindhani, B. K., Thatoi, H., & Panda, C. R. (2019). Active and passive biosorption of Pb(II) using live and dead biomass of marine bacterium Bacillus xiamenensis PbRPSD202: Kinetics and isotherm studies. *Journal of Environmental Management*, *247*, 121–134. https://doi.org/10.1016/J.JENVMAN.2019.06.073

Murphy, V., Hughes, H., & McLoughlin, P. (2008). Comparative study of chromium biosorption by red, green and brown seaweed biomass. *Chemosphere*, *70*(6), 1128–1134. https://doi.org/10.1016/j.chemosphere.2007.08.015

Navarro, A. E., Portales, R. F., Sun-Kou, M. R., & Llanos, B. P. (2008). Effect of pH on phenol biosorption by marine seaweeds. *Journal of Hazardous Materials*, *156*(1–3), 405–411. https://doi.org/10.1016/j.jhazmat.2007.12.039

Nguyen, T. A. H., Ngo, H. H., Guo, W. S., Zhang, J., Liang, S., Yue, Q. Y., Li, Q., & Nguyen, T. V. (2013). Applicability of agricultural waste and by-products for adsorptive removal of heavy metals from wastewater. *Bioresource Technology*, *148*, 574–585. https://doi.org/10.1016/j.biortech.2013.08.124

Ofudje, E. A., Akiode, O. K., Oladipo, G. O., Adedapo, A. E., Adebayo, L. O., & Awotula, A. O. (2015). Application of raw and alkaline-modified coconut shaft as a biosorbent for Pb2+ removal. *BioResources*, *10*(2), 3462–3480. https://doi.org/10.15376/biores.10.2.3462-3480

Oh, S. E., Hassan, S. H. A., & Joo, J. H. (2009). Biosorption of heavy metals by lyophilized cells of Pseudomonas stutzeri. *World Journal of Microbiology and Biotechnology*, *25*(10), 1771–1778. https://doi.org/10.1007/S11274-009-0075-6/FIGURES/7

Omorogie, M. O., Babalola, J. O., & Unuabonah, E. I. (2014). Regeneration strategies for spent solid matrices used in adsorption of organic pollutants from surface water: A critical review. *New Pub: Balaban*, *57*(2), 518–544. https://doi.org/10.1080/19443994.2014.967726

Ozdemir, G., Ozturk, T., Ceyhan, N., Isler, R., & Cosar, T. (2003). Heavy metal biosorption by biomass of Ochrobactrum anthropi producing exopolysaccharide in activated sludge. *Bioresource Technology*, *90*(1), 71–74. https://doi.org/10.1016/S0960-8524(03)00088-9

Özer, A., & Özer, D. (2003). Comparative study of the biosorption of Pb(II), Ni(II) and Cr(VI) ions onto S. cerevisiae: Determination of biosorption heats. *Journal of Hazardous Materials*, *100*(1–3), 219–229. https://doi.org/10.1016/S0304-3894(03)00109-2

Pala, M., Kantarli, I. C., Buyukisik, H. B., & Yanik, J. (2014). Hydrothermal carbonization and torrefaction of grape pomace: A comparative evaluation. *Bioresource Technology*, *161*, 255–262. https://doi.org/10.1016/J.BIORTECH.2014.03.052

Rajapaksha, A. U., Chen, S. S., Tsang, D. C. W., Zhang, M., Vithanage, M., Mandal, S., Gao, B., Bolan, N. S., & Ok, Y. S. (2016). Engineered/designer biochar for contaminant removal/immobilization from soil and water: Potential and implication of biochar modification. In *Chemosphere* (Vol. 148, pp. 276–291). Elsevier Ltd. https://doi.org/10.1016/j.chemosphere.2016.01.043

Rajapaksha, A. U., Dilrukshi Premarathna, K. S., Gunarathne, V., Ahmed, A., & Vithanage, M. (2019). Sorptive removal of pharmaceutical and personal care products from water and wastewater. In *Pharmaceuticals and personal care products: Waste management and treatment technology emerging contaminants and micro pollutants* (pp. 213–238). Elsevier. https://doi.org/10.1016/B978-0-12-816189-0.00009-3

Reguyal, F., Sarmah, A. K., & Gao, W. (2017). Synthesis of magnetic biochar from pine sawdust via oxidative hydrolysis of FeCl2 for the removal sulfamethoxazole from aqueous solution. *Journal of Hazardous Materials*, *321*, 868–878. https://doi.org/10.1016/J.JHAZMAT.2016.10.006

Rehman, A., Park, M., & Park, S. J. (2019). Current progress on the surface chemical modification of carbonaceous materials. *Coatings*, *9*(2), 103. https://doi.org/10.3390/COATINGS9020103

Rial, J. B., & Ferreira, M. L. (2022). Potential applications of spent adsorbents and catalysts: Re-valorization of waste. *Science of the Total Environment*, *823*, 153370. https://doi.org/10.1016/J.SCITOTENV.2022.153370

Rocha, L. S., Sousa, É. M. L., Pereira, D., Gil, M. V., Otero-Irurueta, G., Hortigüela Gallo, M. J., Otero, M., Esteves, V. I., & Calisto, V. (2021). Sustainable and recoverable waste-based magnetic nanocomposites used for the removal of pharmaceuticals from wastewater. *Chemical Engineering Journal*, *426*, 129974. https://doi.org/10.1016/J.CEJ.2021.129974

Rodríguez-Tirado, V., Green-Ruiz, C., & Gómez-Gil, B. (2012). Cu and Pb biosorption on Bacillus thioparans strain U3 in aqueous solution: Kinetic and equilibrium studies. *Chemical Engineering Journal*, *181–182*, 352–359. https://doi.org/10.1016/J.CEJ.2011.11.091

Sağ, Y. (2001). Biosorption of heavy metals by fungal biomass and modeling of fungal biosorption: A review. *Separation and Purification Methods*, *30*(1), 1–48. https://doi.org/10.1081/SPM-100102984

Sahmoune, M. N. (2018). Performance of *Streptomyces rimosus* biomass in biosorption of heavy metals from aqueous solutions. *Microchemical Journal*, *141*, 87–95. https://doi.org/10.1016/j.microc.2018.05.009

Saka, C., Şahin, Ö., & Küçük, M. M. (2012). Applications on agricultural and forest waste adsorbents for the removal of lead (II) from contaminated waters. *International Journal of Environmental Science and Technology*, *9*(2), 379–394. https://doi.org/10.1007/S13762-012-0041-Y

Schreiter, I. J., Schmidt, W., Kumar, A., Graber, E. R., & Schüth, C. (2020). Effect of water leaching on biochar properties and its impact on organic contaminant sorption. *Environmental Science and Pollution Research*, *27*(1), 691–703. https://doi.org/10.1007/s11356-019-06904-2

Silva, A., Delerue-Matos, C., Figueiredo, S. A., & Freitas, O. M. (2019). The use of algae and fungi for removal of pharmaceuticals by bioremediation and biosorption processes: A review. *Water*, *11*(8), 1555. https://doi.org/10.3390/W11081555

Torres, E. (2020). Biosorption: A review of the latest advances. *Processes*, *8*(12), 1584. https://doi.org/10.3390/PR8121584

Tsezos, M., & Volesky, B. (1981). Biosorption of uranium and thorium. *Biotechnology and Bioengineering*, *23*(3), 583–604. https://doi.org/10.1002/bit.260230309

Tsezos, M., & Volesky, B. (1982). The mechanism of uranium biosorption by Rhizopus arrhizus. *Biotechnology and Bioengineering*, *24*(2), 385–401. https://doi.org/10.1002/bit.260240211

Vélez, J. M. B., Martínez, J. G., Ospina, J. T., & Agudelo, S. O. (2021). Bioremediation potential of Pseudomonas genus isolates from residual water, capable of tolerating lead through mechanisms of exopolysaccharide production and biosorption. *Biotechnology Reports*, *32*, e00685. https://doi.org/10.1016/J.BTRE.2021.E00685

Vieira, R. H., & Volesky, B. (2000). Biosorption: A solution to pollution? *International Microbiology: The Official Journal of the Spanish Society for Microbiology*, *3*(1), 17–24. www.ncbi.nlm.nih.gov/pubmed/10963329

Vijayaraghavan, K., & Yun, Y.-S. (2008). Bacterial biosorbents and biosorption. *Biotechnology Advances*, *26*(3), 266–291. https://doi.org/10.1016/j.biotechadv.2008.02.002

Vimala, R., & Das, N. (2009). Biosorption of cadmium (II) and lead (II) from aqueous solutions using mushrooms: A comparative study. *Journal of Hazardous Materials*, *168*(1), 376–382. https://doi.org/10.1016/J.JHAZMAT.2009.02.062

Wan Ngah, W. S., & Hanafiah, M. A. K. M. (2008). Removal of heavy metal ions from wastewater by chemically modified plant wastes as adsorbents: A review. *Bioresource Technology*, *99*(10), 3935–3948. https://doi.org/10.1016/J.BIORTECH.2007.06.011

Wang, B., Zhang, W., Li, H., Fu, H., Qu, X., & Zhu, D. (2017). Micropore clogging by leachable pyrogenic organic carbon: A new perspective on sorption irreversibility and kinetics of hydrophobic organic contaminants to black carbon. *Environmental Pollution*, *220*, 1349–1358. https://doi.org/10.1016/j.envpol.2016.10.100

Wu, K., Liu, T., Lei, C., & Zhang, F. (2017). Evaluation of Al-based nanoparticle-impregnated sawdust as an adsorbent from byproduct for the removal of arsenic(V) from aqueous solutions. *Environmental Progress & Sustainable Energy*, *36*(5), 1314–1322. https://doi.org/10.1002/EP.12583

Xin-hui, D., Srinivasakannan, C., Jin-hui, P., Li-bo, Z., & Zheng-yong, Z. (2011). Comparison of activated carbon prepared from Jatropha hull by conventional heating and microwave heating. *Biomass and Bioenergy*, *35*(9), 3920–3926. https://doi.org/10.1016/J.BIOMBIOE.2011.06.010

Xu, J., Zhang, Y., Li, B., Fan, S., Xu, H., & Guan, D. X. (2022). Improved adsorption properties of tetracycline on KOH/KMnO4 modified biochar derived from wheat straw. *Chemosphere*, *296*, 133981. https://doi.org/10.1016/J.CHEMOSPHERE.2022.133981

Yan, J., Zuo, X., Yang, S., Chen, R., Cai, T., & Ding, D. (2022). Evaluation of potassium ferrate activated biochar for the simultaneous adsorption of copper and sulfadiazine: Competitive versus synergistic. *Journal of Hazardous Materials*, *424*, 127435. https://doi.org/10.1016/J.JHAZMAT.2021.127435

Zhou, M., Liu, Y., Zeng, G., Li, X., Xu, W., & Fan, T. (2007). Kinetic and equilibrium studies of Cr(VI) biosorption by dead Bacillus licheniformis biomass. *World Journal of Microbiology and Biotechnology*, *23*(1), 43–48. https://doi.org/10.1007/s11274-006-9191-8

Zoroufchi Benis, K., Motalebi Damuchali, A., Soltan, J., & McPhedran, K. N. (2020). Treatment of aqueous arsenic—a review of biochar modification methods. *Science of The Total Environment*, *739*, 139750. https://doi.org/10.1016/J.SCITOTENV.2020.139750

6 Carbon-Based Hybrid Metal Composites for the Degradation of Toxic Chemicals

Vallari R. Chourasia, Nidhi Kushwaha, Aisha Noor, Neethu Thoppil Joshikumar and K. K. Pant

LIST OF ABBREVIATIONS

MB: methylene blue
MG: malachite green
GO: graphene oxide
rGO: reduced graphene oxide
NP: nanoparticle
AB: amido black
VB: valence band
CB: conduction band
RhB: rhodamine B
BET: Brunauer Emmett Teller
CD: carbon nanodot
MR: methyl red
WS: water soluble
SPR: surface plasmon resonance
MOF: metal-organic frameworks
UV: ultraviolet
PL: photoluminescence
DRS: differential reflectance spectroscopy
LMCT: ligand to metal charge transfer
LSPR: localized surface plasmon resonance
CNT: carbon nanotube
SWCNT: single-walled carbon nanotube
MWCNT: multi-walled carbon nanotube
CNF: carbon nanofiber
TC: tetracycline
MO: methyl orange

6.1 INTRODUCTION

Life quality has increased beyond all recognition, together with the rapid advancement of the Industrial Revolution. The ensuing issues with the environment and natural resources, however, have only become worse. Around 80% of all industrial and municipal wastewater are discharged into nature without any prior treatment, which either directly or indirectly results in the degradation

DOI: 10.1201/9781003391487-6

of water quality, according to data from the United Nations World Water Development (Huang et al. 2019). These effluent streams often include polyaromatic pollutants, the majority of which are synthetic dyestuffs, and their residues are resistant to traditional wastewater treatment methods such as activated carbon adsorption, coagulation, or biotreatment (Singh et al. 2022). Currently, advanced oxidation methods, such as ozonation, sonochemical oxidation, persulfate oxidation, Fenton processes, photocatalysis, and UV/H_2O_2 oxidation, have shown potential for the effective degradation of dye pollutants, mineralization of organic pollutants, and removal of toxic metal ions from wastewater (Madima et al. 2020; Singh et al. 2022). Photocatalysis has drawn the most attention among several advanced oxidation processing techniques in wastewater because of its low setup cost, stability, ease of design, and complete mineralization of contaminants into benign byproducts (Madima et al. 2020).

Typically, in photocatalysis, photocatalysts increase the pace of the chemical transformation while being illuminated by a light source without altering the structure of the molecule (Singh et al. 2022). The photocatalysts, which are often metal oxides, or photocatalytic semiconductors, absorb light energy to create an electron (e^-) and hole (h^+) pair (Ramalingam et al. 2022). An e^- in the valence band (VB) of the semiconductor absorbs light, moves to the conduction band (CB), and then engages in a series of redox processes to produce a degraded product creating a h^+ in the CB (Singh et al. 2022). Generally, photocatalysts may utilize ultraviolet (UV) or visible light for catalysis, depending on the electrical structure and energy bandgap (E_g) of the materials (Madima et al. 2020). These photo-generated holes (h^+) from the VB then react with water producing OH radicals. Depending on the structure and stability of the pollutants, these hydroxyl radicals, which are powerful antioxidants, may mineralize wastewater by eradicating pathogens via photocatalytic oxidation processes (Ramalingam et al. 2022). Meanwhile, the electrons in the CB produce anionic superoxide radical ($^\bullet O_2^-$) by reacting with oxygen. These superoxide ions then participate in the oxidization process, while reducing electron-hole recombination. The protonation of $^\bullet O_2^-$ produces a hydroperoxyl radical ($HO_2^\bullet$), followed by the formation of H_2O_2, which is then dissociated into extremely reactive hydroxyl radicals ($OH^\bullet$) that are primarily responsible for the decomposition of organic compounds. Figure 6.1

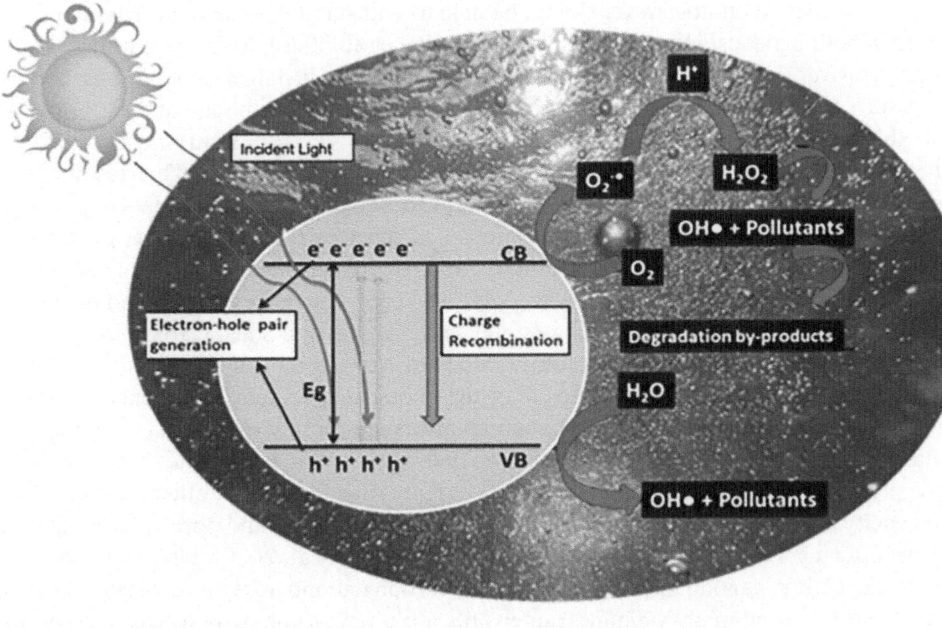

FIGURE 6.1 Mechanism of photodegradation of water pollutants (adapted from (Madima et al. 2020).

illustrates the photocatalytic mechanism, while the basic photocatalytic processes described here are as follows (Madima et al. 2020; Ramalingam et al. 2022):

$$\text{Photocatalyst} + h\nu\ (\text{UV/Visible light}) \rightarrow \text{Photocatalyst}\ (\text{CB}\ (e^-) + \text{VB}\ (h^+))$$

$$h^+\ (\text{VB}) + H_2O \rightarrow OH^\bullet + H^+$$

$$e^-\ (\text{CB}) + O_2 \rightarrow {}^\bullet O_2^-$$

$${}^\bullet O_2^- + H^+ \leftrightarrow {}^\bullet OOH$$

$$2{}^\bullet OOH \rightarrow H_2O_2 + O_2$$

$$H_2O_2 + e^- \rightarrow {}^\bullet OH + OH^-$$

$$\text{Pollutant} + h^+ \rightarrow \text{Oxidation Products}$$

$$\text{Pollutant} + e^- \rightarrow \text{Reduction Products}$$

$$\text{Pollutant} + {}^\bullet OH \rightarrow \text{Degradation Products/Pollutant intermediates}$$

In principle, photocatalysis is a heterogeneous process that takes place at the solid-liquid interface to remove contaminants from water (Madima et al. 2020). In light of this, the critical elements of the photocatalytic reaction are the quantity of active sites, surface area, and semiconductor heterojunctions. Doping is one method that has been used to slow down electron-hole recombination (Ramalingam et al. 2022). The most popular method for enhancing photocatalytic activity is bulk doping of semiconductors, which typically uses metals, non-metals, noble metals, metal oxides, or metal hydroxides. Doping causes the conduction and valence bands of semiconductors to generate acceptable electronic states. This indicates that, compared to band-to-band transitions, the electrical transition between these states and semiconductor bands takes less energy (Ramalingam et al. 2022). Titanium dioxide (TiO_2), tin dioxide (SnO_2), zinc oxide (ZnO), tungsten dioxide (WO_3), cerium dioxide (CeO_2), cadmium sulfide (CdS), iron oxide (Fe_2O_3), and other metal oxides are examples of metal oxide semiconductors (Ramalingam et al. 2022), while TiO_2 and ZnO catalysts are more often utilized, since their photocatalytic activity is found to be greater than that of other catalysts.

However, an effective photocatalyst should be able to withstand intense light irradiation as well as be stable in both aqueous and organic solvents (Zango et al. 2020). Moreover, it must be highly porous, inexpensive, simple to use, and readily regenerable. For instance, graphitic carbon nitride (g-C_3N_4), which has semiconductor characteristics, has been used as a photocatalyst for the degradation of RhB (Ajiboye et al. 2021). To improve its photocatalytic capabilities, it is often coupled with other semiconductor photocatalysts to create heterojunction systems. This results in better charge separation and a decrease in the rate at which produced charges recombine. Graphene oxide, carbon nanotubes, and activated carbon are examples of other carbon-based materials that are not semiconductors but might still be used as a basis or substrate for other semiconductor photocatalysts (Ajiboye et al. 2021). To increase the longevity of a charge carrier, carbon-based materials like carbon quantum dots are used as co-catalysts and serve as reaction sites. Carbon nanotubes have the capacity to collect photoinduced electrons as a conductor, and they also lower the rate at which the charge carriers recombine. The composite of the photocatalysts may experience the bandgap-narrowing effect if any of these carbon-based components are present (Madima et al. 2020). These carbon-based compounds are polymeric; therefore, their inclusion in composites increases the photocatalytic process by increasing the number of adsorption sites for dyes or other pollutants. Due to the homogeneity of their surface morphology, increased particle size, and pore volume, these composites have drawn a lot of interest among researchers (Zango et al. 2020). Graphene and reduced graphene oxide, carbon nanodots, carbon nanotubes, carbon nanofibers with associated metal composites, and carbon-based metal organic frameworks are a few of the carbon-based materials that have been reviewed in this study.

6.1.1 Bibliometric Analysis

An in-depth research strategy involving critical analysis, bibliometric analysis focuses on identifying the academic value and characteristics of publications in a given area of study. In this study, the Topic field tag with the query ("photocatalytic degradation" and "carbon" and "wastewater") was used to conduct a thorough review of the scientific and academic literature available in Web of Science (WOS), an online, scientific citation-indexing database maintained by Clarivate Analytics. The database produced 2,830 articles that matched this search query, with environmental sciences, engineering chemical, engineering environmental, physical interdisciplinary, and water resources ranking as the top five fields of study. The operators "or" and "and" were used to make sure that the final search included at least one word within the brackets. The quotation marks ("") were necessary to search the articles for particular phrases while eliminating synonyms. Additionally, publications from 2000 to April 2022 produced 2,578 research papers and 252 review articles. With more than 100 articles each year and 71 already published in 2023, the key publishing years for this discipline have been from 2015 to 2022. This demonstrates the recent increase in research interest of this field.

A network map was also created after using VOS viewer to analyze and illustrate the co-occurrence of terms. The circles in the network map, as shown in Figure 6.2, denote the frequency of the keywords in the chosen dataset; hence, the bigger the bubble, the more often the keyword occurs. In addition, the links between the keywords are depicted by the lines joining the bubbles, showing their lineage; hence, the closer the circles, the shorter the lines, and the stronger the association. By default, only the 1,000 strongest connections are shown, represented by no more than 1,000 lines.

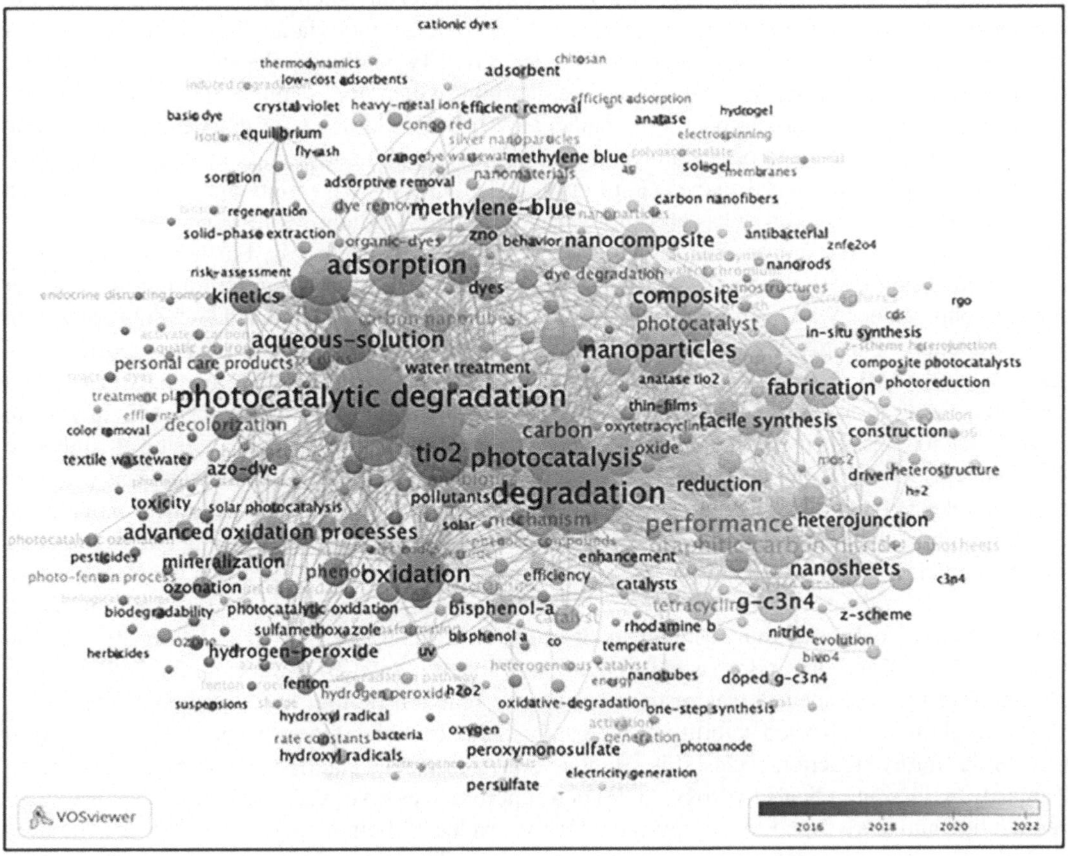

FIGURE 6.2 Network map of co-occurrence of keywords.

By identifying terms that occurred at least 10 times, the keyword co-occurrence map was produced. Out of 8,674 keywords, 485 met the criteria and were further grouped into five core clusters, as seen by the color gradient running from left to right of the Figure 6.2, which represents the publishing chronology from 2015 to 2023. Although the keywords photocatalytic degradation, TiO_2, adsorption, etc., are used to identify the largest bubbles in the central portion of the figure also demonstrating the most association, it is interesting to note that the yellow cluster, which is located on the rightmost side and includes 165 items, represents the most recent keywords used. These primarily highlight the terms nanoparticles, nanosheets, nanorods, carbon nanofibers, composite photocatalysts, etc. As a result, it amply underlines the importance of a thorough review in this still-emerging field of photocatalytic degradation of wastewater using carbon based composite materials.

6.2 CARBON-BASED METAL HYBRIDS FOR PHOTODEGRADATION

6.2.1 Graphene-Based Metal Composites

Nanomaterials such as metal oxides have wide applications in catalysis, electrochemistry, storage devices, sensors, and, importantly, environmental remediation. Herein, several transition metal oxides have been used for photodegradation of pollutants due to their superior characteristics such as crystalline structure, stability in water, wide band gap, and non-toxic behavior (Rauf et al. 2011). As explained earlier, the photodegradation activity of semiconductor material is generally due to generation of holes and electrons via photoexcitation. However, these reactive species are prone to recombination, thereby decreasing the activity of the material. Therefore, modification of these nanomaterials can be done in order to increase the catalytic activity via loading them over carbon-based material such as graphene and graphene oxide (GO). Typically, graphene-based materials are 2D, layered structures possessing sp^2 hybridized carbon, which could enhance electron transport, thus making it an excellent photocatalyst (Zhu et al. 2012). In addition to graphene, graphene oxide has also been developed as a photocatalyst due to the presence of additional oxygenated functional groups on the edge and basal planes (Kim et al. 2012). GO displays high surface area and high charge conduction characteristics, preventing charge recombination. Moreover, it provides resistance to corrosion of nanomaterial during the photodegradation process. Further, the composite also provided stability to the catalyst by inhibiting the graphene-based layers from joining together due to strong van der Waals attraction. In this regard, Atchudan et al., reported GO/ZnO composite for degradation of methylene blue (MB) (Atchudan et al. 2016). The authors have reported ZnO metal oxide for MB degradation, and the degradation efficiency was found to be nearly 13% for a reaction time of 15 min (Table 6.1, entry-1); meanwhile, GO/ZnO degraded 98.5% MB in 15 min (Table 6.1, entry-2). The increased efficiency of composite, as compared to ZnO, can be attributed to the reduced electron hole recombination due to the presence of GO. The excited electrons (e^-) from the ZnO valence band (VB) are transported to conduction band (CB) of GO under light irradiation, which promotes the formation of hydroxyl (·OH) and superoxide radical ($\cdot O_2^-$), thereby interacting with MB and initiating degradation process. The interaction of metals and carbon material has also shown activity in the form of ternary complexes. For example, Rout et al. synthesized GO-ZnO/Mn_2O_3 ternary composite that resulted in degradation efficiency of 98.75% within 30 min and showed enhanced activity under visible irradiation (Table 6.1, entry-3) (Rout et al. 2022). Primarily, formation of reactive species was responsible for the reaction taking place. However, addition of Mn_2O_3 might have improved stability of the catalyst, due to which the performance of the catalyst was similar for five reaction cycles; this might be due to enhanced stable structure of the composite. Primarily, oxidizing radicals hydroxyl (·OH) and superoxide ($\cdot O_2^-$) of the catalyst are responsible for degradation and were found to be recyclable for five cycles without losing its activity, as compared to GO/ZnO, which has not been considered for recycling. The recyclability of the catalyst is crucial in terms of scaling up the process. The authors compared the ternary catalyst with Mn_2O_3, GO,

TABLE 6.1
Graphene Oxide-Metal Composite for Degradation of Various Pollutants

S. no.	Pollutant	Catalyst	Time (min)	Catalyst Quantity (mg)	Pollutant conc. mg/L	Light spectra	Efficiency %	Ref.
1.	MB	ZnO	15	10	-	UV	~13	(Atchudan et al. 2016)
2.	MB	GO/ZnO	15	10	-	UV	98.5	(Atchudan et al. 2016)
3.	MG	GO/ZnO/ Mn_2O_3	30	20	-	Visible	98.75	(Rout et al. 2022)
4.	MG	rGO/ZnO/ Ni	60	-	10	UV	~100	(Q. Wang et al. 2018)
5.	MB	rGO/ $ZnFe_2O_4$	60	20	40	Visible	94.2	(Qureshi et al. 2022)
6.	MB	rGO/ $ZnFe_2O_4$	30	0.02	10	Visible	95	(Baynosa et al. 2020)
7.	MB	rGO/TiO_2/ AuNP	240	10	-	Visible	~100	(Yang et al. 2016)
8.	MG	rGO/S-TiO_2	60	-	20	UV	99.86	(Giang et al. 2022)
9.	AB	rGO/NiS	360	10	-	Visible	99	(Muninathan & Arumugam 2021)
10.	MG	rGO/CuS	90	-	10	Visible	97.6	(El-Hout et al. 2020)
11.	MB	rGO/SnO_2	60	30	-	Visible	~90	(Tuan et al. 2022)
12.	MG	rGO/Al/Ag	5	-	-	-	99.8	(Yang et al. 2021)
13.	MB	rGO/Fe_3O_4	12	10	-	-	95.9	(Vinothkannan et al. 2015)
14.	MB	rGO/Fe_3O_4/ TiO_2	6	-	1	UV	99	(Banerjee et al. 2018)

and GO-ZnO and found that the ternary complex showed superior activity. This was mainly due to high electron-hole generation, high surface area, and less re-combination rate. Incorporating a metal alone to rGO/ZnO composite can also enhance the photocatalytic activity. For example, Q. Wang et al. synthesized rGO/ZnO/Ni, and it showed complete degradation of MG in seawater under UV light irradiation after 15 min (Table 6.1, entry-4) (Q. Wang et al. 2018). This high degradation efficiency was attributed to the smaller size of ZnO nanorod on composite, which provides more photoactive surface, thereby increasing the degradation efficiency, besides lowering the recombination rate.

Some of the carbon composites have carboxyl and hydroxyl groups present on the surface that feasibly form bonds with water molecules, thereby promoting separability issues. Thus, some studies show the effect of addition of magnetic material in the composite in order to separate them without any complex processing. In this regard, Qureshi et al. reported reduced graphene oxide and $ZnFe_2O_4$ composite prepared via in situ dispersion of $ZnSO_4$ and $FeCl_3$ exhibiting excellent catalytic activity with a short reaction time of 30 min, resulting in 95% degradation (Table 6.1, entry-5) (Baynosa et al. 2020). The recyclability of the catalyst was seen up to six cycles for MB degradation at a pH of 10. The enhanced activity of the composite can be due to the increase in defects and the elongated doted strands on the surface of the composite. Interestingly, the variation in catalytic activity also depended on the synthesis procedure using different precursor. For example, Baynosa et al. synthesized a similar composite via a hydrothermal route at moderate temperatures and using non-toxic chemicals (Baynosa et al. 2020). The catalyst reported 94.2% MB degradation within 60 min under solar irradiation (Table 6.1, entry-6) (Yang et al. 2016). Moreover, this catalyst was also utilized for degradation of Rhodamine B (RhB) and the degradation efficiency after 60 min of solar irradiation was 49%, and this could be mainly due to complex and large structure of the RhB molecule. The large size and structure of the RhB dye might have hindered efficient adsorption of the molecules on the surface of the catalyst, which would have further inhibited photocatalytic degradation. Hence, the catalyst synthesis approach is crucial in order to tune the catalyst for achieving higher efficiency and also providing a cost-effective and environmentally benign process.

In general, TiO_2 nanoparticles are used for the photodegradation of toxic pollutants. However, it can only absorb UV light due to its high band gap of 3.2eV (Kim & Kwak 2007). Hence, the properties of TiO_2 can be modified by incorporating graphene-based materials as well as doping with metals or other compounds such as nitrogen and sulfur. For instance, Yang et al. synthesized rGO/meso-TiO_2/AuNPs ternary catalyst, which was highly efficient for MB degradation, achieving 100% degradation of MB after 240 min of irradiation under visible light (Yang et al. 2016). The degradation efficiency was higher than that of meso-TiO_2/AuNPs and meso-TiO_2 (Table 6.1, entry-7). In another study, carried out by Giang et al., degradation of Malachite Green (MG) utilizing sulfur doped rGO/TiO_2 (S-TiO_2) showed 99.6% degradation in 60 min under UV irradiation (Table 6.1, entry-8), but the degradation efficiency was only 78.88% using S-TiO_2 (Giang et al. 2022). The enhanced catalytic activity could be attributed to the higher crystallinity of the composite which came from calcination of S-TiO_2 at various temperatures. As a matter of fact, photocatalytic activity of the catalyst increases with increase in crystallinity, which, in turn, is increased by the calcination process at various temperatures. Additionally, metal and non-metal doping on rGO composite has been reported in the literature for photocatalytic degradation. For instance, rGO/NiS was synthesized by S. Muninathan et al, which showed ~99% degradation of Amido Black (AB) dye under visible light irradiation within 360 min (Table 6.1, entry-9) (Muninathan & Arumugam 2021). The high degradation efficiency can be attributed to high BET surface area of 142.68 $m^2\ g^{-1}$. However, the high degradation time was due to visible light irradiation, since it has low energy-intensity. Also, in another study by Tuan et al., rGO/CuS was used for photocatalytic degradation of MG under direct sunlight irradiation (Tuan et al. 2022). The authors found that the degradation efficiency of 97.6% was achieved within 90 min under solar radiation for catalyst with 7wt% of GO (Table 6.1, entry-10).

The superior catalytic activity for 7wt% GO composite might be due to the high BET surface area and pore volume, in addition to the high crystallinity. Another combination of metal oxide and carbon material was given as rGO/SnO_2, reported by P. Van Tuan et al. (Tuan et al. 2022). The authors showed degradation efficiency of more than 90% for MB dye within 60 min for SnO_2/rGO with 2wt% rGO (Table 6.1, entry-11). The increase in I_D/I_G ratio, as observed from the Raman spectra of the composite suggests presence of defects, which could have been the reason for decrease in bang gap of the composite compared to metal oxide. Another alternate way to improve photocatalytic activity is to dope metals on rGO. Y. Yang et al. synthesized a rGO/Al/Ag sheet via a layer-by-layer technique, which was used for detection and degradation of MG dye in water (Yang et al. 2021). The authors found that MG dye degradation efficiency was 99.8% within 5 min (Table 6.1, entry-12). The high degradation efficiency can be attributed to the π-π interaction of rGO and MG.

However, there are limitations to using metal oxides such as Fe_3O_4 nanoparticles as a catalyst for dye removal, as they lose their active site in the presence of organic dyes due to agglomeration. This can be prevented by supporting Fe_3O_4 nanoparticle on support materials such as graphene, graphene oxide, and carbon nanotube. In a study carried out by M. Vinothkannan et al., they had synthesized rGO/Fe_3O_4 using reducing agent solanum trilobatum extract via a one-pot process (Vinothkannan et al. 2015). The study reported complete degradation of MB dye within 12 min (Table 6.1, entry-13). The main factors which contributed to high degradation efficiency was the increased number of structural defects and disorders, which can be concluded from the increased

I_D/I_G of the composite evaluated from the Raman spectra and the increased surface to volume ratio of the composite. Further increase in degradation efficiency and decrease in reaction time can be achieved by using ternary catalyst. For example, rGO/TiO_2/Fe_3O_4 catalyst degraded 99% MB dye in just 6 min (Table 6.1, entry-14) (Banerjee et al. 2018). The increased photocatalytic activity was due to significant increase in surface area, which was due to Fe_3O_4 and TiO_2 anchoring over rGO sheets and high synergetic effect among rGO, TiO_2 and Fe_3O_4.

6.2.2 Carbon-Nanodot-Based Metal Composites

Carbon nanodots (CD) are another class of carbon compounds that are highly biocompatible, photoluminescent, have photostability, possess definite optical energy gap and dispersibility in water, and have a band gap of about 3.64eV (Duarah & Karak 2017) (Hazarika & Karak 2016). CD or metal oxide individually does not provide high efficiency due to the limitations, as explained in the previous section. Hence, it is important to dope metals, non-metals, or metal oxide on CD to enhance its photocatalytic activity for degradation of dyes. In this regard, A. Thakur et al. synthesized plasmonic nanostructure CD loaded on TiO_2 (TiO_2-P_{Au}/-CDs and TiO_2-P_{Ag}-CDs) using *Citrus limetta* waste pulp-derived CD for photocatalytic degradation of MB (Table 6.2, entry-1,2). The dye was completely degraded within 20 min and 30 min by Au composite and Ag composite, respectively (Thakur et al. 2020). Highly efficient degradation using visible light in a short duration was mainly due to the well-dispersed structure of the composite, and also, the higher degradation by Au composite can be attributed to the boosted surface plasmon resonance (SPR) effect and water stability of Au nanostructures. Another CD composite immobilized over electro-spun TiO_2 nanofibers (TiO_2-wsCD) were synthesized by A. Tyagi et al., where the CD was prepared from waste lemon peel via hydrothermal route (Table 6.2, entry-3) (Tyagi et al. 2016). On comparing with TiO_2 nanofibers, the kinetic rate constant of composite was 2.5 times higher, which might be due to the improved structural morphology of the composite.

To further enhance photocatalytic degradation, the CD can be doped with non-metal. For example, R. Atchudan et al. synthesized TiO_2-N-CD (Table 6.2, entry-4), which showed 90% MB degradation within 40 min. The main factor that contributed to enhanced degradation was the synergistic effect between TiO_2 nanoparticle and CD (Atchudan et al. 2018). Incorporating both metal and non-metal on CD could also enhance the photocatalytic degradation. In a study conducted by B. Reddy et al., the authors synthesized silver nanoparticle/nitrogen-doped carbon dots (Ag-N-CD) (Table 6.2, entry-5,6) composite, which showed very high phototcatalytic degradation efficiency of 98.83 % of RhB dye and 97.14 % methyl red (MD) dye. Notably, the catalyst exhibited recyclability up to five cycles without significant lose in activity. The high degradation and stability can be attributed to the well-dispersed and spherical shape of the composite (Bhagavanth et al. 2021). A similar catalyst with red emitting Mg metal doping was synthesized by A. Bhati et al. (r-Mg-N-CD) (Table 6.2, entry-7), and it completely degraded MB dye within 120 min. The factors that influenced complete degradation of dye are high quantum yield, solubility in aqueous media, and emission in red wavelength region (Bhati et al. 2018).

ZnO and Tungsten oxide are another metal oxide that is used for degradation of dyes, but, as mentioned earlier, it has certain limitations. Hence, doping CD on ZnO can be another alternative to improve its photocatalytic degradation activity. A. Sekar and R. Yadav et al. synthesized ZnO supported CD (ZnO-CD) and found that 1 wt% ZnO-CD showed 94.8% degradation of MG dye in 60 min under visible light irradiation (Table 6.2, entry-8). The degradation efficiency was higher than the individual components, which might be due to the increase in surface roughness of the carbon in the composite (Sekar & Yadav 2021). Another study done by S.P. Smrithi et al. utilizing monoclinic Tungsten oxide modified carbon nanodot (m-WO-g-CD) (Table 6.2, entry-9) showed 95% degradation of crystal violet dye in 120 min under visible light irradiation. The DRS measurement showed a slight decrease in band gap energy for the composite compared to metal oxide, which was attributed to an increase in degradation of the dye (Smrithi et al. 2020).

TABLE 6.2
Carbon Nanodot-Metal Composite for Degradation of Various Pollutants

S. no.	Pollutant	Catalyst	Time (min)	Catalyst Quantity (mg)	Light Spectra	Efficiency %	Ref.
1.	MB	TiO_2-P_{Au}-CDs	20	-	Visible	~100	(Thakur et al. 2020)
2.	MB	TiO_2-P_{Ag}-CDs	30	-	Visible	~100	(Thakur et al. 2020)
3.	MB	TiO_2–wsCD	120	10	UV	~95	(Tyagi et al. 2016)
4.	MB	TiO_2-N-CD	40	-	UV	40	(Atchudan et al. 2018)
5.	RhB	Ag-N-CD	-	-	-	98.83	(Bhagavanth et al. 2021)
6.	MR	Ag-N-CD	-	-	-	97.14	(Bhati et al. 2018)
7.	MB	r-Mg-N-CD	120	-	Visible	~100	(Bhati et al. 2018)
8.	MG	ZnO-CD	60	1.5	Visible	94.8	(Sekar & Yadav 2021)
9.	Crystal violet dye	m-WO-g-CD	120	50	Visible	95	(Smrithi et al. 2020)

6.2.3 Carbon-Nanotube-Based Metal Composites

Recently, carbon nanotubes are gaining attention for the removal of toxic chemicals. Notably, they are capable of degrading three types of contaminants; namely, organic, inorganic, and biological pollutants. The inherent characteristics such as a higher length-to-width ratio of nanotubes and high surface-area are responsible for high activity in the degradation process. Moreover, carbon-based materials are cost effective, less energy-intensive, and environmentally benign. CNTs have honeycomb-like crystal structures, where graphitic sheets get folded into cylindrical tube shapes. The number of layering of graphitic sheets during wrapping classifies nanotubes as single-walled (SWCNTs) and multi-walled nanotubes (MWCNTs). To increase the efficiency of CNTs in photo-degradation of toxic pollutants, modifications such as functionalization with carbonyl and hydroxyl groups were introduced. CNTs also provide potential space to adhere metal or metal oxide inside the tube, thereby enhancing its photo degradability. However, high discharge of waste and massive toxic compounds accumulation in water have surpassed the efficiency of CNTs. Metal CNT composite can be prepared by loading metals such as ferrous, cobalt, nickel, and titanium in the tubular structure of CNTs. Interestingly, various CNT metal composites have been developed such as CNT/CuS, MgO@CNT, ZFO@CNT, GO-CNT, TiO_2/CNT to degrade dye and heavy metals produced from various industrial sectors in waste discharge. For instance, Isari et al. (2020) developed CNT/WO_3, which significantly degraded tetracycline (TC) 60 mg/L in 60 minutes with 100% efficiency under the influence of visible light and ultrasonic irradiation (Table 6.3, entry-1). The ultrasonic irradiation has created several cavities in the solution, which further broke down in the presence of light and caused the explosive collapse of a cavity-generating radical; for example, OH, $O_2^{\cdot}$. In addition, many electron-hole pairs were also formed, but the recombination of radicals and electron hole pair were inhibited by the presence of CNT, prolonging its charge shell life. A schematic figure of photodegradation of pollutants over CNT metal composite has been shown in Figure 6.3. J. Zhang et al. (2022) reported that TiO_2-based CNT composites have exceptional potential for removal of heavy metals like mercury to a final extent of around 86% in 100 minutes (Table 6.3, entry-2). In addition, charge transfer properties, active species, and radical generation has further accelerated the degradation activity. Similarly, when Pd was loaded with ZnO-CNT (CNT/Pd/ZnO), it stimulated visible

TABLE 6.3
Carbon Nanotube-Metal Composite for Degradation of Various Pollutants

S. no.	Pollutant	Catalyst	Time (min)	Pollutant Amount (mg/L)	Light Spectra	Efficiency %	Ref.
1.	TC	CNT/WO_3	60	60	Visible	100	(Isari et al. 2020)
2.	Hg vapor	CNT/TiO_2	100	60 μg/m^3	Visible	86	(J. Zhang et al. 2022)
3.	Hg	CNT/Pd/ZnO	30	100	Visible	100	(Mohamed & Abdel Salam 2014)
4.	Aniline	$CNT/ZnO/MoS_2$	120	100	Visible	76	(Ghasemipour et al. 2020)
5.	RhB	CNT/CuS-MD	120	5	Visible	93	(Y. Wang et al. 2020)
6.	Erythrosine	CNT-$CuFe_{12}O_{19}$	50	20	UV	89	(Mahdiani et al. 2018)
7.	Ampicillin	MWCNT-$CuNiFe_2O_4$	60	-	UV	100	(Al-Musawi et al. 2022)
8.	TC	FCNT-$Bi_2O_2CO_3$/G-C_3N_4	160	-	Visible	97	(J. Zhang et al. 2022)
9.	MG	Ag_3PO_4@MWCNT@Cr:$SrTiO_3$	10	25	Visible	100	(Deonikar et al. 2020)
10.	Phenol	Ag_3PO_4@CNT@PANI	20	25	Visible	100	(Jo et al. 2019)

light response, aiding in heavy metal degradation. The authors have reported 100% degradation of Hg metal in 30 minutes (Mohamed & Abdel Salam 2014) (Table 6.3, entry-3). Ghasemipour et al. (2020) has prepared a similar composite $CNT/ZnO/MoS_2$ by ligating MoS_2 in place of Pd exhibiting excellent activity because of MoS_2 low band gap, light driven excitation ability, and high surface area. The composite showed 76% aniline degradation in 120 minutes (Table 6.3, entry-4). The surface area and morphology display a meticulous role in the degradation rate. One similar study is reported by Y. Wang et al. (2020) on CNT/CuS-MD, where CuS mixed and dried with CNT aids in nanoflakes formation and thus provides higher surface area for degradation, which exhibited 93% degradation of RhB in 120 minutes (Table 6.3, entry-5). On the other hand, CNT composite of copper and iron oxide were developed showing outstanding degradation of dyes such as MB, MO, and erythrosine. Metal oxide such as iron is widely used in forming magnetic CNT composites. Herein, the unpaired spin electron in carbon nanotubes creates line defects and vacancy in tubular structures, which supports the development of magnetic compounds. Due to the vacancy defect in CNTs, it gets arranged into a pentagonal structure, making it ferromagnetic, which degrades impurities, thereby enhancing its degradation efficiency. For instance, Mahdiani et al. (2018) showed CNT-$CuFe_{12}O_{19}$ composite that resulted in ~89% of erythrosine dye degradation within 50 minutes (Table 6.3, entry-6). This composite has facilitated better adsorption and electrochemical activity due to high wavelength utilization of UV-visible, which supports improved degradation. In the same way, Al-Musawi et al. (2022) modified iron composite with the introduction of nickel in it – i.e., MWCNT-$CuNiFe_2O_4$ – which has achieved 100% degradation of ampicillin in less than 60 minutes (Table 6.3, entry-7). In here, multiple layers of CNT are acting as a shield to prevent the

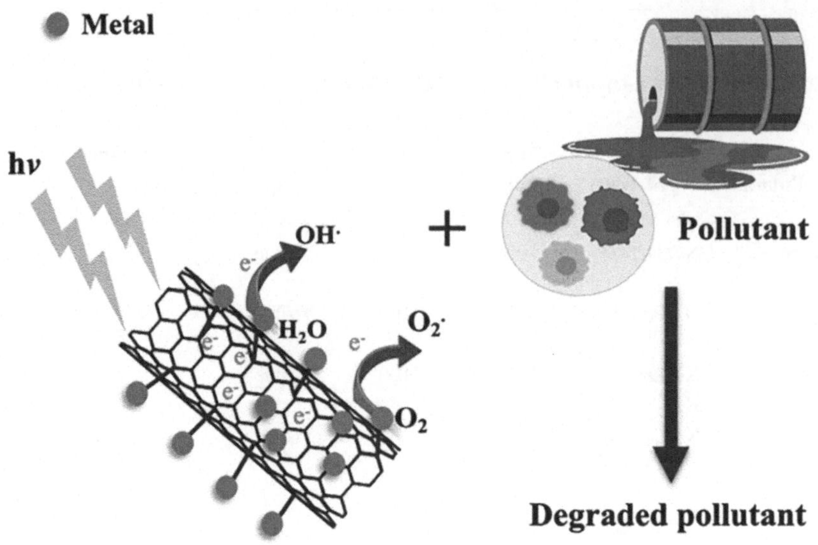

FIGURE 6.3 Schematic of photo-degradation mechanism of pollutants on CNT-metal composite.

recombination of generated electrons and thereby ensuring continuous production and supply of reactive species like O_2^- from the composites, which, in turn, expediates the photodegradation rate. The reduced time and excellent degradation of CNT metal composites are primarily because of homogenous supply of radicals. In addition, another composite FCNT-$Bi_2O_2CO_3$/G-C_3N_4 was developed by J. Zhang et al. (2022), where its tunable morphology and heterostructure arrangement encourages continuous and extended charge supply, eventually accelerating radical performance, thereby resulting in 97% degradation of TC in 160 minutes (Table 6.3, entry-8). Besides, various ternary CNT composites such as Ag_3PO_4@MWCNT@Cr:$SrTiO_3$ and Ag_3PO_4@CNT@PANI were developed which displayed high degradation performance in the removal of MG and phenol, respectively. Surprisingly, the rate of photocatalytic degradation with ternary CNT composites Ag_3PO_4@MWCNT@Cr:$SrTiO_3$ were found to be 10 times higher than single compounds Ag_3PO_4 efficiency (Deonikar et al. 2020). Owing to electron hole pair recombination inhibition by MWCNT, ternary composites are showing profound degradation effects. The heterojunction interface management is swiftly attained by ternary CNT; for instance, in Ag_3PO_4@MWCNT@Cr:$SrTiO_3$, a highly stable heterojunction photocatalytic degradation is supported by light irradiation. Here, multilayer CNT encourages charge transfer, which alters the morphology of Ag_3PO_4 from polyhedron to spherical due to electrostatic forces between positively charge Ag^+ and MWCNT. Here, Ag acts as a hole trapper and photogenerated electron, enabling the charge transfer between CB of Ag_3PO_4 and VB of Cr:$SrTiO_3$. Also, when the size of Ag decreases due to conformational modification, higher surface area is achieved that supports a better photocatalytic activity. Therefore, Ag_3PO_4@MWCNT@Cr:$SrTiO_3$ was capable of displaying degradation activity of 100% in 10 minutes for MG dye (Table 6.3, entry-9), and the activity was similar up to five cycles without any significant loss in the quantity. Similarly, another ternary composite, Ag_3PO_4@CNT@PANI, showed 100% degradation of phenol in 20 minutes (Table 6.3, entry-10) (Jo et al. 2019).

6.2.4 Carbon-Nanofiber-Based Metal Composites

In recent years, carbon nanofibers (CNFs) have been acquiring attention because of their ability to meticulously cope with emerging pollutants or toxic chemicals (Gong, J. et al.), owing to its ease of handling, process, low cost, and better performance, crediting it to be a suitable candidate to

TABLE 6.4
Carbon Nanofiber-Metal Composite for Degradation of Various Pollutants

S. no.	Pollutant	Catalyst	Time (min)	Light spectra	Efficiency %	Ref.
1.	MB	CNF/ZnO	30	Visible	95	(Gadisa et al. 2019)
2.	MB	CNF/Ag/ZnO	60	UV	95	(Pant et al. 2016)
3.	RhB	CNF- TiO_2	180	UV	97.4	(H. Wang et al. 2018)
4.	MB	CNF/Mo/N/TiO_2	180	Visible	80	(Qiu et al. 2019)
5.	MB	CNF/$ZnFe_2O_4$@TiO_2	40	Visible	80	(Al-Meer et al. 2017)

eliminate organic and inorganic contaminants and toxic pollutants (Mehrizad & Gharbani, 2017). The π-π stacking in CNFs, its hexagonal arrangement, mesoporous structure, and chemical inertness makes it potential and stable for a photodegradation reaction. The π-conjugates facilitate the adsorption of dye on the surface of CNF, which are followed by photocatalytic degradation (Gong et al., 2022). To augment the efficacy of CNFs, several composites have been synthesized for toxic pollutant degradation. Mehrizad and Ghabrani have reported zinc oxide doped carbon nanofibers (CNF/ZnO) toward degradation of MB, RhB, and indigo carmine dye (Mehrizad & Gharbani 2017). The doping of ZnO on CNFs assists in lowering the band energy of ZnO, thereby making its excitation facile to produce oxygen and hydroxyl radicals. The composite showed 95% degradation of MB in 30 minutes under visible light (Table 6.4, entry-1) (Gadisa et al. 2019). In another study, the effect of addition of Ag nanoparticle onto the CNF/ZnO composite was studied, which elevated the efficiency of composite resulting in the degradation of MB up to 95% in 60 minutes under UV light (Table 6.4, entry-2) (Pant et al. 2016). Wang et al. showed degradation activity of CNF/TiO_2 toward RhB, which was found effective for five cycles. The composite degraded 97.4% RhB in 180 minutes (Table 6.4, entry-3) (H. Wang et al. 2018). In addition, several hetero-conjugate composites were developed; for instance, Mo/N co-doped TiO_2 CNF (CNF/Mo/N/TiO_2) were recognized as excellent scavengers of MB dye (Qiu et al. 2019). The remarkable increase in photodecomposition activity is due to the generation of H_+ species from the metal conjugate surface, encouraging a high degradation rate of 80% MB in 180 minutes (Table 6.4, entry-4). Magnetic CNFs composites such as CNF/$ZnFe_2O_4$@TiO_2 exhibited high potential in selectively degrading dyes like MB up to 80% in 40 minutes (Table 6.4, entry-5) (Al-Meer et al. 2017). The synergistic effect of iron oxide magnetism, Zn low band gap, and presence of CNF enhanced charge transfer and prevented electron hole pair recombination, which further boosted its photodegradation activity.

6.2.5 Carbon-Based Material Metal-Organic Frameworks Composites

In addition to carbon metal composites such as graphene-oxide-based, carbon-nanodots-based, CNT-based and CNF-based metal composites, porous metal organic frameworks (MOF) with intermixing of carbon-based material has recently emerged as a potential pollutant degradation material. MOF are a crystalline inorganic-organic hybrid material with high porosity, surface area, and three-dimensional structure, which can be tailored specifically to the application. These properties expand its application to gas storage, drug delivery, separation, catalysis, chemical sensors, and adsorbent (Liu et al. 2016). Owing to these properties, MOFs have been recently used in degradation of pollutant via electrocatalysis, sulfate-radical-based reaction, Fenton-like oxidation, and photocatalysis. Amongst them, photocatalytic degradation is the most widely accepted process due to its ability to convert solar energy into chemical energy with no residue production, thereby associating sustainable development. Nevertheless, limited metallic centers in MOF structure and larger void space enhances diffusion coefficient, which leads to lower diffusion capacity, and weak dispersive forces reduce the retention of smaller adsorbents, thereby not fully utilizing the porosity of MOF

(Zhu et al., 2019). Moreover, harsh conditions used for photocatalytic degradation puts a condition to stability of MOF. The photocatalytic degradation activity of MOF occurs due to the generation of hydroxyl radicals ($^{\bullet}OH$), superoxide radical ($^{\bullet}O^{2-}$), and holes (h^+); however, MOF are very susceptible to charge recombination, which leads to low efficiency (Zango et al. 2020). To circumvent these issues, MOF-based composites, especially with carbon-based compounds such as graphene, graphene oxide, (GO) and reduced graphene oxide (rGO), are emerging as suitable photocatalytic material. The composites possess synergy of high surface area, charge mobility, electrical conductivity, reduced bandgap, lower photoluminescence, and photocurrent response (Yang et al. 2017a; Z. Yang et al. 2017; Alfonso-Herrera et al. 2018). The synergistic mechanism for photodegradation using carbon MOF composite is shown in Figure 6.4.

Herein, Zhang et al. illustrated the application of MIL-53(Fe) for the degradation of cationic Methylene Blue (MB) dye pollutant. The authors reported 82% degradation of MB using MIL-53(Fe) in 80 minutes under ultraviolet (UV) light (Table 6.5, entry-1) whereas only 32% removal after 420 min under the visible light (Table 6.5, entry-2) (Zhang et al. 2014). The difference in the removal efficiency under different light spectra might be due to higher energy intensity of UV light than visible light (Ahmad et al. 2006). The degradation activity enhanced on introducing rGO in the crystal structure of MOF. Herein, the activity of MB degradation via MIL-53(Fe)/rGO composite increased to 95% in UV light (Table 6.5, entry-3) and 49% in visible light (Table 6.5, entry-4) (Zhang et al. 2014). This can be attributed to electron withdrawing behavior of rGO, which eliminates electron hole recombination, thereby increasing the effective absorbance of light for MB degradation (Xu et al. 2015). The subsequent generation of strong oxidizing radicals $^{\bullet}OH$ and $^{\bullet}O^{2-}$ degrade MB. However, it was observed that increase in rGO beyond optimal range decreased the degradation, possibly due to covering of active sites of MOF for electron generation. Besides the light spectra, the defect created in the lattice of MOF due to incorporation of rGO also plays

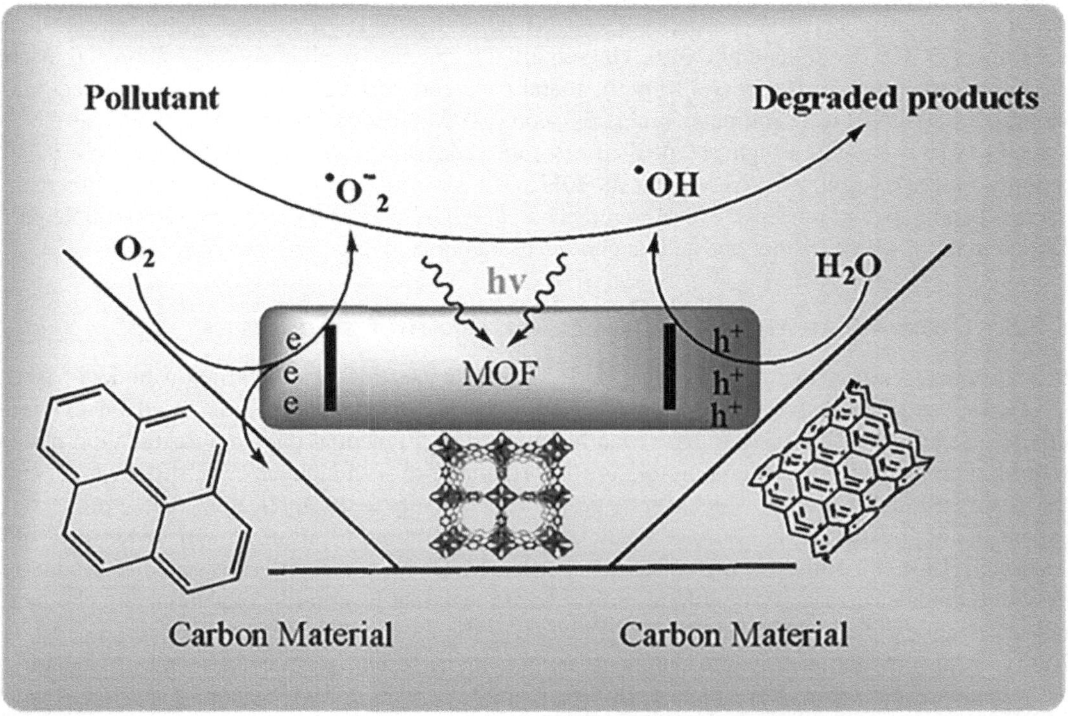

FIGURE 6.4 General photo-degradation mechanism of pollutants on carbon-MOF composite.

TABLE 6.5
Carbon-Based Material Metal-Organic Frameworks Composite for Degradation of Various Pollutants

S. no.	PollutaZnt	Catalyst	Time (min)	Catalyst quantity (mg)	Light spectra	Efficiency %	Ref.
1.	MB	MIL-53 (Fe)	80	100	UV	82	(Zhang et al. 2014)
2.	MB	MIL-53 (Fe)/ rGO	80	100	UV	95	(Zhang et al. 2014)
3.	MB	MIL-53 (Fe)	80	100	Visible	32	(Zhang et al. 2014)
4.	MB	MIL-53 (Fe)/ rGO	80	100	Visible	49	(Zhang et al. 2014)
5.	MB	NH_2-MIL-53 (Al)/rGO	210	0.01	Visible	59	(Y. Yang et al. 2017)
6.	Cr(VI)	MIL-53 (Fe)/ rGO	80	40	Visible	100	(Liang et al. 2015)
7.	RhB	MIL-125 (Ti)	50	40	Visible	~86	(Yuan et al. 2016)
8.	RhB	MIL-125 (Ti)/ rGO	50	40	Visible	~93.5	(Yuan et al. 2016)
9.	RhB	Ag/MIL-125 (Ti)/rGO	50	40	Visible	95.7	(Yuan et al. 2016)
10.	Methyl Blue	NH_2-MIL-125 (Ti)/rGO	30	30	Visible	~98%	(Huang & Liu 2016)
11.	MB	NH_2-MIL-53 (Al)/rGO	210	10	Visible	59%	(Y. Yang et al. 2017)
12.	MB	CuCd-BMOF/ GO	40	30	Visible	92%	(Chen et al. 2022)
13.	RhB	Bi_2O_3/Cu-MOF/ GO	60	50	Visible	84%	(Chen et al. 2020)

a crucial role in promoting the photocatalytic efficiency of the material. For example, Yang et al. demonstrated in a study that NH_2-MIL-53 (Al)/rGO composite shows a red shift of 57 nm, which is indicative of the internal and external defects, which aids the prevention of electron hole recombination. This was also supported from PL spectra showing less emission intensity. The authors reported 59% degradation of MB in 210 min in visible light (Table 6.5, entry-5) (Y. Yang et al. 2017).

In addition to dyes, MIL-53 (Fe)/rGO was also potential for the degradation of heavy metal such as Cr(VI). Liang et al. reported a synergistic process where reduction of Cr(VI) over MIL-53 (Fe)/rGO composite also fulfilled the oxidation of Malachite Green (MG) and Rhodamine B (RhB) simultaneously. Primarily, the activity of MIL-53 (Fe)/rGO for degradation of Cr(VI) was observed, which was almost 100% in 80 minutes (Table 6.5, entry-6) in visible light (Liang et al. 2015). From UV-vis DRS spectra, it was noted that MIL-53 (Fe)/rGO composite absorbs visible light; therefore, ligand to metal charge transfer (LMCT) mechanism directed for carboxylate oxygen to metal bonding must be followed for the photocatalytic activity (Fu et al. 2012; Ai et al. 2014). The effect of metal dopants on the surface of MOF has showed enhanced photo-degradation of pollutant dyes. For example, Yuan et al. illustrated application of MIL-125 (Ti) MOF, which possess high density of Ti sites immobilized in the porous structure and can support the introduction of dopant metal. The degradation of RhB dye under visible light using MIL-125 was nearly 86% (Table 6.5, entry-7); however, charge recombination was readily possible, and the absorption of visible light was suppressed due to dye-sensitization effect of RhB (Yuan et al. 2016). Therefore, rGO and Ag metal were introduced into the crystal to primarily understand the effect of dopant metal. The photoactivity for RhB degradation increased for MIL-125 (Ti)/rGO and reached ~93.5% in 50 minutes (Table 6.5, entry-8), whereas for Ag/MIL-125 (Ti)/rGO, the activity was 95.7% under similar conditions (Table 6.5, entry-9) (Yuan et al. 2016). The authors attributed the development in the activity after the addition of rGO due to effective charge separation channel in a planar conjugation rGO structure, which prevents electron-hole recombination. On the other hand, the ternary composite consisting of Ag metal showed superior activity due to localized surface plasmon resonance (LSPR) absorption, increasing the light absorption band from 400–750 nm in the visible range. Thus, absorption in a wider solar spectrum gives more energy for electron excitation (Arif Sher Shah et al. 2013; Sim et al. 2014). Moreover, the synergistic effect of Ag metal and rGO structure simultaneously working for charge transfer to electron acceptors like oxygen molecules, thereby reducing into superoxide ions ($^{\bullet}O^{2-}$), and subsequently, back photo-oxidizing Ti^{3+} cluster into Ti^{4+} and the generation of reactive hole species, which altogether oxidize the pollutant and degrade them to lower molecules (Dan-Hardi et al. 2009; Yuan et al. 2016). However, the ternary composite includes Ag, which belongs to precious metals and is quite expensive. The metal dopant was therefore replaced by metal free molecules, which is shown in the study of Huang and Liu for the photodegradation of Methyl Blue (MB). The introduction of NH_2 functional group led to an extension in absorption band range from 250–350 nm to 550 nm, since the amine group showed absorption at 400 nm. Moreover, the organic groups NH_2 and rGO showed strong π-π interaction with each other, bringing distortion in the idea octahedral square grid crystal structure of MIL-125 (Ti) (Huang & Liu 2016). These reasons can possibly be attributed to the higher activity of NH_2/MIL-125(Ti)/rGO in photodegradation of MB in visible light. The photodegradation efficiency of NH_2/MIL-125(Ti)/rGO was thus ~98% in 30 minutes (Table 6.5, entry-10) (Huang & Liu 2016). Similar functional groups were incorporated into another MOF, MIL-53 (Al) for the photodegradation of Methylene Blue. Yang et al. studied the activity of NH_2-MIL-53(Al)/rGO achieving 59% degradation of Methylene Blue in 210 minutes under visible light (Table 6.5, entry-11) (Y. Yang et al. 2017). The absorption spectrum of NH_2-MIL-53 (Al) was extended up to greater than 450 nm due to the presence of rGO with an absorption edge of NH_2-MIL-53(Al)/rGO was 517 nm, which corresponded to a lower bandgap energy of 2.40 eV than 2.70 eV of NH_2-MIL-53(Al). Similarly, Su et al. has also corroborated the energy shift to a lower band gap energy (Su et al. 2008). The virtue of lower Fermi level (0 V) of GO has made it a strong photocatalyst, since it provides a channel for easy movement of photogenerated electrons toward the support. Besides a lower Fermi level and light spectra, pH of the medium also has an effect of the photoactivity and degradation curve of MB (Banerjee et al. 2022). In this regard, Chen et al. showed the use of phosphonates-based CuCd-BMOF/GO as a promising photocatalyst for the degradation of MB dye by 92% in 40 min at pH = 9 (Table 6.5, entry-12) (Chen et al. 2022). The metal phosphonates are thermally stable and not soluble in highly acidic media. The process of degradation of MB under dark conditions in alkaline conditions was accelerated, contrary to sequestration of the degradation under acidic conditions. This was due to a positive charge acquired by the surface of CuCd-BMOF/GO under basic pH of 7–11 (above the 0 zeta potential of 6.8 pH), which interacts strongly with the negatively charged MB, thereby expediting the process (W. Zhang et al. 2022). However, in acidic media of pH 2–5, the negatively charged surface and MB result in repulsive behavior and decelerate the process. In addition to the medium of reaction, as explained earlier, the higher activity of the CuCd-BMOF/GO is due to the synergistic effect between CuCd-BMOF and GO; however, optimum GO is crucial, since higher GO concentration in the photocatalyst wraps the superficial active sites and blocks the channel for charge transfer (Luo & Wang 2018). Interestingly, visible light is crucial for the degradation of the pollutant. This is due to allowing the adsorption process in the porous MOF structure to take place, thereby reducing the concentration of the pollutant; however, their degradation will still be on the table. For example, Chen et al. illustrated the activity of Bi_2O_3 semiconductor having very small bandgap of 2.4 to 2.8 eV. The semiconductor was used to make a ternary composite along with Cu-MOF and GO to produce a highly porous, large surface area and pore size possessing material for photocatalytic degradation of RhB by 84% in 60 minutes (Table 6.5, entry-13) (Chen et al. 2020). The high surface area (117.339 m^2/g) and pore size (4.9 nm) of the catalyst allowed higher accessibility of active sites for photoactivity in visible light. However, greater size of pore size than RhB (1.8 x 0.7 nm) facilitates the adsorption of RhB first in dark mode, thereafter benefiting in its degradation in visible light by redox reactions.

6.2.6 Graphitic Carbon Nitride-Metal Composites

Graphitic carbon nitrides (g-C_3N_4) are another class of compounds that have received significant attention recently as a visible-light-responsive photocatalyst. They have found a wide range of applications for use as photocatalyst for organic transformation, dye and antibiotic degradation, heavy metal removal, and CO_2 reduction. However, similar to the previously discussed carbon-based hybrid materials, fast recombination and low surface area limit the application (Meena et al. 2023). This can be overcome by incorporating other metal or elemental compound doping or formation of heterojunction by a semiconductor material. For example, N. Venkatesh et al. synthesized silver-decorated polymeric graphitic carbon nitride with cerium oxide nanocomposite Ag/g-C_3N_4/CeO_2 (Table 6.6, entry-1,2), which degraded 100% acid yellow dye in 150 min. The high degradation efficiency can be attributed to the presence of crystalline phases of Ag, g-C_3N_4 and CeO_2 and due to the small size and porous structure (Venkatesh et al. 2023). However, the use of precious metals in catalyst would increase the cost of the process; hence, it would be ideal to incorporate other metals. In this regard, Das et al. synthesized manganese (III) acetylacetonate over graphitic carbon nitride (2-Mn/g-C_3N_4) (Table 6.6, entry-3,4), which showed 99.59% degradation efficiency of RhB dye within 55min and 97.6% degradation of metronidazole within 40 min. The improved degradation efficiency was found to be due to enhanced properties of the composite such as high BET surface area and a smooth and layered structure, as seen from the FE-SEM image (Das et al. 2023). The π-π stacking of the metal complex on g-C_3N_4 sheet enhanced the roughness and corrugation in the complex, which thereby improved the degradation efficiency. As mentioned in the previous sections, metal oxide can also be mounted on g-C_3N_4 to improve photocatalytic activity. For example, H. An et al. studied the degradation efficiency of MB by zinc oxide decorated graphitic carbon nitride (ZnO/g-C_3N_4) (Table 6.6, entry-5) composite (An et al. 2023). The catalyst took 180 min to degrade MB, whereas another catalyst, Ti/Ga/ZnO/g-C_3N_4 (Table 6.6, entry-6), synthesized by J. Guo, took only 105 min to degrade MB dye with lower catalyst quantity, but the degradation efficiency was only 95.4%. The improved degradation within less time was due to Z-scheme heterostructure of the

TABLE 6.6
Graphitic Carbon Nitride Composite for Degradation of Various Pollutants

S. no.	Pollutant	Catalyst	Time (min)	Catalyst quantity (mg)	Pollutant conc. mg/L	Light spectra	Efficiency %	Ref.
1.	Acid yellow-36	Ag/g-C_3N_4/ CeO_2	150	25	20	Visible	~100	(Venkatesh et al. 2023)
2.	Acid yellow-12	Ag/g-C_3N_4/ CeO_2	150	25	20	Visible	~100	(Venkatesh et al. 2023)
3.	RhB	2-Mn/g-C_3N_4	55	-	20	Visible	99.59	(Das et al. 2023)
4.	Metronidazole	2-Mn/g-C_3N_4	40	-	20	Visible	97.6	(Das et al. 2023)
5.	MB	ZnO/g-C_3N_4	180	50	10	Visible	99.16	(An et al. 2023)
6.	MB	Ti/Ga/ ZnO/g-C_3N_4	105	30	10	Visible	95.4	(Guo et al. 2023)

catalyst. It is noteworthy that the XRD pattern of the catalyst did not generate any additional impure crystal compounds, and the structure of the composite was lamellar, both of which have enhanced the degradation of the dye (Guo et al. 2023).

6.3 CARBON-BASED METAL HYBRIDS FOR PHOTOCATALYTIC REMOVAL OF PHARMACEUTICAL POLLUTANTS

With a dramatic increase in consumption of pharmaceuticals in recent years, it has been almost impossible to practice modern medicine without the use of pharmaceutical drugs. While global population growth, urbanization, increased investment in the health-care sector, and advances in research have been the primary causes of the accumulation of these compounds in our environment, their excessive consumption has led to environmental contamination, particularly of water bodies. Antibiotics, antipyretics, analgesics, chemotherapeutics, steroids, β-blockers, hormones, anti-inflammatory agents, and antidepressants are among the most prominent pharmaceutical contaminants found in wastewater (Javaid et al. 2022). They enter the wastewater via the feces and urine of humans and animals owing to their inability to be assimilated. Similarly, pharmaceutical industry effluents, hospital wastes, and discarded medications are all significant contributors of these pollutants. Therefore, the elimination of pharmaceuticals from wastewater resources is a critical environmental remediation concern. In view of the vast amount of research on the photocatalytic characteristics of carbon-based hybrid materials for the photodegradation of dyes and organic contaminants, these materials have also emerged as promising candidates for the removal of pharmaceuticals from wastewater. For instance, a novel carbon-metal hybrid composite (MWCNT-TiO_2-SiO_2) was reported in order to remove carbamazepine and bisphenol A from an aqueous solution by photodegradation. The inclusion of merely a small amount of MWCNT (0.15–17.8 wt %) increased the catalytic activity of the synthesized hybrid against the aforementioned pollutants (Buda 2015). Furthermore, tests on the toxicity of these pollutants revealed the development of non-toxic, photodegraded compounds. In addition, TiO_2-rGO materials synthesized by various methods were successfully used to degrade carbamazepine, acetaminophen, atenolol, diclofenac, ibuprofen, sulfamethoxazole, risperidone, and diphenhydramine (Minella et al.). In another study, it was further reported that the coupling of rGO and TiO_2 particles decreased the bandgap and improved the photocatalytic activity of nanohybrid substantially when evaluated against three pharmaceuticals, including carbamazepine, ibuprofen, and sulfamethoxazole. With the addition of about 2.7% rGO, impressive degradation efficiencies of 54% for carbamazepine, 81% for ibuprofen, and 92% for sulfamethoxazole were achieved (Lin et al. 2017). In a separate investigation, the photocatalytic activity of Ag/g-C_3N_4 (5 wt %) for the elimination of sulfamethoxazole was reported to be 99.5% after 60 minutes of irradiation. Photogenerated h^+ and $^{\bullet}O_2$ were identified as the most active species responsible for catalysis (Song et al. 2018). Using a nanohybrid comprised of carbon-doped g-C_3N_4 and TiO_2 nanoparticles, the photodegradation of diclofenac and carbamazepine has been shown to be accelerated with their degradation efficiency reaching up to 98.9 and 99.7 %, respectively (Hu et al. 2019). Furthermore, it was demonstrated that the MIL-68(In)-NH_2/graphene oxide (GrO) was a promising MOF composite for the photocatalytic removal of amoxicillin, with the photodegradation efficiency of the MIL-68(In)-NH_2/GrO (93%) to be significantly higher than the MOF MIL-68(In)-NH_2 without GrO (60%) (Yang et al. 2017b). This study revealed that the presence of graphene oxide encouraged electron movement, decreased the recombination of photoinduced charges, and enhanced visible light absorption, contributing to the increased photoactivity of the carbon composite MOF (Yang et al. 2017b). In a similar investigation by Cao et al. (Cao et al. 2020), g-C_3N_4/MIL-68(In)-NH_2 carbon composite MOF was synthesized for the photodegradation of ibuprofen. The photocatalytic efficiency of the composite was much higher (93%) than individually used g-C_3N_4 (9%) and MIL-68(In)-NH_2 (68%). As a result, owing to the formation of heterojunctions, more reactive species were generated in the composite, which expedited the synthesis of free radicals, hence, enhancing the photocatalytic efficiency and adsorption capacity of ibuprofen.

Similarly, in a work by Wang et al., Z. Wang et al. (2020) anchored $Bi_2O_2CO_3$ in the porous g-C_3N_4 hybrids by calcining CAU-17/melamine precursor in a single step for the removal of tetracycline and sulfonamide. Therefore, $Bi_2O_2CO_3$ particles, thus generated using CAU-17, were uniformly dispersed on the surface and between the layers of g-C_3N_4 sheets, which considerably increased the direct contact at the interface and therefore improved the photocatalytic activity.

6.4 CONCLUSIONS

Various oxidation methods, such as ozonation, sonochemical oxidation, persulfate oxidation, Fenton processes, photocatalysis, and UV/H_2O_2 oxidation, show potential for the effective degradation of dye pollutants, organic pollutants, and toxic metal ions from wastewater. In this comprehensive study, the photocatalytic activity of carbon-based metal composites for wastewater treatment has been thoroughly examined and summarized. The literature review revealed that incorporating carbon-based materials into transition metal oxides, metal ions, and other semiconductor materials can significantly increase the photocatalytic efficiency against the removal of a variety of contaminants in wastewater by increasing their adsorption capacity on photocatalytic surfaces, improving the visible light absorptivity range, and also by minimizing the electron-hole pair recombination after photocatalysis.

Moreover, it was demonstrated that the improvement of carbon-based metal composites is dependent on the synthesis process, the quantity of doping material used, the form of irradiation source along with it, the duration of the irradiation, etc. The quantity of dopant defines the photocatalytic performance, the technique of synthesis determines the shape and surface area, and the source of irradiation and stability controls the material's reusability. It is also shown that these carbon-based semiconductor materials have a wide range of further applications in medical and health sectors, including energy storage systems, medical sensors, medication delivery, cancer treatment, environmental production, biomedicines, and much more. Though the photocatalytic degradation efficiency of organic dyes and other industrial pollutants in wastewater has significantly increased due to the development of these materials, their scalability is still constrained by their production costs, stability, and reusability. This opens the door for further exploration in this field and research into the discovery of newer materials in this regard.

Declaration

Authors declare no conflict of interest.

BIBLIOGRAPHY

Ahmad I, Fasihullah Q, Vaid FHM. 2006. Effect of light intensity and wavelengths on photodegradation reactions of riboflavin in aqueous solution. *J Photochem Photobiol B Biol*. 82(1):21–27.

Ai L, Zhang C, Li L, Jiang J. 2014. Iron terephthalate metal-organic framework: revealing the effective activation of hydrogen peroxide for the degradation of organic dye under visible light irradiation. *Appl Catal B Environ*. 148–149:191–200. http://dx.doi.org/10.1016/j.apcatb.2013.10.056

Ajiboye TO, Oyewo OA, Onwudiwe DC. 2021. Adsorption and photocatalytic removal of Rhodamine B from wastewater using carbon-based materials. *FlatChem*. 29(June):100277. https://doi.org/10.1016/j.flatc.2021.100277

Alfonso-Herrera LA, Huerta-Flores AM, Torres-Martínez LM, Rivera-Villanueva JM, Ramírez-Herrera DJ. 2018. Hybrid SrZrO3-MOF heterostructure: surface assembly and photocatalytic performance for hydrogen evolution and degradation of indigo carmine dye. *J Mater Sci Mater Electron*. 29(12):10395–10410. http://dx.doi.org/10.1007/s10854-018-9096-y

Al-Meer S, Ghouri ZK, Elsaid K, Easa A, Al-Qahtani MT, Shaheer Akhtar M. 2017. Engineering of magnetically separable ZnFe2O4@ TiO2 nanofibers for dye-sensitized solar cells and removal of pollutant from water. *J Alloys Compd*. 723:477–483. http://dx.doi.org/10.1016/j.jallcom.2017.06.211

Al-Musawi TJ, Asgariyan R, Yilmaz M, Mengelizadeh N, Asghari A, Balarak D, Darvishmotevall M. 2022. Synthesis of a doped \$α\$-Fe2O3/g-C3N4 catalyst for high-efficiency degradation of diazinon contaminant from liquid wastes. *Magnetochemistry*. 8(11).

An H, Huong LM, Dat NM, Hai ND, Cong CQ, Nam NTH, Tai LT, Thi DNM, Nghi HB, Huyen NTT, et al. 2023. Photocatalytic degradation of organic dyes using zinc oxide-decorated graphitic carbon nitride composite under visible light. *Diam Relat Mater*. 131(November 2022):109583. https://doi.org/10.1016/j.diamond.2022.109583

Arif Sher Shah MS, Zhang K, Park AR, Kim KS, Park NG, Park JH, Yoo PJ. 2013. Single-step solvothermal synthesis of mesoporous Ag-TiO2-reduced graphene oxide ternary composites with enhanced photocatalytic activity. *Nanoscale*. 5(11):5093–5101.

Atchudan R, Edison TNJI, Perumal S, Karthikeyan D, Lee YR. 2016. Facile synthesis of zinc oxide nanoparticles decorated graphene oxide composite via simple solvothermal route and their photocatalytic activity on methylene blue degradation. *J Photochem Photobiol B Biol*. 162:500–510. http://dx.doi.org/10.1016/j.jphotobiol.2016.07.019

Atchudan R, Edison TNJI, Perumal S, Vinodh R, Lee YR. 2018. In-situ green synthesis of nitrogen-doped carbon dots for bioimaging and TiO2 nanoparticles@nitrogen-doped carbon composite for photocatalytic degradation of organic pollutants. J Alloys Compd. 766:12–24.

Banerjee D, Kushwaha N, Shetti NP, Aminabhavi TM, Ahmad E. 2022. Green hydrogen production via photo-reforming of bio-renewable resources. *Renew Sustain Energy Rev*. 167(December 2021):112827. https://doi.org/10.1016/j.rser.2022.112827

Banerjee S, Benjwal P, Singh M, Kar KK. 2018. Graphene oxide (rGO)-metal oxide (TiO 2/Fe 3 O 4) based nanocomposites for the removal of methylene blue. *Appl Surf Sci*. 439:560–568.

Baynosa ML, Mady AH, Nguyen VQ, Kumar DR, Sayed MS, Tuma D, Shim JJ. 2020. Eco-friendly synthesis of recyclable mesoporous zinc ferrite@reduced graphene oxide nanocomposite for efficient photocatalytic dye degradation under solar radiation. *J Colloid Interface Sci*. 561:459–469. https://doi.org/10.1016/j.jcis.2019.11.018

Bhagavanth RR, Dadigala R, Bandi R, Seku K, Koteswararao D, Girija Mangatayaru K, Shalan AE. 2021. Microwave-assisted preparation of a silver nanoparticles/N-doped carbon dots nanocomposite and its application for catalytic reduction of rhodamine B, methyl red and 4-nitrophenol dyes. *RSC Adv*. 11(9):5139–5148.

Bhati A, Anand SR, Gunture G, Garg AK, Khare P, Sonkar SK. 2018. Sunlight-induced photocatalytic degradation of pollutant dye by highly fluorescent red-emitting Mg-N-embedded carbon dots. *ACS Sustain Chem Eng*. 6(7):9246–9256.

Buda W. 2015. Photocatalytic treatment of pharmaceutical wastewater using new multiwall-carbon nanotubes/TiO 2/SiO 2 nanocomposites. 137:176–184.

Cao W, Yuan Y, Yang C, Wu S, Cheng J. 2020. In-situ fabrication of g-C 3 N 4/MIL-68 (In)-NH 2 heterojunction composites with enhanced visible-light photocatalytic activity for degradation of ibuprofen. *Chem Eng J*. 391(August 2019):123608. https://doi.org/10.1016/j.cej.2019.123608

Chen J, Wei J, Zhang H, Wang X, Fu L, Yang TH. 2022. Construction of CuCd-BMOF/GO composites based on phosphonate and their boosted visible-light photocatalytic degradation. *Appl Surf Sci*. 594(November 2021):153493. https://doi.org/10.1016/j.apsusc.2022.153493

Chen Y, Zhai BY, Liang YN, Li Y. 2020. Hybrid photocatalysts using semiconductor/MOF/graphene oxide for superior photodegradation of organic pollutants under visible light. *Mater Sci Semicond Process*. 107(May 2019):104838. https://doi.org/10.1016/j.mssp.2019.104838

Dan-Hardi M, Serre C, Frot T, Rozes L, Maurin G, Sanchez C, Férey G. 2009. A new photoactive crystalline highly porous titanium(IV) dicarboxylate. *J Am Chem Soc*. 131(31):10857–10859.

Das B, Devi M, Deb S, Dhar SS. 2023. Boosting photocatalytic property of graphitic carbon nitride with metal complex fabrication for efficient degradation of organic pollutants. *Chemosphere*. 323(February):138230. https://doi.org/10.1016/j.chemosphere.2023.138230

Deonikar VG, Rathod P V, Pornea AM, Kim H. 2020. Superior decontamination of toxic organic pollutants under solar light by reduced graphene oxide incorporated tetrapods-like Ag3PO4/MnFe2O4 hierarchical composites. *J Environ Manage*. 256(November 2019):109930. https://doi.org/10.1016/j.jenvman.2019.109930

Duarah R, Karak N. 2017. Facile and ultrafast green approach to synthesize biobased luminescent reduced carbon nanodot: an efficient photocatalyst. *ACS Sustain Chem Eng*. 5(10):9454–9466.

El-Hout SI, El-Sheikh SM, Gaber A, Shawky A, Ahmed AI. 2020. Highly efficient sunlight-driven photocatalytic degradation of malachite green dye over reduced graphene oxide-supported CuS nanoparticles. *J Alloys Compd*. 849:156573. https://doi.org/10.1016/j.jallcom.2020.156573

Fu Y, Sun D, Chen Y, Huang R, Ding Z, Fu X, Li Z. 2012. An amine-functionalized titanium metal-organic framework photocatalyst with visible-light-induced activity for CO_2 reduction. *Angew Chemie—Int Ed*. 51(14):3364–3367.

Gadisa BT, Appiah-Ntiamoah R, Kim H. 2019. In-situ derived hierarchical ZnO/Zn-C nanofiber with high photocatalytic activity and recyclability under solar light. *Appl Surf Sci*. 491(February):350–359. https://doi.org/10.1016/j.apsusc.2019.06.159

Ghasemipour P, Fattahi M, Rasekh B, Yazdian F. 2020. Developing the Ternary ZnO doped MoS2 nanostructures grafted on CNT and reduced graphene oxide (RGO) for photocatalytic degradation of aniline. *Sci Rep*. 10(1):1–16. http://dx.doi.org/10.1038/s41598-020-61367-7

Giang NTH, Hai ND, Thinh NT, Tan NN, Phuong LP, Thinh DB, Van Duc N, Viet VND, Duy HK, Phong MT, Hieu NH. 2022. Enhanced photocatalytic degradation of malachite green by sulfur-doped titanium dioxide/porous reduced graphene oxide. *Diam Relat Mater*. 129.

Guo J, Sun Y, Luo Q, Zhang J, Fang L. 2023. Construction of Z-scheme Ti/Ga co-doped ZnO heterostructure photocatalyst with graphitic carbon nitride for efficient visible-light-driven dye degradation. *Environ Sci Pollut Res*. 30, 43702–43713. https://doi.org/10.1007/s11356-023-25276-2

Hazarika D, Karak N. 2016. Photocatalytic degradation of organic contaminants under solar light using carbon dot/titanium dioxide nanohybrid, obtained through a facile approach. *Appl Surf Sci*. 376:276–285.

Hu Z, Cai X, Wang Z, Li S, Wang Z, Xie X. 2019. Construction of carbon-doped supramolecule-based g-C 3 N 4/TiO 2 composites for removal of diclofenac and carbamazepine : a comparative study of operating parameters, mechanisms, degradation pathways. *J Hazard Mater*. 380(July):120812. https://doi.org/10.1016/j.jhazmat.2019.120812

Huang L, Liu B. 2016. Synthesis of a novel and stable reduced graphene oxide/MOF hybrid nanocomposite and photocatalytic performance for the degradation of dyes. *RSC Adv*. 6(22):17873–17879.

Huang Q, Song S, Chen Z, Hu B, Chen J, Wang X. 2019. Biochar-based materials and their applications in removal of organic contaminants from wastewater: state-of-the-art review. *Biochar*. 1(1):45–73. https://doi.org/10.1007/s42773-019-00006-5

Isari AA, Mehregan M, Mehregan S, Hayati F, Rezaei Kalantary R, Kakavandi B. 2020. Sono-photocatalytic degradation of tetracycline and pharmaceutical wastewater using WO3/CNT heterojunction nanocomposite under US and visible light irradiations: a novel hybrid system. *J Hazard Mater*. 390(January):122050. https://doi.org/10.1016/j.jhazmat.2020.122050

Javaid A, Latif S, Imran M, Hussain N, Shahid M, Rajoka R, Iqbal HMN, Bilal M. 2022. Chemosphere nanohybrids-assisted photocatalytic removal of pharmaceutical pollutants to abate their toxicological effects—a review. *Chemosphere*. 291(P2):133056. https://doi.org/10.1016/j.chemosphere.2021.133056

Jo YK, Lee JM, Son S, Hwang SJ. 2019. 2D inorganic nanosheet-based hybrid photocatalysts: design, applications, and perspectives. *J Photochem Photobiol C Photochem Rev*. 40:150–190. https://doi.org/10.1016/j.jphotochemrev.2018.03.002

Kim DS, Kwak SY. 2007. The hydrothermal synthesis of mesoporous TiO2 with high crystallinity, thermal stability, large surface area, and enhanced photocatalytic activity. *Appl Catal A Gen*. 323:110–118.

Kim J, Cote LJ, Huang J. 2012. Two dimensional soft material: new faces of graphene oxide. *Acc Chem Res*. 45(8):1356–1364.

Liang R, Shen L, Jing F, Qin N, Wu L. 2015. Preparation of MIL-53(Fe)-reduced graphene oxide nanocomposites by a simple self-assembly strategy for increasing interfacial contact: efficient visible-light photocatalysts. *ACS Appl Mater Interfaces*. 9507–9515.

Lin L, Wang H, Xu P. 2017. Immobilized TiO_2-reduced graphene oxide nanocomposites on optical fibers as high performance photocatalysts for degradation of pharmaceuticals. *Chem Eng J*. 310:389–398. http://dx.doi.org/10.1016/j.cej.2016.04.024

Liu XW, Sun TJ, Hu JL, Wang SD. 2016. Composites of metal-organic frameworks and carbon-based materials: preparations, functionalities and applications. *J Mater Chem A*. 4(10):3584–3616.

Luo S, Wang J. 2018. MOF/graphene oxide composite as an efficient adsorbent for the removal of organic dyes from aqueous solution. *Environ Sci Pollut Res*. 25(6):5521–5528.

Madima N, Mishra SB, Inamuddin I, Mishra AK. 2020. Carbon-based nanomaterials for remediation of organic and inorganic pollutants from wastewater. a review. *Environ Chem Lett*. 18(4):1169–1191. https://doi.org/10.1007/s10311-020-01001-0

Mahdiani M, Soofivand F, Ansari F, Salavati-Niasari M. 2018. Grafting of CuFe12O19 nanoparticles on CNT and graphene: eco-friendly synthesis, characterization and photocatalytic activity. *J Clean Prod*. 176:1185–1197.

Meena PL, Poswal K, Surela AK, Saini JK. 2023. Synthesis of graphitic carbon nitride/zinc oxide (g-C3N4/ZnO) hybrid nanostructures and investigation of the effect of ZnO on the photodegradation activity of g-C3N4 against the brilliant cresyl blue (BCB) dye under visible light irradiation. *Adv Compos Hybrid Mater*. 6(1):1–17. https://doi.org/10.1007/s42114-022-00577-1

Mehrizad A, Gharbani P. 2017. Novel ZnS/carbon nanofiber photocatalyst for degradation of rhodamine 6G: kinetics tracking of operational parameters and development of a kinetics model. *Photochem Photobiol.* 93(5):1178–1186.

Minella M, Fabbri D, Calza P, Minero C. ScienceDirect selected hybrid photocatalytic materials for the removal of drugs from water. *Curr Opin Green Sustain Chem.* 6:11–17. http://dx.doi.org/10.1016/j.cogsc.2017.05.002

Mohamed RM, Abdel Salam M. 2014. Photocatalytic reduction of aqueous mercury(II) using multi-walled carbon nanotubes/Pd-ZnO nanocomposite. *Mater Res Bull.* 50:85–90. http://dx.doi.org/10.1016/j.materresbull.2013.10.031

Muninathan S, Arumugam S. 2021. Enhanced photocatalytic activities of NiS decorated reduced graphene oxide for hydrogen production and toxic dye degradation under visible light irradiation. *Int J Hydrogen Energy.* 46(9):6532–6546. https://doi.org/10.1016/j.ijhydene.2020.11.178

Pant B, Park M, Kim HY, Park SJ. 2016. Ag-ZnO photocatalyst anchored on carbon nanofibers: synthesis, characterization, and photocatalytic activities. *Synth Met.* 220:533–537. http://dx.doi.org/10.1016/j.synthmet.2016.07.027

Qiu J, Liu F, Yue C, Ling C, Li A. 2019. A recyclable nanosheet of Mo/N-doped TiO2 nanorods decorated on carbon nanofibers for organic pollutants degradation under simulated sunlight irradiation. *Chemosphere.* 215:280–293. https://doi.org/10.1016/j.chemosphere.2018.09.182

Qureshi I, Khan S, Shifa MS, Wazir AH. 2022. Graphene oxide-based ZnFe2O4 catalyst for efficient adsorption and degradation of methylene blue from water. *J Dispers Sci Technol.* 43(2):282–288.

Ramalingam G, Perumal N, Priya AK, Rajendran S. 2022. A review of graphene-based semiconductors for photocatalytic degradation of pollutants in wastewater. *Chemosphere.* 300(March):134391. https://doi.org/10.1016/j.chemosphere.2022.134391

Rauf MA, Meetani MA, Hisaindee S. 2011. An overview on the photocatalytic degradation of azo dyes in the presence of TiO2 doped with selective transition metals. *Desalination.* 276(1–3):13–27. http://dx.doi.org/10.1016/j.desal.2011.03.071

Rout DR, Chaurasia S, Jena HM. 2022. Enhanced photocatalytic degradation of malachite green using manganese oxide doped graphene oxide/zinc oxide (GO-ZnO/Mn2O3) ternary composite under sunlight irradiation. *J Environ Manage.* 318.

Sekar A, Yadav R. 2021. Green fabrication of zinc oxide supported carbon dots for visible light-responsive photocatalytic decolourization of malachite green dye: optimization and kinetic studies. *Optik (Stuttg).* 242.

Sim LC, Leong KH, Ibrahim S, Saravanan P. 2014. Graphene oxide and Ag engulfed TiO2 nanotube arrays for enhanced electron mobility and visible-light-driven photocatalytic performance. *J Mater Chem A.* 2(15):5315–5322.

Singh AR, Dhumal PS, Bhakare MA, Lokhande KD, Bondarde MP, Some S. 2022. In-situ synthesis of metal oxide and polymer decorated activated carbon-based photocatalyst for organic pollutants degradation. *Sep Purif Technol.* 286(December 2021):120380. https://doi.org/10.1016/j.seppur.2021.120380

Smrithi SP, Kottam N, Arpitha V, Narula A, Anilkumar GN, Subramanian KRV. 2020. Tungsten oxide modified with carbon nanodots: integrating adsorptive and photocatalytic functionalities for water remediation. *J Sci Adv Mater Devices.* 5(1):73–83.

Song Y, Qi J, Tian J, Gao S, Cui F. 2018. Construction of Ag/g-C 3 N 4 photocatalysts with visible-light photocatalytic activity for sulfamethoxazole degradation. *Chem Eng J.* 341(December 2017):547–555. https://doi.org/10.1016/j.cej.2018.02.063

Su Y, Chen S, Quan X, Zhao H, Zhang Y. 2008. A silicon-doped TiO2 nanotube arrays electrode with enhanced photoelectrocatalytic activity. *Appl Surf Sci.* 255(5 PART 1):2167–2172.

Thakur A, Kumar P, Kaur D, Devunuri N, Sinha RK, Devi P. 2020. TiO2 nanofibres decorated with green-synthesized PAu/Ag@CQDs for the efficient photocatalytic degradation of organic dyes and pharmaceutical drugs. *RSC Adv.* 10(15):8941–8948.

Tuan P Van, Tuong HB, Tan VT, Thu LH, Khoang ND, Khiem TN. 2022. SnO2/reduced graphene oxide nanocomposites for highly efficient photocatalytic degradation of methylene blue. *Opt Mater (Amst).* 123.

Tyagi A, Tripathi KM, Singh N, Choudhary S, Gupta RK. 2016. Green synthesis of carbon quantum dots from lemon peel waste: applications in sensing and photocatalysis. *RSC Adv.* 6(76):72423–72432.

Venkatesh N, Mohankumar A, Murugadoss G, Sundararaj P, Hatamleh AA, Alnafisi BK, Kumar MR, Gouse Peera S, Sakthivel P. 2023. Visible light active hybrid silver decorated g-C3N4–CeO2 nanocomposite for ultrafast photocatalytic activity and toxicity evaluation. *Environ Res.* 216(P3):114749. https://doi.org/10.1016/j.envres.2022.114749

Vinothkannan M, Karthikeyan C, Gnana Kumar G, Kim AR, Yoo DJ. 2015. One-pot green synthesis of reduced graphene oxide (RGO)/Fe3O4 nanocomposites and its catalytic activity toward methylene blue dye degradation. *Spectrochim Acta—Part A Mol Biomol Spectrosc*. 136(PB):256–264. http://dx.doi.org/10.1016/j.saa.2014.09.031

Wang H, Huang X, Li W, Gao J, Xue H, Li RKY, Mai YW. 2018. TiO2 nanoparticle decorated carbon nanofibers for removal of organic dyes. *Colloids Surfaces A Physicochem Eng Asp*. 549(March):205–211.

Wang Q, Cai C, Wang M, Guo Q, Wang B, Luo W, Wang Y, Zhang C, Zhou L, Zhang D, et al. 2018. Efficient photocatalytic degradation of malachite green in seawater by the hybrid of zinc-oxide nanorods grown on three-dimensional (3D) reduced graphene oxide (RGO)/Ni foam. *Materials (Basel)*. 11(6).

Wang Y, Jiang F, Chen J, Sun X, Xian T, Yang H. 2020. In situ construction of CNT/cus hybrids and their application in photodegradation for removing organic dyes. *Nanomaterials*. 10(1).

Wang Z, Wang H, Zeng Z, Zeng G, Xu P, Xiao R, Huang D, Chen X, He L, Zhou C, et al. 2020. Applied catalysis B : environmental metal-organic frameworks derived Bi 2 O 2 CO 3/porous carbon nitride : a nanosized Z-scheme systems with enhanced photocatalytic activity. *Appl Catal B Environ*. 267(November 2019):118700. https://doi.org/10.1016/j.apcatb.2020.118700

Xu J, He S, Zhang H, Huang J, Lin H, Wang X, Long J. 2015. Layered metal-organic framework/graphene nanoarchitectures for organic photosynthesis under visible light. *J Mater Chem A*. 3(48):24261–24271.

Yang C, You X, Cheng J, Zheng H, Chen Y. 2017a. A novel visible-light-driven In-based MOF/graphene oxide composite photocatalyst with enhanced photocatalytic activity toward the degradation of amoxicillin. *Appl Catal B Environ*. 200:673–680. http://dx.doi.org/10.1016/j.apcatb.2016.07.057

Yang C, You X, Cheng J, Zheng H, Chen Y. 2017b. Applied catalysis B: environmental a novel visible-light-driven in-based MOF/graphene oxide composite photocatalyst with enhanced photocatalytic activity toward the degradation of amoxicillin. *Applied Catal B, Environ*. 200:673–680. http://dx.doi.org/10.1016/j.apcatb.2016.07.057

Yang Y, Cao L, Wu S, Qin L, Kang S, Li X. 2021. A patterned aluminum/reduced graphene oxide/silver sheet for detection and degradation of malachite green in water. *Sep Purif Technol*. 272.

Yang Y, Ma Z, Xu L, Wang H, Fu N. 2016. Preparation of reduced graphene oxide/meso-TiO2/AuNPs ternary composites and their visible-light-induced photocatalytic degradation n of methylene blue. *Appl Surf Sci*. 369:576–583.

Yang Y, Wang W, Li H, Jin X, Wang H, Zhang L, Zhang Y. 2017. NH2-MIL-53(Al) nanocrystals anchored on the surface of RGO hollow spheres and its visible light degradation of methylene blue. *Mater Lett*. 197:17–20. http://dx.doi.org/10.1016/j.matlet.2017.03.041

Yang Z, Xu X, Liang X, Lei C, Gao L, Hao R, Lu D, Lei Z. 2017. Fabrication of Ce doped UiO-66/graphene nanocomposites with enhanced visible light driven photoactivity for reduction of nitroaromatic compounds. *Appl Surf Sci*. 420:276–285. http://dx.doi.org/10.1016/j.apsusc.2017.05.158

Yuan X, Wang H, Wu Y, Zeng G, Chen X, Leng L, Wu Z, Li H. 2016. One-pot self-assembly and photoreduction synthesis of silver nanoparticle-decorated reduced graphene oxide/MIL-125(Ti) photocatalyst with improved visible light photocatalytic activity. *Appl Organomet Chem*. 30(5):289–296.

Zango ZU, Jumbri K, Sambudi NS, Ramli A, Bakar NHHA, Saad B, Rozaini MNH, Isiyaka HA, Jagaba AH, Aldaghri O, Sulieman A. 2020. A critical review on metal-organic frameworks and their composites as advanced materials for adsorption and photocatalytic degradation of emerging organic pollutants from wastewater. *Polymers (Basel)*. 12(11):1–42.

Zhang J, Dai M, Zhang S, Dai M, Zhang P, Wang S, He Z. 2022. Recent progress on carbon-nanotube-based materials for photocatalytic applications: a review. *Sol RRL*. 6(9):1–31.

Zhang W, Zhang YZ, Yang JM. 2022. MIL-100(Fe)@GO composites with superior adsorptive removal of cationic and anionic dyes from aqueous solutions. *J Mol Struct*. 1265:133365. https://doi.org/10.1016/j.molstruc.2022.133365

Zhang Y, Li G, Lu H, Lv Q, Sun Z. 2014. Synthesis, characterization and photocatalytic properties of MIL-53(Fe)-graphene hybrid materials. *RSC Adv*. 4(15):7594–7600.

Zhu L, Meng L, Shi J, Li J, Zhang X, Feng M. 2019. Metal-organic frameworks/carbon-based materials for environmental remediation: a state-of-the-art mini-review. *J Environ Manage*. 232(August 2018):964–977. https://doi.org/10.1016/j.jenvman.2018.12.004

Zhu Y, James DK, Tour JM. 2012. New routes to graphene, graphene oxide and their related applications. *Adv Mater*. 24(36):4924–4955.

7 Microbial Biofilms in Biodegradation of Toxic Chemicals

Antonina Nikolayeva

7.1 INTRODUCTION

Biofilms are a form of microbial communities that can attach themselves to various abiotic and biotic surfaces. Biofilms are composed of microbial cells and their associated extracellular matrix, which consists of polysaccharides, proteins, and DNA. Biofilms are ubiquitous in nature. They also are formed in catheters, implants, and contact lenses. Everyone knows such an example of a biofilm as plaque—a thin layer that forms on the teeth. Bacterial biofilms are resistant to various stresses, including antibiotics and disinfectants. In biofilms, bacteria are protected from phagocytosis and other components of innate and adaptive immunity.

In nature, as a rule, several types of microorganisms exist in biofilms as a single community. It has been shown that the numerous physiological processes occur, including the production of metabolites and biologically active substances, in a microbial biofilm, in comparison with pure planktonic cultures of bacteria. The response of microorganisms to changing environmental conditions in a biofilm is significantly different from the response of each individual species in a monoculture. This organization ensures its physiological and functional stability and, therefore, is the key to competitive survival in the ecological niche.

Biofilms can also be very dangerous, as they are often formed in various infectious pathologies. The course of infectious diseases can proceed with complications because of the formation of microbial biofilms in the body. Many chronic diseases are associated with biofilm infections—cystic fibrosis pneumonia, otitis media, pathology of the teeth and periodontal tissues, osteomyelitis, urinary tract infections, and others. It is believed that up to 5–80% of all human bacterial infections are associated with the formation of biofilms.

The study of biofilms is currently of great interest to researchers, mainly due to the fact that this mode of existence of bacteria creates great problems in medical practice. This chapter describes biofilms that are formed at the interface between liquid and solid media.

Bacteria that form communities are able to exhibit new properties, such as resistance to antimicrobial agents.

At the moment, the ability of microorganisms (fungi and bacteria) to switch between free-living and attached states under various conditions is well known, using biofilm as a factor of colonization, pathogenicity, virulence, and protection from environmental conditions.

Thus, the accumulation of microorganisms attached to the surface—and specific polymeric substances released by them into the environment—is called a biofilm. The surface on which the biofilm is formed is conventionally called the substrate (Mardanova et al., 2016).

7.1.1 Formation of Microbial Biofilms

In the process of its development, a biofilm goes through a number of stages (Figure 7.1). First, the adhesion of microorganisms to the surface of the substrate from the environment (most often a liquid medium) occurs. This stage is reversible, since adherent cells can return to the planktonic form

DOI: 10.1201/9781003391487-7

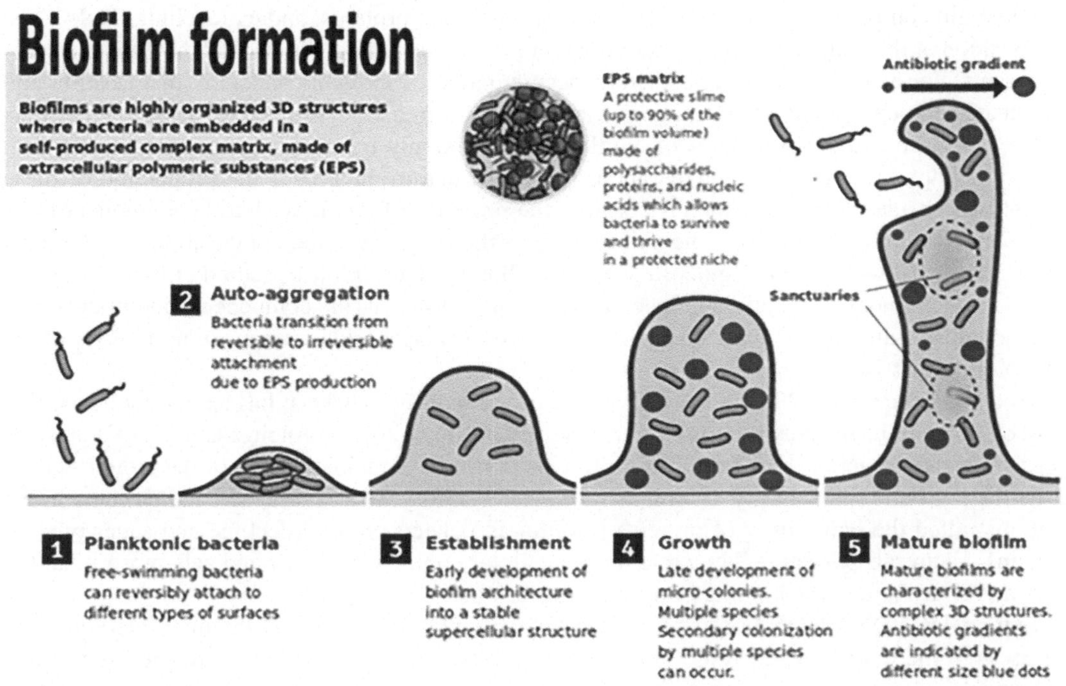

FIGURE 7.1 Schematic representation of the different steps required for the formation of a mature biofilm (Meroni et al., 2021).

of existence. Then, the final attachment of cells to the surface occurs, and this is called fixation. At this stage, microbes secrete extracellular polymers that provide strong adhesion. Then microcolonies (separate clusters of adherent cells) are formed. At this stage, the cells are actively dividing, and the secreted matrix holds the entire colony together. Finally, the microcolonies merge and form a mature biofilm with a complex three-dimensional structure. It can change its size and shape. The extracellular matrix protects it from external threats. Also, for many bacteria with developed motility, the stage of a cell monolayer is distinguished. In this case, the cells are able to propagate along the substrate, using various surface elements of the cells (pili, fimbria) and forming schwermers—elongated cells with several chromosomes. Also, especially at low nutrient levels, cells are able to leave the biofilm and become planktonic, which is called dispersion. As a result of dispersion, individual cells are periodically detached from the biofilm, which, after some time, can attach themselves to the surface and form a new colony. Inside mature biofilms, a population of persisters (cells with particular resistance to antibiotics) is isolated (Tolker-Nielsen, 2015; Mardanova et al., 2016).

7.1.2 Composition and Structure of Biofilms

Biofilms are composed of microbial cells and a matrix. According to modern concepts, a biofilm is a formation formed by microcolonies of microorganisms in the form of towers or mushrooms (15–20% of the volume) and an exopolymer matrix (75–85% of the volume). Polymeric substances secreted by adherent cells into the external environment are called matrix or extracellular polymeric substance (EPS).

The matrix is hydrated and consists of 97% water. This matrix is similar to a "sponge" or porous structure that allows small molecules to pass through but retains large molecules and proteins. In the structure of a biofilm, water channels are often found, through which substances dissolved in the medium enter the underlying layers of cells.

The main components of the matrix are polysaccharides, proteins, and extracellular DNA. The composition of the matrix may vary, depending on environmental conditions and the type of microorganisms. For example, in urinal catheters, accumulation of inorganic calcium, magnesium, and phosphate compounds with the formation of crystallins is often observed on the surface of biofilms.

The polysaccharide components of biofilms are commonly referred to as exopolysaccharides. The most important condition for the appearance of a mature biofilm is the production of these exopolysaccharides. Some researchers consider the presence of a polysaccharide component to be necessary for biofilm formation. Sugars can vary in their structure and composition, for the most part being beta-1,6-N-acetyl-D-glucosamine and cellulose. The simplest method of their visualization is immunofluorescent staining. Specific stains for polysaccharide components and microscopy with cellulose staining (e.g., calcofluor) are used. Exopolysaccharides can make up to 80% of the total biofilm matrix.

Proteins make up to 40% of the total volume of the biofilm matrix. It has been established that most of the proteins in many biofilms are represented by amyloid-like proteins: CsgA/CsgB proteins in gram-negative bacteria and TasA/TapA proteins in bacilli. Amyloid groups of the proteins determine the high thermal stability as well as its adhesive ability. In addition, specific proteins in the biofilm, called the bap-family of proteins (biofilm-associated protein), lectins, and sugar-binding proteins—autotransporters—play a role in intercellular contacts, cell-cell and cell-substrate adhesion, and contribute to the binding of polysaccharides in the biofilm structure.

Extracellular DNA (eDNA) can be excreted by vesicular transport, but the main method of its release into the matrix is cell lysis. This DNA is involved in horizontal gene transfer in biofilms and signaling between cells. Extracellular DNA can play a structural role and serve as a target for exonucleases. The total content of DNA in the matrix is low (Hobley et al., 2015).

The ratio of the matrix and cells in a biofilm can be different. It is believed that, in some cases, cells can occupy only 10% of the biofilm. Cells in biofilms can be at different stages of the cell cycle (Figure 7.1). Thus, dead and destroyed cells are also part of the biofilm.

Biofilms formed by several types of microorganisms are called mixed (multispecies). Biofilms formed by fungal filaments are called fungal. Most often, in the natural environment, microorganisms coexist in mixed biofilms, jointly producing exopolysaccharides and proteins (Mardanova et al., 2016).

7.1.3 Quorum Sensing in Bacteria

Many genetically determined programs regulate the processes of biofilm formation at different stages. The essential links of these processes at the early stages are interactions: cell-environment and cell-surface. In the later stages, this situation is changed. The driving force in the development of the bacterial community is self-organization and cooperation between cells, not classical, "competitive" natural selection of individual microorganisms.

The most important mechanism of such regulation in the cell community is quorum sensing. Quorum sensing is a special type of regulation of bacterial gene expression, depending on their population density. Specialized substances, called autoinductors, are excreted by cells, and they are accumulated in the medium during cell growth populations. The response of the cell depends on the concentration of the inductor in the medium, and the genetic program will turn on when inductor reaches the critical importance of each cell of the community. The result is a high degree of coordination of gene expression and adaptation mechanisms (Miller and Bassler, 2001). The role of autoinducer messengers in gram-negative bacteria is performed by N-acetylhomoserine lactones (AHL) and in gram-positive bacteria—oligopeptides. AHL molecules can move between the environment and the cell with the help of membrane-bounded proteins. Messengers, also called autoinductors-2 (AI-2), responsible for signal transduction between gram-positive and gram-negative microorganisms, most often represented by homologues of furanosyl borate diester or tetrahydroxy furan (Mardanova et al., 2016).

7.2 BIOREMEDIATION

7.2.1 Different Types of Bioremediation

Bioremediation can be classified according to the organisms that are used: application of bacteria, application of fungi, application of algae, and combined method. Bioremediation is a safe process of neutralization of harmful pollutants of soil, water, and air using microorganisms (Prasad and Prasad, 2012). Bioremediation does not use aggressive chemicals, and this fact reduces the harmful impact of toxic substances on the environment (Vidali, 2001). Bioremediation involves no or minimal disruption of environmental objects, and bioremediation is an economical process, compared to conventional decontamination methods. The bioremediation process uses a variety of microbes to decompose or process xenobiotic compounds, aromatic hydrocarbons, pesticides, herbicides, heavy metals, and radionuclides (Gaur et al., 2014).

Microbes, especially bacteria, can be easily grown and genetically manipulated, making them suitable for bioremediation.

In addition to bacteria, fungi are also used for pollution removal (Mishra and Malik, 2014). Importantly, microbes are extremely efficient at degrading waste pollutants through a variety of catabolic pathways, as microbes adapt to survive in different conditions (Bruins et al., 2000). Some microbes are extremophiles that can withstand acidic or heavy metal-contaminated or radioactive environment (Prince, 2000).

The bioremediation depends on the enzymatic activity of microbes to decompose environmental pollutants into less toxic or harmless components such as carbon dioxide and water (Das and Dash, 2014). The metabolic pathways require the transmission of electrons from electron donors to electron acceptors. Electron donors serve as nutrition for microbes, which are usually limited in an uncontaminated site. The release of an organic electron donor may encourage microbes to compete for available acceptors to rebalance the system. Microorganisms can decompose pollutants without oxygen and with oxygen in anaerobic and aerobic conditions, respectively. In aerobic degradation, microbes use oxygen as the final electron acceptor to convert organic and inorganic pollutants into harmless products—often carbon dioxide and water. In anaerobic decomposition, microbes use other electron acceptors such as nitrate, iron, manganese, and sulfate to break down organic compounds—often into carbon dioxide and methane. As a rule, aerobic microbes are capable of more rapid decomposition of pollutants than anaerobic microbes. Sometimes, some microbes can break down contaminants by fermentation. Pollutants that are electron donors can be easily degraded in the presence of oxygen by aerobic microbes, while those pollutants that are poor electron donors can degrade pollutants under anaerobic conditions. Depending on the contaminant, both electron donors and acceptors can be supplied to facilitate the degradation process.

Fungi have a significant impact on humans and environmental components due to their ability to mineralize and decompose toxins. These microorganisms can destroy any natural biopolymer, aromatic and phenolic compounds, pesticides, and other substances that enter the soil with industrial waste (Young Darcy, 2012). The bioremediation properties of fungi are associated with rapid colonization and their deep penetration into the substrate, and they can survive and develop in a wide range of environmental conditions. The fungus *Phanerochaete chrysosporium* is a bioremediation organism. Preparations based on this microorganism are able to remove a large number of pollutants from the soil: organochlorine compounds (chlorophenol, polychlorinated biphenyl), aromatic hydrocarbons (benzene, toluene), dyes, pesticides, and many synthetic polymers. Many fungal species, such as basidiomycetes and ascomycetes, are capable of degrading cellulose materials. The natural components of the oil—i.e. polycyclic aromatic hydrocarbons—are degraded by enzyme laccase, which is present in cells of basidiomycetes. Fungi *Pleurotus pulmonarius* are known for their ability to degrade oil compounds (Olusola and Anslem, 2010). Various types of yeasts are used to bioremediate soil contaminated with metals. The most effective actions to remove organic pollutants are biological products based on *Saccharomyces cerevisiae* and *Leurotus ostreatus*. These microorganisms are basidiomycetes that secrete enzymes and acids that break down organic polymers. The use of fungi is an effective method of soil bioremediation (Xu, 2015).

It has been established that application of green algae as biopreparation for soil bioremediation increases the effectiveness of these preparations. *Chlamydomonas reinhardtii* has a high efficiency in the accumulation and degradation of herbicides. This is due to the metabolic processes of the algae, which leads to rapid degradation and removal of herbicides from the soil. *Monoraphidium braunii* is applied for soil phytoremediation from phenol, the algae converting phenol into its monoglucoside (Jin, 2012). The role of the algae *Chlorococcum sp.* and *Scenedesmus sp.* in the degradation of insecticides has been demonstrated in many experiments (Sethunathan, 2015).

Communities of algae and cyanobacteria with other aerobic or anaerobic microorganisms form microbial groups that live as symbiotic organisms. Consortiums of algae and bacteria increase significantly the biodegradation of organic and inorganic pollutants compared to single microorganisms. Benefits of consortiums: resistance to environmental fluctuations, the ability to survive during a lack of nutrients, the ability to share metabolites, and increased resistance to invasion by other species. Algae accumulate toxic chemicals on their cell wall and form a surface zone that is degraded by bacteria (Subashchandrabose, 2013). Bacteria protect the algae from the toxic effects of oil products, while cyanobacteria provide the bacteria with oxygen and other extracellular substances (Abed, 2010).

7.2.2 Role of Microbial Biofilms in Bioremediation

A biofilm is a microbial community in which bacterial cells are attached to a substrate or to each other and are immersed in a matrix of an extracellular polymeric substance. Microorganisms in the composition of biofilms are protected from the effects of environmental factors and are able to transform organic pollutants. Pollutants such as polycyclic aromatic hydrocarbons, glyphosate, and hydrocarbons are the most common environmental pollutants in the world. Understanding the basic principles of the bacterial biofilms is important for a more effective solution of the problems of bioremediation, medical microbiology, and other areas in which it is necessary to stimulate or suppress the formation of microbial biofilms. Therefore, the study of the formation, metabolism, and functional state of bacterial biofilms under the influence of organic pollutants is actual.

Various types of microbes that are present in biofilms are able to degrade several pollutants, individually or collectively (Gieg et al., 2014). Biofilm-forming bacteria are adapted to survive and are suitable for bioremediation because they compete with nutrients and oxygen. Biofilm is an economical way for removing environmental contaminants. The use of biofilms is effective for bioremediation, since biofilms absorb, immobilize, and decompose various environmental pollutants. Bacterial biofilms exist near contaminated sites to persist and survive extremal environmental conditions. Gene expression in biofilms differs from free planktonic cells. Differential gene expression in biofilms is due to different local concentrations of nutrients and oxygen in the biofilm matrix. Such variable gene expression may be important for the degradation of various pollutants by numerous metabolic pathways. An important factor in the formation of biofilms in microorganisms is chemotaxis and flagella-dependent motility (Pratt and Kolter, 1999). Motility, chemotaxis, and quorum sensing help to coordinate the movement of microbes toward the pollutant and improve biodegradation (Lacal et al., 2013).

Most bacteria persist as a biofilm enclosed in a matrix, which helps microbes to participate in bioremediation (More et al., 2014). The matrix consists of polysaccharides, proteins, lipids, nucleic acids, and water (Branda et al., 2005). The composition of the matrix varies in different microbes and depends on growth conditions and stress situations. Bacterial biofilms and matrix production depend on environmental conditions in which microbes are located (Kreft and Wimpenny, 2001; Miqueleto et al., 2010). For example, bacteria often adapt to the presence of protozoa in the form of large colonies that are responsible for survival and resistance to extreme conditions (Matz and Kjelleberg, 2005). The biofilm matrix provides greater resistance than planktonic cells to environmental stress, antimicrobial agents, UV damage, biocides, high concentrations of toxic chemicals, and pollutants (Davey and O'Toole, 2000; Mah and O'Toole, 2001). In contrast, planktonic cells deactivate environmental pollutants through metabolic activity, but such cells are not adapted to survive under extreme conditions (Sutherland, 2001). Biofilm microbes receive limited nutrients and oxygen compared to planktonic cells. The three-dimensional structure of the matrix with a low oxygen content connects aerobes and anaerobes and sulfate reducers with sulfate oxidizers, which promotes degradation of various pollutants (Field et al., 1995). The cyanobacterial matrix acts as a sorbent and removes heavy metals (De Philippis et al., 2011). The presence of the matrix is involved in the exchange of nutrients and the removal of dangerous products in the biofilm. Extracellular enzymes within the biofilm matrix decontaminate substances such as heavy metals and organic compounds (Flemming and Wingender, 2010). The matrix binds a wide range of metals, including copper, magnesium, zinc, cadmium, iron, and nickel (Ferris et al., 1989).

In nature, mixed biofilms of various types are usually formed, which have the ability to withstand environmental stress (Burmolle et al., 2006). Natural ecological biofilms exist in soil, plants, lakes and rivers, and swamps. Biofilms in the environment can destroy and remove pollutants from the soil and river. Biofilms that are produced on the surface of the water are composed of bacteria, protozoa, fungi, and algae. Microbial consortiums of phototrophic bacteria and algae are formed under extreme conditions to survive stress (Logsdon et al., 2002).

Biofilms are often used as an indicator for assessing heavy metal pollution in rivers, since physiological changes in biofilms occur in the presence of toxic substances (Fuchs et al., 1997). Biofilms interact with nutrients and pollutants in aquatic systems, and therefore, such biofilms can be used as environmental monitoring in water resources (Peacock et al., 2004). Severe environmental pollution, such as zinc and cadmium, can affect species diversity within biofilms, and therefore, the assessment of microbial biofilms can indicate environmental pollution (Ramírez et al., 2009). Modern molecular biology techniques such as denaturing gradient gel electrophoresis, ribosomal spacer analysis, and terminal restriction fragment length polymorphism have been used to assess species diversity in biofilms (Kostanjsek et al., 2005; Bricheux et al., 2013). Changes in algae biofilms under the influence of heavy metals and herbicides can be used as an indicator of environmental pollution (Navarro et al., 2002). Analysis and samples of pigment composition can serve as a biomarker because pigment composition among biofilm members can be changed after exposure to toxic chemicals (Dewez et al., 2008).

7.2.3 Factors Affecting Microbial Remediation

Biological factors include the presence of microorganisms, the diversity of the microbial population, the presence of specific metabolic pathways, the intensity of metabolism, and the concentration of microorganisms (Vasudevan and Rajaram, 2001).

Physicochemical factors can be divided into several groups (Boopathy, 2000): substrate (chemical characteristics, molecular structure, concentration, availability, and toxicity of pollutants), environmental (physical and chemical properties of the soil, nutrient concentration, redox potential (Eh), pH, humidity, and temperature).

7.2.3.1 Presence of Microorganisms

The presence of bacteria with decomposing abilities is the main requirement for remediation (Bundy et al., 2002). It has been established that hydrocarbons are decomposed by various types of heterotrophic microorganisms under aerobic and anaerobic conditions (Ramírez et al., 2009). The ability to decompose substances is observed in fungi—*Candida*, *Saccharomyces*, *Cryptococcus*, *Rhodotorula*, *Rhodospondium*, *Sporobolomyces*, *Torulopsis*, *Trichosporon* (Prenafeta-Boldu et al., 2002)—some cyanobacteria—*Oscillatoria*, *Anabaena*, *Nostoc*, *Microcoleus*, *Chlamydomonas*, *Scenedesmus*, *Phormidium*—and green algae—*Chlorella*, *Microcoleus*, *Chlamydomonas*, *Ulva*, *Scenedesmus*, *Phormidium* (Antizar-Ladislao et al., 2004). However, bacteria are mainly used for soil bioremediation because they are characterized by rapid growth and a wide range of used hydrocarbons (Wolicka et al., 2009). The environment usually contains microorganisms that degrade various hydrocarbons using various metabolic pathways: alk (for alkanes), xyl (for xylene, toluene, aromatic hydrocarbons) and nah (for polycyclic aromatic hydrocarbons) (Whyte et al., 1997). Rarely, microorganisms possess both alk and nah catabolic pathways. Whyte, Bourbonniere, and Greer showed that both pathways located in different plasmids and can coexist in the same bacterium. Also, alk and xyl pathways for the degradation of aliphatic and aromatic hydrocarbons can be found in the same organism.

7.2.3.2 Properties and Concentrations of Pollutants

Biological degradation of pollutants can differ significantly, depending on their physical and chemical properties (Rojas-Avelizapa et al., 2007). The intensity of decomposition of hydrocarbons is related to the size of their molecules: small molecules (from methane to butane C1-C4) and large molecules > C44 are poorly destroyed (Schaefer and Filser, 2007). Simple hydrocarbons C5-C15, alcohols, phenols, amines, acids, esters, and amides are very easily decomposed by microorganisms (Huesemann et al., 2004). The most easily destroyed hydrocarbons are from decane to hexadecane (C10-C16); high molecular weight paraffins (C17-C24) are poorly destroyed. Substances with a large molecular weight are hydrophobic, and therefore, more resistant to biological degradation (Schaefer and Filser, 2007). Polychlorinated biphenyls, polycyclic aromatic hydrocarbons, and pesticides are very difficult to decompose during bioremediation (Gianfreda and Rao, 2004). The most difficult thing is the biological decomposition of a mixture of various chemical compounds. The intensity of decomposition of polycyclic aromatic hydrocarbons by microorganisms depends on their chemical structure, mainly on the number of rings (Margesin et al., 2000). Low molecular weight polycyclic aromatic hydrocarbons with < four rings (naphthalene, phenanthrene, and anthracene) are moderately toxic and have a high rate of biodegradation. High-molecular-weight polycyclic aromatic hydrocarbons with more than four rings (for example, pyrene, benzo(a)anthracene, and benzo(a) pyrene) exhibit a toxic, mutagenic effect, and their decomposition is possible only through combined metabolic pathways (Silva et al., 2009).

Organic matter is consumed by microorganisms if it can be used by them to obtain carbon and energy. Using hydrocarbons as a carbon source, microorganisms are able to decompose alkanes in three stages: first, turning them into alkyl alcohols; then into aldehydes; and finally, into aliphatic carboxylic acids (Schaefer and Filser, 2007). Soils contaminated with hydrocarbons often contain mixtures of different substances, which makes them more difficult to decompose. Metabolic cooperation of several microorganisms can lead to an increase in the use of hydrocarbons because the metabolic products produced by some organisms can serve as substrates for other microorganisms (Silva et al., 2009). For individual microorganisms and microbial communities, it has been shown that fungi and bacteria are capable of partial or complete decomposition of high-molecular-weight polycyclic aromatic hydrocarbons (Boonchan et al., 2000). The ability to biodegradation depends on the concentration of the pollutant. Low concentrations of oil are easily destroyed, while high concentrations have an acute toxic effect on microorganisms.

7.2.3.3 Soil Types

Bundy, Paton, and Campbell (Bundy et al., 2002) showed that different microbial communities are formed in different soil types. Differences in the composition of soil organic matter affect the structure and density of the microbial community (Cookson et al., 2005). Microbial communities are able to use high levels of soil organic matter to increase their abundance, diversity, and ability to resist pollution. Insufficient reproduction and diversity of microorganisms are typical for soils with a sandy texture and low content of organic carbon (Scherr et al., 2007).

The soils with heavy mechanical composition are a real threat of excessive accumulation of pollutants. Light soils are less prone to pollution by oil and products of its processing. The distribution of the pollutant and its availability depends on various characteristics of the soil: the content of organic matter, the presence of dissolved organic matter, and the mechanical composition of the soil (Gianfreda and Rao, 2004).

7.2.3.4 Molecular Oxygen Concentration

The rate of bioremediation of deep soil layers in situ can be increased by providing the access of microorganisms and oxygen to oil hydrocarbons (Hyun et al., 2008). Aerobic conditions and microorganisms are necessary for the optimal rate of bioremediation of soils contaminated with petroleum hydrocarbons (Vasudevan and Rajaram, 2001). Aerobic decomposition of hydrocarbons occurs much faster than anaerobic processes. The oxygen concentration in soils depends on microbial activity, the mechanical composition of the soil, and water content. It has been shown that the low oxygen content in soils limits the bioremediation of soils contaminated with oil hydrocarbons. It has been established that the mineralization of soil hydrocarbons is very limited when the oxygen concentration is below 10% (Freijer, 1986). Mechanical tillage improves soil aeration, enhances the contact between oil and bacteria, and reduces toxicity of soil (Wang and Bartha, 1990).

7.2.3.5 Humidity

The water content has a significant effect on the rate of decomposition of hydrocarbons in the soil (Ramírez et al., 2009). The optimal humidity level for aerobic biological decomposition of hydrocarbons is 50–80%. At a lower water content, the metabolism in microorganisms decreases. Some researchers (Admon et al., 2001) believe that the liquid solution of the soil creates the most favorable conditions for the biological decomposition of hydrocarbons. Liu, Banks, and Schwab (Liu et al., 2001) found that the microbial population increased with increased water content. However, other studies showed that the number of microorganisms increased insignificantly (Breedveld and Sparrevik, 2000). This effect is more important in clay soils than in sandy soils because clays form aggregates that make it difficult to biodegrade. The water content also increases the mobility of microorganisms, which facilitates the contact of microorganisms and pollutants (Ramírez et al., 2009).

7.2.3.6 Temperature

Temperature has a significant effect on the biological decomposition of hydrocarbons. Low temperatures increase the adaptation period and limit the rate and degree of biological decomposition of oil hydrocarbons in polluted soils (Sanscartier et al., 2009). At low temperatures, the viscosity of oil is increased, the evaporation of short chain alkanes is reduced, and thus their toxicity is increased (Atlas and Bartha, 1992). An increase in temperature leads to a decrease in viscosity, higher solubility, and faster diffusion of hydrophobic contaminants to the cell, thus enhancing biodegradation rates. Elevated temperatures can activate communities of thermophilic microorganisms, increasing the intensity of bioremediation (Perfumo et al., 2007). Among the various methods of bioremediation, the most preferable method is the bioremediation of soil contaminated with oil and oil products, based on the use of the metabolic capabilities of the local microflora, which has a minimal impact on the environment.

7.2.4 Advantages and Disadvantages of Microbial Remediation

Advantages of bioremediation:

- Bioremediation is a natural process for the recycling of contaminated materials such as soil. The number of microbes that can destroy pollutants increases in the presence of a pollutant.
- Bioremediation requires less energy than other technologies.
- Bioremediation may be less expensive than other technologies used to treat hazardous waste.

- Bioremediation is useful for the destruction of a wide range of contaminants. It does not use any dangerous chemicals. Bioremediation changes harmful chemicals into water and harmless gases; the harmful chemicals are completely destroyed (Kulshreshtha et al., 2014).

Disadvantages of bioremediation:

- Bioremediation is limited for biodegradable compounds. Not all compounds are disposed to a quick and complete degradation process. Bioremediation does not work well on clay soils where oxygen or nutrients are difficult to introduce into the soil.
- There are some new products of biodegradation that may be more toxic than the initial compounds.
- The bioremediation process takes much longer than other methods, such as dredging and burning.
- Further migration of the pollutant and its transfer to the environment is possible (Iosob et al., 2016).

7.3 CONCLUSIONS

This chapter describes the usage of microbial biofilms in the degradation of toxic chemicals. Biofilm is the accumulation of microbial communities that can attach themselves to various abiotic and biotic surfaces. These are composed of microbial cells and their associated extracellular matrix, and their development process occurs in multiple stages. New methods are constantly being tested in the field of bioremediation. Bioremediation involves no or minimal disruption of environmental objects, and bioremediation is an economical process compared to conventional decontamination methods. Microbial strains are being designed that use some pollutants as preferred substrates over other available compounds for effective bioremediation. New bioremediation methods use a process of transformation process within biofilms using DNA-containing genes that promote the biodegradation of selected contaminants. Genetically modified microorganisms can degrade various pollutants, including halogenated aromatic compounds. Horizontal transfer of biodegradable genes from genetically modified microorganisms to biofilm members can further enhance the biodegradation process. Reengineering of secreted proteins in the biofilm matrix is also an area of further development in the field of bioremediation. Intercellular interactions between members of the biofilm community can be further studied to improve the bioremediation process. In some cases, biofilm bioremediation can be used in combination with phytoremediation or chemical treatment for the destruction of persistent pollutants. Microbial consortia of bacteria-fungi can be used to decompose xenobiotic compounds. Overall, bioremediation based on biofilm communities remains an attractive choice for reducing environmental pollution. There are factors such as the presence of microorganisms, the diversity of the microbial population, specific metabolic pathways and the intensity of metabolism, and the concentration of microorganisms that affect microbial bioremediation.

BIBLIOGRAPHY

Abed R.M. (2010) Interaction between cyanobacteria and aerobic heterotrophic bacteria in the degradation of hydrocarbons. *International Biodeterioration & Biodegradation*, 60.

Admon S., Green M., Avnimelech Y. (2001) Biodegradation kinetics of hydrocarbons in soil during land treatment of oily sludge. *Bioremediation Journal* 5: 193–209.

Antizar-Ladislao B., Lopez-Real J.M., Beck A.J. (2004) Bioremediation of polycyclic aromatic hydrocarbon (PAH)—contaminated waste using composting approaches. *Critical Reviews in Environmental Science and Technology* 34: 249–289.

Atlas R.M., Bartha R. (1992) Hydrocarbon biodegradation and oil spill bioremediation. *Advances in Microbial Ecology* 12: 287–338.

Boonchan S., Britz M.L., Stanley G.A. (2000) Degradation and mineralization of high-molecular-weight polycyclic aromatic hydrocarbons by defined fungal-bacterial cocultures. *Applied and Environmental Microbiology* 66: 1007–1019.

Boopathy R. (2000) Factors limiting bioremediation technologies. *Bioresource Technology* 74: 63–67.

Branda S.S., Vik S., Friedman L., et al. (2005) Biofilms: the matrix revisited. *Trends in Microbiology* 13: 20–26. doi: 10.1016/j.tim.2004.11.006

Breedveld D.G., Sparrevik M. (2000) Nutrient-limited biodegradation of PAH in various soil strata at a creosote contaminated site. *Biodegradation* 11: 391–399.

Bricheux G., Le Moal G., Hennequin C., et al. (2013) Characterization and evolution of natural aquatic biofilm communities exposed in vitro to herbicides. *Ecotoxicology and Environmental Safety* 88: 126–134. doi: 10.1016/j.ecoenv.2012.11.003

Bruins M.R., Kapil S., Oehme F.W. (2000) Microbial resistance to metals in the environment. *Ecotoxicology and Environmental Safety* 45: 198–207. doi: 10.1006/eesa.1999.1860

Bundy J.G., Paton G.I., Campbell C.D. (2002) Microbial communities in different soil types do not converge after diesel contamination. *Journal of Applied Microbiology* 92: 276–288.

Burmolle M., Webb J.S., Rao D., et al. (2006) Enhanced biofilm formation and increased resistance to antimicrobial agents and bacterial invasion are caused by synergistic interactions in multispecies biofilms. *Applied and Environmental Microbiology* 72: 3916–3923. doi: 10.1128/AEM.03022-05

Cookson W.R., Abayeb D.A., Marschner P., Murphy D.V., Stockdaleb E.A., Goulding K.W.T. (2005) The contribution of soil organic matter fractions to carbon and nitrogen mineralization and microbial community size and structure. *Soil Biology and Biochemistry* 37: 1726–1737.

Das S., Dash H.R. (2014) Microbial bioremediation: a potential tool for restoration of contaminated areas. In: Das S., editor. *Microbial Biodegradation and Bioremediation* (1st edition). Oxford: Elsevier, pp. 1–21.

Davey M.E., O'Toole G.A. (2000) Microbial biofilms: from ecology to molecular genetics. *Microbiology and Molecular Biology Reviews* 64: 847–867. doi: 10.1128/MMBR.64.4.847-867.2000

De Philippis R., Colica G., Micheletti E. (2011) Exopolysaccharide-producing cyanobacteria in heavy metal removal from water: molecular basis and practical applicability of the biosorption process. *Applied Microbiology and Biotechnology* 92: 697–708. doi: 10.1007/s00253-011-3601-z

Dewez D., Didur O., Vincent-Héroux J., et al. (2008) Validation of photosynthetic-fluorescence parameters as biomarkers for isoproturon toxic effect on alga Scenedesmus obliquus. *Environmental Pollution* 151: 93–100. doi: 10.1016/j.envpol.2007.03.002

Ferris F.G., Schultze S., Witten T.C., et al. (1989) Metal interactions with microbial biofilms in acidic and neutral pH environments. *Applied and Environmental Microbiology* 55: 1249–1257.

Field J.A., Stams A.J., Kato M., et al. (1995) Enhanced biodegradation of aromatic pollutants in cocultures of anaerobic and aerobic bacterial consortia. *Antonie Van Leeuwenhoek* 67: 47–77. doi: 10.1007/BF00872195

Flemming H.C., Wingender J. (2010) The biofilm matrix. *Nature Reviews Microbiology* 8: 623–633.

Freijer J.I. (1986) Mineralization of hydrocarbons in soil under decreasing oxygen availability. *Journal of Environmental Quality* 25: 296–304.

Fuchs S., Haritopoulou T., Schäfer M., et al. (1997) Heavy metals in freshwater ecosystems introduced by urban rainwater runoff—monitoring of suspended solids, river sediments and biofilms. *Water Science and Technology* 36: 277–282.

Gaur N., Flora G., Yadav M., et al. (2014) A review with recent advancements on bioremediation-based abolition of heavy metals. *Environmental Science: Processes & Impacts* 16: 180–193. doi: 10.1039/C3EM00491K

Gianfreda L., Rao M.A. (2004) Potential of extra cellular enzymes in remediation of polluted soils: a review. *Enzyme and Microbial Technology* 35: 339–354.

Gieg L.M., Fowler S.J., Berdugo-Clavijo C. (2014) Syntrophic biodegradation of hydrocarbon contaminants. *Current Opinion in Biotechnology* 27: 21–29. doi: 10.1016/j.copbio.2013.09.002

Hobley L., Harkins C., MacPhee C.E., Stanley-Wall N.R. (2015, September) Giving structure to the biofilm matrix: an overview of individual strategies and emerging common themes. *FEMS Microbiology Reviews* 39(5): 649–669. doi: 10.1093/femsre/fuv015; Epub 2015, April 22; PMID: 25907113; PMCID: PMC4551309.

Huesemann M.H., Hausmann T.S., Fortman T.J. (2004) Does bioavailability limit biodegradability? A comparison of hydrocarbon biodegradation and desorption rates in aged soils. *Biodegradation* 15: 261–274.

Hyun S., Ahn M.-Y., Zimmerman A.R., Kim M., Kim J.-G. (2008) Implication of hydraulic properties of bioremediated diesel-contaminated soil. *Chemosphere* 71: 1646–1653.

Iosob, G.-A., Prisecaru, M., Stoica, I., Călin, M., Cristea, T.O. (2016) Biological remediation of soil polluted with oil products: an overview of available technologies. *Research Gate*, 91.

Jin Z.P. (2012) Bioaccumulation and catabolism of prometryne in green algae. In: Jin Z.P., Luo K., Zhang S., et al., editors. *Chemospere*, 143–147.

Kostanjsek R., Lapanje A., Drobne D., et al. (2005) Bacterial community structure analyses to assess pollution of water and sediments in the Lake Shkodra/Skadar, Balkan Peninsula. *Environmental Science and Pollution Research International* 12: 361–368. doi: 10.1065/espr2005.07.271

Kreft J.U., Wimpenny J.W. (2001) Effect of EPS on biofilm structure and function as revealed by an individual-based model of biofilm growth. *Water Science and Technology* 43: 135–141.

Kulshreshtha A., Agrawal R., Barar M., Saxena S. (2014) A review on bioremediation of heavy metals in contaminated water. *IOSR Journal of Environmental Science* 8(7): 47.

Lacal J., Reyes-Darias J.A., García-Fontana C., et al. (2013) Tactic responses to pollutants and their potential to increase biodegradation efficiency. *Journal of Applied Microbiology* 114: 923–933. doi: 10.1111/jam.12076

Liu B., Banks M., Schwab P. (2001) Effects of soil water content on biodegradation of phenanthrene in a mixture of organic contaminants. *Soil and Sediment Contamination* 10: 633–658.

Logsdon G.S., Kohne R., Abel S., et al. (2002) Slow sand filtration for small water systems. *Journal of Environmental Engineering* 1: 339–348. doi: 10.1139/s02-025

Mah T.F., O'Toole G.A. (2001) Mechanisms of biofilm resistance to antimicrobial agents. *Trends in Microbiology* 9: 34–39. doi: 10.1016/S0966-842X(00)01913-2

Mardanova A.M., Kabanov D.A., Rudakova N.L., Sharipova M.R. (2016) *Bioplenki: osnovnye printsipy organizatsii i metody issledovaniya: uchebnoe posobie* [Biofilms: Basic Principles of Organization and Research Methods: Study Guide]. Kazan, Russian, p. 42. https://kpfu.ru/portal/docs/F1250326711/posobie._.Bioplenki._.Mardanova.AM.Kabanov.D.A..Sharipova.M.R.pdf

Margesin R., Walder G., Schinner F. (2000) The impact of hydrocarbon remediation (diesel oil and polycyclic aromatic hydrocarbons) on enzyme activities and microbial properties of soil. *Acta Biotechnologica* 20: 313–333.

Matz C., Kjelleberg S. (2005) Off the hook-how bacteria survive protozoan grazing. *Trends in Microbiology* 13: 302–307. doi: 10.1016/j.tim.2005.05.009

Meroni G., Panelli S., Zuccotti G., Bandi C., Drago L., Pistone D. (2021) Probiotics as therapeutic tools against pathogenic biofilms: have we found the perfect weapon? *Microbiology Research* 12: 916–937. doi: 10.3390/microbiolres12040068

Miller M.B., Bassler B.L. (2001) Quorum sensing in bacteria. *Annual Review of Microbiology* 55: 165–199. doi: 10.1146/annurev.micro.55.1.165; PMID: 11544353.

Miqueleto A.P., Dolosic C.C., Pozzi E., et al. (2010) Influence of carbon sources and C/N ratio on EPS production in anaerobic sequencing batch biofilm reactors for wastewater treatment. *Bioresource Technology* 101: 1324–1330. doi: 10.1016/j.biortech.2009.09.026

Mishra A., Malik A. (2014) Novel fungal consortium for bioremediation of metals and dyes from mixed waste stream. *Bioresource Technology* 171: 217–226. doi: 10.1016/j.biortech.2014.08.047

More T.T., Yadav J.S.S., Yan S., et al. (2014) Extracellular polymeric substances of bacteria and their potential environmental applications. *Journal of Environmental Management* 144: 1–25. doi: 10.1016/j.jenvman.2014.05.010

Navarro E., Guasch H., Sabater S. (2002) Use of microbenthic algal communities in ecotoxicological tests for the assessment of water quality: the Ter river case study. *Journal of Applied Phycology* 14: 41–48. doi: 10.1023/A:1015242301451

Olusola S.A., Anslem E.E. (2010) Bioremeditation of a crude oil polluted soil with Pleurotus pulmonarius and Glomus mosseae using Amaranthus hybridus as a test plant. In: Olusola S.A., Anslem E.E., editors. *Bioremediation & Biodegradation*, 115.

O'Toole G.A. (2011, January 30) Microtiter dish biofilm formation assay. *Journal of Visualized Experiments* (47): 2437. doi: 10.3791/2437; PMID: 21307833; PMCID: PMC3182663.

Peacock A.D., Chang Y.J., Istok J.D., et al. (2004) Utilization of microbial biofilms as monitors of bioremediation. *Microbial Ecology* 47: 284–292.

Perfumo A., Banat I.M., Marchant R., Vezzulli L. (2007) Thermally enhanced approaches for bioremediation of hydrocarbon-contaminated soils. *Chemosphere* 66: 179–184.

Prasad M.N., Prasad R. (2012) Nature's cure for cleanup of contaminated environment—a review of bioremediation strategies. *Reviews on Environmental Health* 27: 181–189.

Pratt L.A., Kolter R. (1999) Genetic analyses of bacterial biofilm formation. *Current Opinion in Microbiology* 2: 598–603. doi: 10.1016/S1369-5274(99)00028-4

Prenafeta-Boldu F.X., Vervoort J., Grotenhuis J.T., Van Groenestijn J.W. (2002) Substrate interactions during the biodegradation of benzene, toluene, ethylbenzene and xylene (BTEX) hydrocarbons by the fungus Cladophialophora sp. strain T1. *Applied and Environmental Microbiology* 68: 2660–2665.

Prince R.C. (2000) Bioremediation. In: *Kirk-Othmer Encyclopedia of Chemical Technology*. John Wiley & Sons, Inc.

Ramírez M.E., Zapién B., Zegarra H.G., Rojas N.G., Fernández L.C. (2009) Assessment of hydrocarbon biodegradability in clayed and weathered polluted soils. *International Biodeterioration and Biodegradation* 63: 347–353.

Rojas-Avelizapa N.G., Roldán-Carrillo T., Zegarra-Martínez H., Muñoz-Colunga A.M., Fernández-Linares L.C. (2007) A field trial for an ex-situ bioremediation of a drilling mud-polluted site. *Chemosphere* 66: 1595–1600.

Sanscartier D., Zeeb B., Koch I., Reimer K. (2009) Bioremediation of diesel-contaminated soil by heated and humidified biopile system in cold climates. *Cold Regions Science and Technology* 55: 167–173.

Schaefer M., Filser J. (2007) The influence of earthworms and organic additives on the biodegradation of oil contaminated soil. *Applied Soil Ecology* 36: 53–62.

Scherr K., Aichberger H., Braun R., Loibner A.P. (2007) Influence of soil fractions on microbial degradation behavior of mineral hydrocarbons. *European Journal of Soil Biology* 43: 341–350.

Sethunathan N.R. (2015) Algal degradation of a known endocrine disrupting insecticide, alpha-endosulfan, and its metabolite, endosulfan sulfate, in liquid medium and soil. In: Sethunathan N.R., Chen Z.L., Williams B.D., et al., editors. *Agricultural and Food Chemistry* 52: 3030–3035.

Silva Í.S., dos Santos E., de Menezes C.R., de Faria A.F., Franciscon E., Grossman M., Regina L. (2009) Durrant bioremediation of a poly aromatic hydrocarbon contaminated soil by native soil microbiota and bioaugmentation with isolated microbial consortia. *Bioresource Technology* 100: 4669–4675.

Subashchandrabose S. (2013) Mixotrophic cyanobacteria and microalgae as distinctive biological agents for organic pollutant degradation. In: Subashchandrabose S., Ramakrishnan B., Megharaj M., Venkateswarlu K., editors. *Environment International* 51: 59–72.

Sutherland I.W. (2001) The biofilm matrix-an immobilized but dynamic microbial environment. *Trends in Microbiology* 9: 222–227. doi: 10.1016/S0966-842X(01)02012-1

Tolker-Nielsen, T. (2015) Biofilm development. *Microbiology Spectrum* 3. MB-0001–2014.

Vasudevan N., Rajaram P. (2001) Bioremediation of oil sludge-contaminated soil. *Environment International* 26: 409–411.

Vidali M. (2001) Bioremediation. an overview. *Pure and Applied Chemistry* 73: 1163–1172. doi: 10.1351/pac200173071163

Wang X., Bartha R. (1990) Effects of bioremediation on residues, activity and toxicity in soil contaminated by fuel spills. *Soil Biology and Biochemistry* 22: 501–505.

Whyte L.G., Bourbonnière L., Greer C.W. (1997) Biodegradation of petroleum hydrocarbons by psychrotrophic Pseudomonas strains possessing both alkane (alk) and naphthalene (nah) catabolic pathways. *Applied and Environmental Microbiology* 63: 3719–3723.

Wolicka D., Suszek A., Borkowski A., Bielecka A. (2009) Application of aerobic microorganisms in bioremediation in situ of soil contaminated by petroleum products. *Bioresource Technology* 100: 3221–3227.

Xu X. (2015) Role of Penicillinum crysogenum XJ-1 in the detoxication and bioremediation of cadmium. In: Xu X., Xia L., Zhu W., Zhang Z., editors. *Frontiers in Microbiology* (6): 1–10.

Young Darcy M.A. (2012) Bioremediation with white-rot fungi at Fisherville Mill: analyses of gene expression and number 6 fuel oil degradation. *Mosakowski Institute for Public Enterprise* 12. https://commons.clarku.edu/mosakowskiinstitute/12

8 Microbial Degradation of Endocrine-Disrupting Chemicals

Punit Kumar, Ravindra Kumar and Shalini Sharma

8.1 INTRODUCTION

The growth and development of different organs is performed in a coordinated way, not in isolation. This coordination is regulated by hormones and signals that regulate the simultaneous development of all organs. This coordination also regulates tissue-specific developmental programs (Oliveira et al. 2021). The network of glands and organs that create, store, and secrete different types of hormones is known as the endocrine system. This system collaborates with other systems and regulates the growth, development, and function of the human body throughout life. Hormones are chemical messengers that regulate bodily function such as growth and development, homeostasis, maintenance of pregnancy, metabolism, and reproduction also. Moreover, hormones (for example thyroid hormone, estrogens and progesterone, glucocorticoids, and insulin growth factor, etc.) also play a key role in fetal growth and development by regulating the process of differentiation and development. These hormones also regulate the metabolism by regulating the transportation of nutrients across the placenta (Baud & Berkane 2019). Thus, deficiency of hormones or impaired hormonal activities during pregnancy may lead development defects in embryo. Overall, it can be suggested that hormones and the endocrine system play a key regulatory role in during fatal development, postnatal health, postnatal neurodevelopment, and health during the entire life of an individual.

Now-a-days, humans use different types of chemical products for a variety of applications, from household to industrial. These include natural and synthetic chemicals. Thus, humans are constantly exposed to these chemicals. Out of these, many chemicals (both synthetic and natural) interfere with the normal functioning of endocrine system and are termed as endocrine disrupting chemicals (EDCs).

EDCs are those substances that mimic, block, or interfere with activities of hormones and their production by the endocrine system. EDCs are heterogeneous groups of chemicals that are exogenous and act like "hormone mimics". These chemicals interfere with any aspect of naturally present hormones in our body. These chemicals affect the pathways of natural hormones, affecting the biosynthesis by endocrine system, metabolism, or can be stored in body tissues (Papalou et al. 2019). EDCs target different physiological mechanisms in humans, like metabolism, epigenetic changes, and chromosomal alterations. EDCs may interfere with the hormonal homeostasis of individuals of any age, but fetuses, infants, and children may have enhanced sensitivity to EDCs (Braun 2017).

EDCs are a global problem for human health, animal health, and the environment. Moreover, EDCs are also defined as "exogenous chemical, or mixture of chemicals, that can interfere with any aspect of hormone action". It is assumed that approximately 1,000 different types of chemicals exist having endocrine-acting properties. In the broader picture, EDCs comprise diverse categories of chemicals such as fungicides, industrial chemicals, metals, nonylphenols, pollutants, plasticizers, pesticides, pharmaceutical agents, and phytoestrogens (Yilmaz et al. 2020). These are present in products of regular use, including pharmaceuticals, food products, cosmetics, and packaging materials, etc. (Dhiman & Dureja 2020). Industrialized areas are typically characterized by contamination

DOI: 10.1201/9781003391487-8

from a wide range of industrial chemicals that may leach into soil and groundwater. Furthermore, the exponential growth of pollutants entering the environment resulting from enhanced agricultural and industrial activities is a mounting concern for human health and environmental health globally (Kumar et al. 2020).

Around the world, there are numerous and diverse sources of EDC exposure. Even a very low concentration of EDC exposure has the potential to cause endocrine or reproductive problems. Interestingly, it is found that EDCs exhibit broad structural diversity, and apparently, don't share any structural similarity, but usually, these molecules have small molecular mass—i.e., approximately less than 1,000 Dalton. Moreover, many EDCs frequently contain a phenolic moiety, which is suggested to imitate natural steroid hormones and allow for analogous or antagonistic interactions with steroid hormone receptors. Even metals and metalloids may show estrogenic action, indicating EDCs and more widespread toxicants. In addition to acting as thyroid hormone receptor agonists or antagonists and antiandrogens, other groups of EDCs have also been found to exhibit androgenic properties (Diamanti-Kandarakis et al. 2009).

EDCs induce many adverse consequences such as androgenic and estrogenic effects on human health, thus disturbing male and female reproductive system. Adverse effects inflected by these chemicals are associated with issues related to fertility, abnormalities in sex organs, endometriosis, breast development, and early puberty, etc. Other adverse effects are noticed on nervous system, immune function, and respiration. Metabolic disorders (such as diabetes, obesity, cardiovascular problems), problems associated with growth and development, and certain type of cancers (prostrate cancers, uterine and ovarian cancer) are also associated with EDCs (Endocrine Disrupting Chemicals 2022; Diamanti-Kandarakis et al. 2009; Papalou et al. 2019). Some EDCs are also found to interference with the hypothalamo-pituitary-thyroid and adrenal axes (Yilmaz et al. 2020) Though EDCs have been found to be associated with cancer progression and cancer regulation, their role in tumor microenvironment is not well elucidated (Buoso et al. 2020).

Endocrine-disrupting chemicals are found in air, soil, water, food sources, and also in personal care products. From these sources, EDCs enter into the body via skin, inhalation, drinking water, and consumption of food products. These EDCs are exposed to humans and enter into systemic circulation (Diamanti-Kandarakis et al. 2009). Some naturally occurring, endocrine-disrupting chemicals, called phytoestrogens, present in soy and soy products, exhibit beneficial effects (Bar-El & Reifen 2010).

EDCs may affect not only the exposed individual but also children and subsequent generations. Studies suggest that this transmission may involve the germline and may be nongenomic (Anway & Skinner 2006). That is, effects may be transmitted not due to mutation of the DNA sequence, but rather, through modifications to factors that regulate gene expression such as DNA methylation and histone acetylation.

Some EDCs have extended half-lives, which is advantageous for industrial use but harmful to humans and wildlife. Even substances that were outlawed decades ago still exist in the environment, and it is possible that these may be found in many tested individual animals or humans because they are persistent and are not decomposed.

The degradation and removal of EDCs from the environment requires alternate approaches because traditional water treatment procedures do not completely remove the EDCs. These approaches utilize physical, chemical, and biological process, alone or in combination.

8.2 TYPES OF ENDOCRINE-DISRUPTING CHEMICALS

Endocrine disrupting chemicals are man-made and natural (Figure 8.1). These are found in everyday products including cosmetics, plastic bottles, food containers, metal cans, pesticides, and flame retardants (Endocrine Disrupting Chemicals 2022).

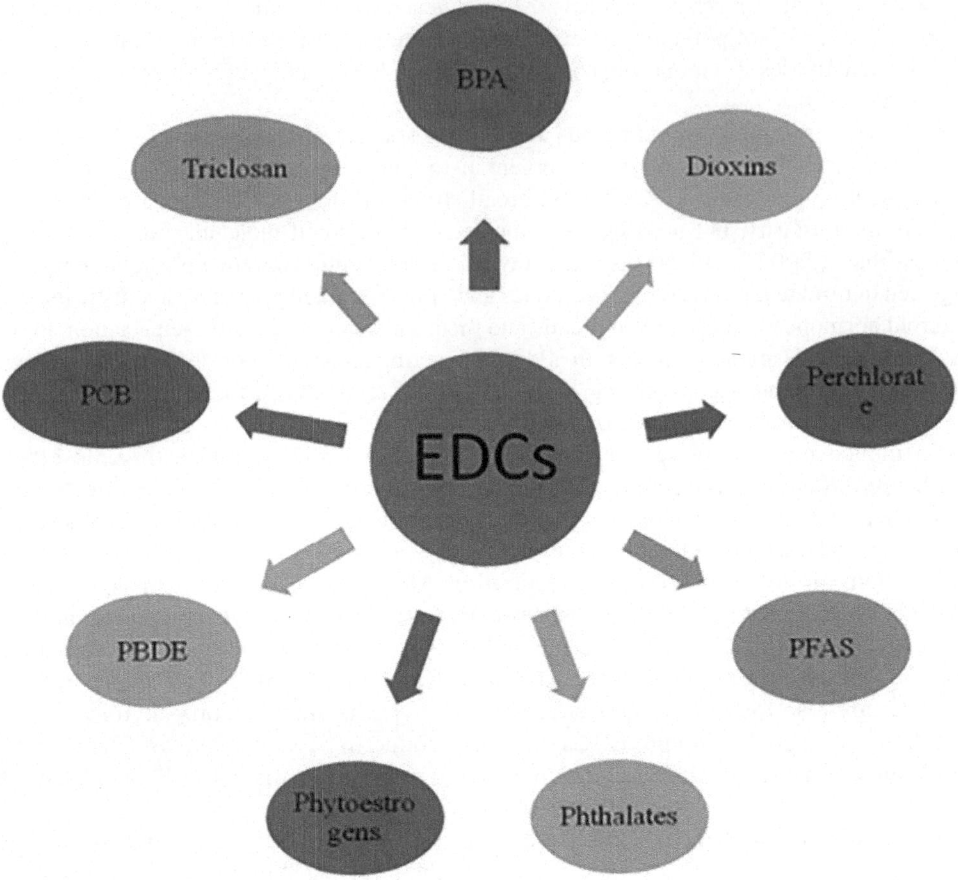

FIGURE 8.1 Different types of EDCs present in nature.

8.2.1 Bisphenol A (BPA)

It is found in a large number of consumer products. It is used to make polycarbonate plastics and epoxy resins, which are found mainly in plastic products, including surface tiles, and food storage containers. It is a developmental, reproductive, and systemic toxicant. The consumption of BPA leads to metabolic disorders such as immune toxicity, neurotoxicity, low sex-specific neurodevelopment, and interference of cellular pathways. Bisphenol S is an analogue used as a replacement for bisphenol A, though very little is known about the toxicity of bisphenol S towards living organisms (Frenzilli et al. 2021).

8.2.2 Dioxins

Dioxins are commonly found environmental toxicants. Collectively, about 210 types of compounds are termed as dioxins, and the majority of them are dibenzofurans. Polychlorinated dibenzo-p-dioxins (PCDDs) and polychlorinated dibenzofurans (PCDFs) are well-studied dioxins. Moreover, 2,3,7,8-tetrachlorodibenzo-p-dioxin (2378-TCDD) has one of the minimum known LD50 (lethal dose 50%) values. Frequently, it is also termed as one of the most toxic chemical compounds made by humans. Dioxins are unwanted products of the industrial processes. These are produced as byproducts in industrial and combustion processes (Hites 2011). As endocrine disruptor, TCDD has been found associated with disruption of hormone signaling, interference with embryonic development, effects on fertility, and miscarriage in humans (Chen et al. 2010).

8.2.3 Perchlorate

It is a byproduct of aerospace, weapon, and pharmaceutical industries. It is also present in drinking water and fireworks. Perchlorate contaminates many types of foodstuffs. It is associated with many types of health disorders—mainly thyroid dysfunction. A study conducted with $NaClO_4$ revealed the toxicity of perchlorate on thyroid gland (Liu et al. 2022).

8.2.4 Perfluoroalkyl and Polyfluoroalkyl Substances (PFAS)

It is mainly used in firefighting foams and non-stick pans, paper, and the textile industry. These are present in foods, drinking water, food packaging materials, and many other consumer products. These compounds are suggested to interfere with ovarian function. In experimental models, PFAS have shown adverse effects on ovarian folliculogenesis and steroidogenesis (Ding et al. 2020).

8.2.5 Phthalates

Phthalates are water-insoluble plasticizers that are used to make plastic more flexible. Such plastics are used in medical devices, the pharmaceutical industry, clothing, and food packagings. They are also found in cosmetics and children's toys. These compounds are commonly found in the environment and exposed to humans via food, personal care products, clothes, inhalation, medication, and nutritional supplements. These compounds are also exposed to infants. Phthalates cause a number of abnormalities, behavioral problems, and diseases such as low IQ and attention deficit, impaired neurodevelopment, sex anomalies, early puberty, and fertility problems and effects on reproductive development, etc. These compounds also perturb sugar and fat homeostasis and cause overweightness, obesity, insulin resistance, and type II diabetes (Farooq et al. 2021).

8.2.6 Phytoestrogens

These are naturally present in many plants and have hormone-like properties such as genistein and daidzein, which are present in soy products. These compounds act as weak estrogen mimics or antiestrogens. These are also considered as EDCs and exhibit some beneficial effects on health, decreased risk of breast cancer, and improvement in metabolism. Though lifetime exposure to estrogenic substances, mainly during the crucial developmental periods, is found associated with formation of malignancies and several anomalies of reproductive systems (Bar-El & Reifen 2010), more study is required to investigate the adverse health effect of phytoestrogens.

8.2.7 Polychlorinated Biphenyls (PCB)

These are used to make electrical equipment like transformers and in hydraulic fluids, heat transfer fluids, lubricants, and plasticizers. These compounds are one of the most studied EDCs for their effects on neurodevelopment and cognitive performance. PCBs also found to disrupt reproductive functions (Bell 2014). Though the production of polychlorinated biphenyls (PCB) was stopped in 1970, these are persistent organic pollutants. These cause an acne-like condition in workers who come into contact with it, increase the risk of cancer, childhood obesity, type II diabetes, and are toxic to the liver and thyroid (Gupta et al. 2018). PCBs are also hydrophobic in nature and have ability to accumulate in the food chain and adipose tissues.

8.2.8 Triclosan

It may be found in some anti-microbial and personal care products, like liquid body wash. Due to excessive use and disposal, triclosan is found in wastewater sources. Triclosan also has potential for accumulation. Triclosan is found to interfere with thyroid homeostasis and association with metabolic disorders, neurodevelopment impairment, cardiotoxicity, and the increased risk for cancer (Milanović et al. 2021).

8.3 TOXIC EFFECTS OF ENDOCRINE-DISRUPTING CHEMICALS

EDCs produce various adverse effects on human health, aquatic life, and the environment. Different EDCs affect human health in a different manner; some of them affect the metabolic system; some, the reproductive system. Some of them produce a carcinogenic effect, childhood obesity, and developmental problems—problems related to neural and cognitive development.

It is suggested that EDCs may follow different mechanisms that interfere with hormonal homeostasis and affect human health (Figure 8.2, Combarnous & Nguyen 2019).

1. EDCs acts as hormones and interact with the hormone receptor. Through this mechanism, EDCs activate hormone receptors and trigger hormone-associated signaling pathways.
2. EDCs interact with the hormone receptor and inhibit the hormone-associated signaling pathways. In this mechanism, EDCs antagonize the hormonal actions.
3. EDCs interact with components of hormone-signaling pathways and interfere with the components of hormone-signaling pathways.
4. EDCs affect endogenous hormone biosynthesis or degradation.

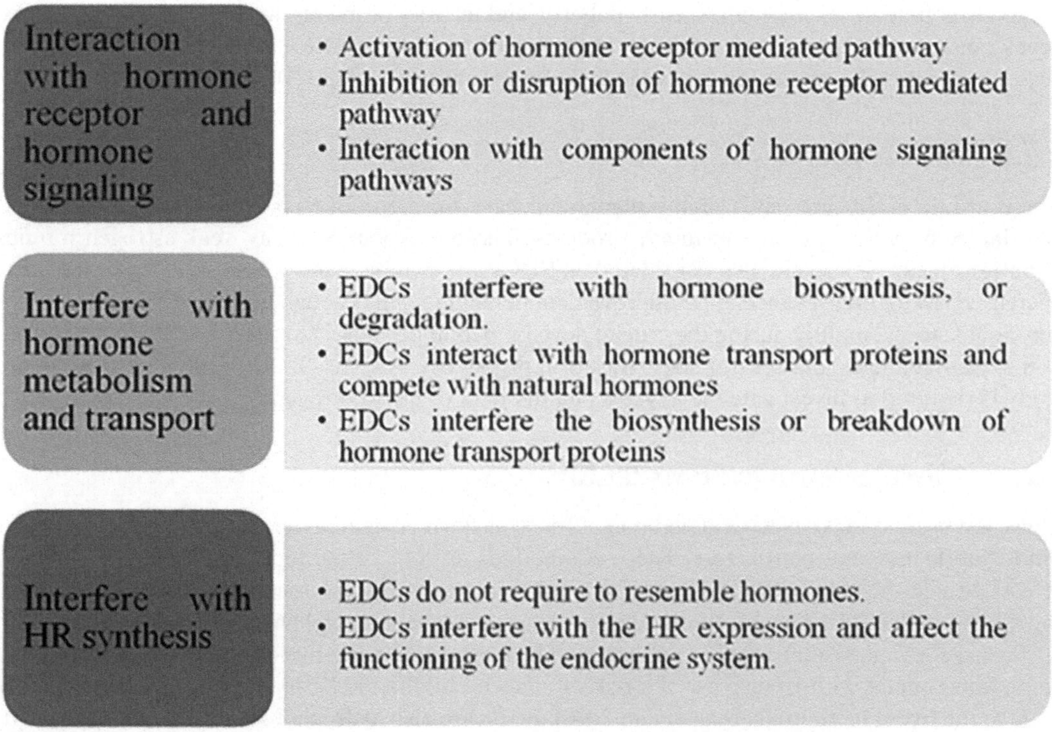

FIGURE 8.2 Different mechanisms by which EDCs interfere with endocrine system.

5. EDCs interfere with hormone-transport proteins and interfere with the biosynthesis/degradation of hormone-binding transport proteins.
6. Many EDCs interfere with synthesis or degradation of hormone receptors.

As EDCs are present in a wide range of products of daily use, people get exposed to EDCs through these products. EDCs are exposed to humans via consumption of food and beverages, pesticides, and use of cosmetics. Through the disposal of these products in the environment, these EDCs are released into environment. Thus, completely avoiding exposure is not possible. Including this, some EDCs are slow to break down in the environment. That nature makes them potentially hazardous over time.

EDCs are present in the environment in water sources, soil, air, contaminated foods, and a variety of sources. Due to health concerns and ecological risks, there is a requirement to develop intervention strategies that can mitigate the risks associated with exposure to EDCs. There are different approaches used to remove EDCs such as adsorption processes, membrane-based filtration, advanced oxidation processes (AOP), phytoremediation, bioremediation, and other integrated approaches.

8.4 APPROACHES TO REMOVE ENDOCRINE-DISRUPTING CHEMICALS

Many approaches are being used to remove the EDCs from the environment—mainly from water sources. These approaches are based on physical, chemical, and biological treatment methods. Some of the important approaches are the following (Figure 8.3).

8.4.1 Adsorption Technology

Adsorption methods are one of the effective and cost-effective removal techniques (Werkneh et al. 2022). It is identified that one of the best techniques for treating wastewater is adsorption, and it mostly relies on the availability of active sites on the sorbent as well as the chemical properties (sorbate pKa, basicity or acidity of the sorbent, etc.) and physical properties (molecular size, pore density, contact area, etc.) of the surface of absorbent. In general, physisorption and chemisorption approaches may be used to perform the adsorption process. The physiosorption method is based formation of physical bonding (van der Waals forces) and chemisorption is based on formation of

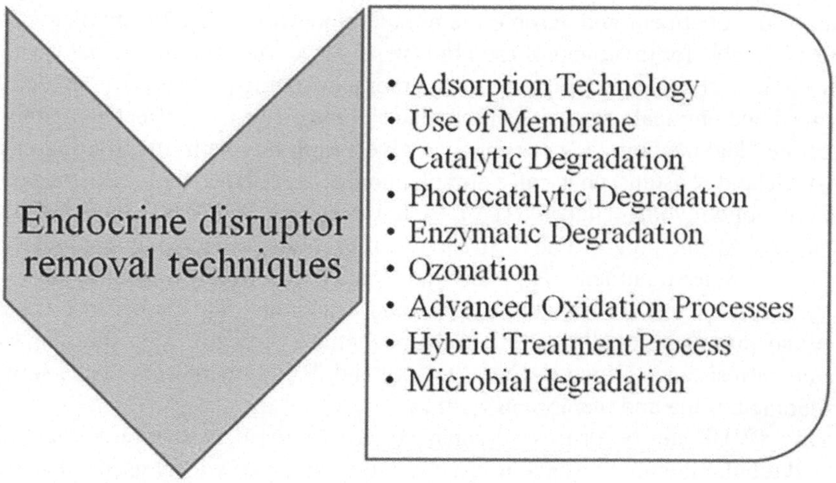

FIGURE 8.3 Different approaches for removal/degradation of EDCs.

chemical bonds. These processes may take place at several interfaces, including solid-liquid and/or solid-gas interfaces (Adeleke et al. 2019; Zhu et al. 2018).

Physisorption is the consequence of the physical interaction between the adsorbed chemicals and the solid surface caused by a weak van der Waals force of attraction. The van der Waal forces are a fundamental result of the interaction between permanent and transient electric dipoles (Malhotra et al. 2018; Zhu et al. 2018).

Adsorbents have excellent potential for regenerating adsorption capacity and releasing high-quality, secure effluent for discharge. Lower temperatures are needed for the adsorption process because of the weak binding energy. The activation energy typically falls between 20 and 40 kJ, which suggests that the active sorbent has a strong inclination to dissolve in an aqueous solution (Dickenson & Drewes 2010; Mailler et al. 2016).

In the chemisorption mechanism, the chemical bonding causes the active plane surface of the sorbent and adsorbate to break and create a new connection. This means that greater adsorption energy and temperature, typically between 200 and 400 kJ/mol, are needed (Dickenson & Drewes 2010).

The activated carbon (AC) adsorption method has been used in the removal of many EDCs and trace contaminants without generation of transformed products. This method offers several advantages such as pore structure, high surface area, cost effective, applicability to remove low concentration of EDCs, hydrophobic nature of surface properties, and possibility of reuse, etc. (Delgado et al. 2012). It has been found that, in AC-based removal of diethyl phthalate (DEP), BPA is dependent on pH due to ionization and hydrolysis (Zhao et al. 2019). Including this, biochar is also gaining the attention for removal of EDCs due to low cost, easy availability, higher condensed structure, and existence of surface oxygen containing functional groups, etc. (Delgado et al. 2012).

8.4.2 Removal of EDCs by Membrane

Membrane technology is the most popular physicochemical separation technique for eliminating salt and germs from water. These techniques have been employed in situations where fresh, clean water is in short supply (Yoon et al. 2006; Yuan & He 2015). This was possible because of its special features, including energy economy, compactness, fast throughput, and cost effectiveness. Based primarily on the pore size and operating pressure employed, pressure-operated membrane processes can be broadly classified into four primary categories: reverse osmosis, microfiltration, ultrafiltration, and nanofiltration (Lin et al. 2013; Zhang et al. 2006). Including this, it had been suggested that anaerobic membrane bioreactor (AnMBR) technology has many advantages over conventional anaerobic treatment and aerobic membrane bioreactor (MBR) technology. It was found that AnMBR is suitable for treatment of food industrial wastewater and municipal wastewater. This approach was found effective in regulating the membrane-fouling problem (Lin et al. 2013).

The chemical and physical properties of the material may have a big effect on membrane performance. Since the ideal membrane is one that can give a high flux with little fouling or clogging, is chemically stable and resistant, physically durable, nonbiodegradable, and inexpensive.

Furthermore, polyvinylidene fluoride (PVDF), polyacrylonitrile (PAN), polyethersulfone (PES), cellulose acetate (CA), and polysulfone (PSF) are the most often utilized polymer materials in membrane filtration for water treatment. The most well-known and widely-used polymeric membrane, PVDF, has growing interest from manufacturers and academics. This is because it has proven to have unique and promising qualities, like high mechanical strength, high thermal stability, and superior chemical resistance (Kang & Cao 2014), though PVDF membranes have some problems, such as membrane fouling and membrane wetting.

Additionally, PVDF can be processed satisfactorily for the construction of flat sheet, hollow fiber (HF), and tubular membranes, and it dissolves in a variety of widely used solvents, such as N, N-dimethyl acetamide (DMAC), dimethylformamide (DMF), and N-methyl-2-pyrrolidone (NMP) (Neale & Schäfer 2012).

Ultrafiltration (UF) and microfiltration (MF) membranes have mostly been viewed as a viable method to lower operating expenses because they can be operated at low pressures (1–3 bar) and produce more permeate in a shorter span of time (Su-Hua et al. 2010). Though Hu et al. (2014) have suggested 50kPa was suitable for UF process operation for EDCs removal and suitable flux, it should be emphasized that dense membranes by size exclusion can remove EDCs more effectively than microporous membranes.

The efficiency of UF and MF membranes in eliminating EDCs could be as effective as that of NF and RO membranes, provided that the adsorption mechanism plays a significant role in retaining EDCs by adsorbing onto the internal pore walls and membrane surface (Hu et al. 2014; Rana et al. 2014; Syahida et al. 2019; Zhao et al. 2015).

Su-Hua et al. (2010) investigated how polysulphone (PS) membranes could be used to reject bisphenol A (BPA). Solution pH has a significant impact on BPA retention, with a rapid drop in retention when the pH rises above its pKa value. It was determined that hydrogen bonding and hydrophobic adsorption, respectively, may both physically and chemically excite sorption onto the membrane. As the polysulphone membrane is particularly vulnerable to adsorptive fouling, several pollutants are deposited, and changes in the feed solution matrix caused leakages, there may be some potential limitations in this work. Therefore, combining several membrane processes, including as MF, UF, RO, and NF, is essential to enhancing the removal of various EDCs (Dharupaneedi et al. 2019; Luo et al. 2014).

When MF and RO were coupled to treat EDCs, they showed elimination effectiveness of 97%. However, after the treatment procedure, BPA at a concentration of 500 ng/L was found in the effluent in the final, recycled water (Al-Rifai et al. 2011).

To integrate the strength of the component materials in the end product, composite membranes made of multiple components (cellulose acetate/zinc oxide-zeolite) have recently been produced. These membranes have shown the efficient removal of benzophenone-3 from water. Typically, one material is used to represent the active surface, and the other is used to represent the support layer. In this membrane, zeolite zinc oxide increases the hydrophilicity of the membranes. Including this, permeate flux and rejection increased with the concentration of zeolite zinc oxide (Rajesha et al. 2019).

Nanoparticles are incorporated into the formulations of membranes to create nano-composite membranes. Most commonly, nanoparticles (NPs) such as iron oxide (Fe3O4), titanium dioxide (TiO2), and silica (SiO2) have been used to enhance the characteristics of UF membrane. Numerous studies have demonstrated that, when nanoparticles were integrated in the membrane matrix, significant removal of BPA and other pollutants was possible. Addition of polyvinylpyrrolidone (PVP) and magnesium oxide (MnO2) nanoparticles to PVDF membranes could significantly enhance the membrane's properties and remove BPA completely (100%). The PVDF-PVP and PVDF-PVP-MnO_2 composites demonstrated the maximum removal of BPA. The PVDF-PVP-MnO_2 membrane composite also exhibited better reusability (Zahari et al. 2018). Including this, Fe-doped TiO_2/polysulfone composite ultrafiltration (UF) membranes caused enhanced photocatalytic degradation of BPA (more than 90%) (Wang et al. 2017). Nasseri et al. (2018) found that polysulfone/graphene oxide nano-composite membranes with graphene oxide (GO) weight ratios of 0.4% exhibited a high BPA-removal efficiency of about 93%.

8.4.3 Photocatalytic Degradation

Photocatalytic degradation processes are considered promising approaches for the treatment of EDCs, but the efficiencies of these photocatalytic processes are dependent on the performance of the photocatalysts (Wang et al. 2020). The elimination of EDCs by photocatalysis depends on many operating conditions that affect the performance and degradation rate of a photocatalytic process. Light intensity, the presence of organic and inorganic compounds, reaction temperature, the amount of catalyst used, the quantity and chemical make-up of the impurities, the starting concentration of the substrate, solution pH, presence of second substrate, and dissolved oxygen are some of these factors (Sin et al. 2013). Moreover, the photocatalysis procedure has been considered to be an effective method for breaking down EDCs because it has minimal secondary contamination, a moderate reaction medium, and better energy-saving properties.

It was found that photocatalytic degradation of phenol and resorcinol was performed generally via the hydroxyl radicals. ZnO photocatalysts are found to be effective photocatalysts, and without considerable loss of activity (Sin et al. 2013). The presence of organic enzymes or UV radiation can accelerate the process of breakdown. As opposed to non-catalyzed mechanisms, the catalyst primarily provides a reaction path option with lower excitation energy. In catalyzed mechanisms, the catalyst often reacts to create a transient intermediate, which then regenerates the original catalyst in a positive feedback-loop. Researchers are trying different types of reactors for catalysis. For example, photocatalytic reactors of the small and scale-up rotating and flat types have been used to perform the treatment of 17α-ethynyl estradiol (EE2), bisphenol A (BPA), and 17β-estradiol (E2) (Kim et al. 2017).

In most cases, photocatalytic degradation of EDC contaminants requires the activation of photoreactions while being influenced by light (solar irradiation or UV) and a catalyst. It takes several phases to complete the process (such as EDC diffusion and adsorption, chemical reactions, intermediate desorption, and product removal from the interface). H_2O and CO_2 are often the produced as byproducts of a photocatalytic degradation process.

8.4.4 Enzymatic Degradation

Enzymatic degradation—and phytoremediation—is another innovative remediation method that shows promise for getting rid of EDCs and other chemical compounds similar to them in wastewater. Enzymatic degradation is also influenced by enzymatic activity, albeit there is a high association between the degree of degradation and a number of environmental variables, including pH, nutrient levels, availability of other contaminants, and temperature.

Among the enzymes used for EDCs biodegradation, oxidative enzymes are being explored for their potential because of their versatility, large scale production, and modification of their properties (Macellaro et al. 2014). Oxidoreductases, such as laccases, tyrosinases, polyphenol oxidases, manganese peroxidases, lignin peroxidases, horseradish peroxidases, and bitter gourd peroxidases, are some of the enzymes explored for degradation for EDCs (Husain & Qayyum 2013). Including this, cytochrome P 450 monooxygenases and hemoglobin have shown the removal of bisphenol A and other related EDCs. Moreover, surfactants, additives, and many redox mediators have also improved the enzymatic oxidation of bisphenol A and other EDCs. Immobilized enzymes have also shown the capacity to perform continuous and enhanced bioprocessing of EDCs.

Out of the enzymes for biodegradation of EDCs, the fungal laccases are one of the most commonly investigated. Haugland et al. (2019) investigated the degradation of 2-chlorophenol and sulfamethoxazole using fungal laccases (commercial product derived from *Myceliphthora thermophile* and a laboratory-generated enzyme mix derived from *Tramates versicolor*) in the presence and absence of mediators (acetosyringone). It was found that commercial enzyme performed the rapid degradation of 2-chlorophenol in the presence of a mediator, while natural enzyme (mixed) degraded without the presence of mediator. Sulfamethoxazole was not removed efficiently in all laboratory-generated chemically buffered solutions, while, in secondary effluent suspensions from municipal water resource recovery facility, it was degraded without the addition of mediators (acetosyringone).

8.4.5 Ozonation and Advanced Oxidation Processes

Advanced oxidation processes (AOPs) and ozonization are more efficient approaches used to remove persistent EDC components from municipal and industrial wastewater. By these approaches, hardly biodegradable or non-biodegradable organic molecules may be removed from wastewater. AOP can perform complete or partial removal of contaminants.

Ozone is a strong oxidative agent, and it converts into hydroxyl radicals (•OH) in water. The process of ozonation is based on mechanisms such as direct selective action of ozone and indirect action of radicals generated by decomposition of ozone in water (Cesaro & Belgiorno 2016). AOPs use potent radical oxidants such ashydroxyl radicals (•OH) to expedite or speed up the oxidation of recalcitrant and non-biodegradable pollutants.

Hydroxy radicals are considered unstable and are highly reactive. These are nonselective in nature and target a large number of organic chemicals and simplify their structures. These are able to target organic compounds through four pathways as electron transfer, hydrogen abstraction, combination, or addition of radicals. Their reaction with organic pollutants generates carbon radicals that are transformed into peroxy radicals with oxygen. Radicals further form other reactive species like superoxide and hydrogen peroxide that further cause chemical destruction and mineralization of water target pollutant (Garrido-Cardenas et al. 2020). AOPs are considered a suitable approach for the treatment of contaminated aqueous media, especially when it is coupled with traditional biological processes (Cesaro & Belgiorno 2016).

Advance oxidation process also does not transform pollutants into another phase; also, it does not produce hazardous sludge in large amounts.

It should be noted that •OH has one of the highest levels of non-selectivity and reactivity of any radical species present in AOPs, with a standard reduction potential of 2.8 V vs. the standard hydrogen electrode. These activities result in the production of highly reactive oxidizing hydroxyl radicals. Including this, the presence of H_2O_2, UV light, and the Fenton reagent may accelerate the creation of •OH (Chen et al. 2011; Egerton & Purnama 2014; Gernjak et al. 2003). It is more advantageous to convert the EDC pollutants into biodegradable aliphatic carboxylic acids, followed by a biological process, rather than completely mineralizing them.

Photochemical oxidation using molecular oxidation is one of the most important photochemical methods, as these methods do not require additional oxidizers like hydrogen peroxide or ozoneor or any extra energy input, as required in the process of UV irradiation or ozonation. As AOP processes are effective in treating EDCs from water, these are not cost effective; thus, immobilization of photocatalyst is assumed to be a good option (Gmurek et al. 2017).

Powerful redox techniques like ozonation and AOPs have many advantages over traditional treatment methods, including less environmental impact, faster rates of degradation, and non-selective elimination of persistent refractory chemicals that are not biodegradable. Additionally, because of direct human interaction, such as in household reclamation applications, these methods enable decontaminating effects that are essential for water reusability applications.

Following successful removal of persistent EDCs organic microcontaminants, research has shown that both ozonation and AOPs are highly effective in removing EDCs from various water matrices with great removal rates of up to 99%. However, insufficient ability to mineralize organic contaminants, ongoing detection of radical scavengers in effluent confining the attack of the radicals to the contaminants, considerable interference of natural organic matter with the ozone decomposition rate, generation of poisonous by-products, oxidation intermediates, solid by-products (mainly unknown), and high operating costs are some of the primary limitations that can limit their widespread adoption at full scale.

8.4.6 Removal of EDCs Using Hybrid Treatment Process

Many researchers have explored the idea of combining multiple treatment procedures (hybrid) to successfully remove these stubborn EDC micropollutants from water.

Sahar et al. (2011) investigated the integration of reverse osmosis (RO) with MBR to remove EDCs from municipal wastewater in order to increase the effectiveness of the traditional treatment method. According to the findings, diclofenac removal was achieved to a level of over 95% and removal of BPA, macrolides, pharmaceuticals to about more than 99%. The major flaw in this study was the excessive energy demand and regular de-fouling requirements that drove up the cost of operating the RO system. When the ozonation method was paired with an ultrafiltration membrane to remove EDCs from drinking water, a limited ozone dose of 2–5 mg/L with permeate flux of 100 L/m^2h also demonstrated significant removal efficiency (64–100%) for all the target pollutants (Fan et al. 2014). Researchers discovered that ozonation in the membrane tank improves the efficiency of multiple-contaminant removal when combined with a UF membrane, putting the emphasis on reducing the fouling rate.

According to Borikar et al. (2015)'s investigation, EDCs were significantly eliminated from water by ozone/H_2O_2, UV/H_2O_2, both with conventional treatment. Furthermore, traditional treatment with strong UV/H_2O_2 showed successful elimination of 92% to 97%. Interestingly, diclofenac, triclosan, and bisphenol A (BPA) were fully eliminated. According to the scientists, using ozone and UF together is an effective way to control EDC contamination and concentration in secondary wastewater. The probable bromate creation and solid by-products of the ozonation process as well as the astronomical cost of ozone generation are this study's main drawbacks.

It is found that effluent organic matters influence the removal of EDCs using UF membrane and ozonation process. It is found that effluent organic matters increased the decay of ozone and thus reduce the degradation efficiency of EDCs by process of ozonation (Si et al. 2019). It was also found that, if EDCs are treated by a combined process of O_3 and UF, then approximately complete (100%) EDC removal may be achieved. In overall, it is suggested that combined usage of O_3 and UF are an attractive effective approach for removal of EDCs in secondary effluent (Si et al. 2018).

Li et al. (2017) investigated the removal of 17-ethynylestradiol (EE2) from water by using a PAC/UF membrane method in conjunction with activated carbon (AC) adsorption. The claimed clearance efficiency ranged from 70 to 80%. The study's main limitations were the high cost and PAC regeneration issues; the coexistence of natural organic matter, which may compete for adsorption sites and clog pores; and the technical difficulties related to separating PAC from the treated water.

In a small-scale experiment, James et al. (2014) investigated the removal of EDCs from secondary urban sewage using an advanced oxidation process (AOP) that relied on UV irradiation together with hydrogen peroxide mixed with MF and RO. AOP successfully reduced EDCs levels and demonstrated a higher removal of more than 90% for all waters. Despite this increased elimination, this study revealed the high-energy requirement. It was also found that AOPs significantly reduce the levels of recalcitrant micropollutants.

Ferreiro et al. (2020) examined the removal of EDCs from wastewater using continuous-UF ultrafiltration, in addition to an existing biological treatment. In the case of biological treatment, the removal of contaminant was dependent on concentration and nature of compound. Thus, some of the contaminants and EDCs (like acetaminophen, caffeine, progesterone, and testosterone) had exceptional removal rates (above 99.5%) while some contaminants (like carbamazepine, diuron, irbesartan, and phenytoin) have removal rate below 20%, while, in the case of UF, the removal rates were about 30%. Thus, biological treatment coupled with continuous-UF allows a higher efficiency in removing the pollutant with respect to the biological process alone.

A combination of powdered activated carbon and submerged membrane-filtration system was used to remove nonylphenol ethoxylates. It is found that initial powdered activated carbon (PAC) dosage and filtration flux exhibit remarkable effects on micropollutants removal efficiency. The high rate of removal was obtained with initial powdered activated carbon of 50 mg/L and filtration flux of 20 $L/m^2.h$ (Nguyen et al. 2021).

Overall, the hybrid procedures revealed astounding removal results, ranging from 64% to 100% for removing of EDCs and micropollutants from waste sources. Researchers are investigating hybrid processes for the removal of EDCs. But these approaches are associated with creation of harmful byproducts, oxidation intermediates, high operating costs, and membrane fouling.

8.5 MICROBIAL DEGRADATION OF ENDOCRINE-DISRUPTING CHEMICALS

Microorganisms are all around us and have a variety of metabolic transformational skills that are crucial for environmental biodegradation of pollutants, which supports ecosystem maintenance and human health. Here, "microbial degradation" refers to the microbial transformation of organic molecules (EDCs) in the environment, frequently those that have a detrimental influence on human health, into less hazardous or more beneficial forms (Figure 8.4).

New techniques for the degradation of these substances are being created in numerous labs with the aid of microorganisms that have a higher potential for the degradation of these substances. Due to its low cost and low energy requirements, biological treatment is a promising technology. However, the key issue is the survival of these microorganisms in the intricate microbial consortia of wastewater treatment plants (Michalska et al. 2020). Therefore, immobilized systems—in which these strains are protected by being bound to a carrier—have received an increasing amount of interest (Fernandez et al. 2017; Liu et al. 2016, 2018; Menashe et al. 2020).

Many scientists have suggested biodegradation approaches for using microorganisms or their enzymes in the treatment of soils or wastewater contaminated with EDCs because standard water treatment procedures are limited. One of the current treatment options for EDCs is biodegradation. This approach offers several benefits for both economic and environmental protection, including minimal input costs, a broad range of application, a lengthy duration, and straightforward equipment and space needs. The pollution of natural water ecosystems is better handled with a biodegradation approach. Numerous studies have demonstrated that specific fungi and bacteria can efficiently break down EDCs. Widespread in sewage treatment, acclimated activated sludge degrades some EDCs in a certain way. Numerous studies have been done on the biodegradation of BPA, demonstrating that it may be successfully broken down in both freshwater and sewage. For addressing the pollution of natural water settings, a biodegradation technique is more appropriate. Numerous studies have demonstrated that many bacteria and fungi are capable of degrading EDCs. Acclimatized, activated

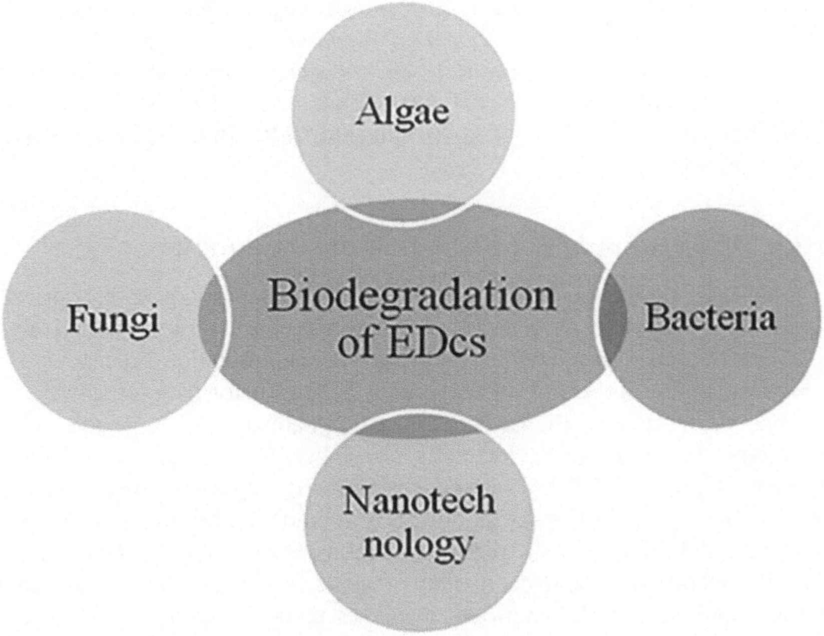

FIGURE 8.4 Use of different types of microorganism and techniques for removal of EDCs.

sludge is frequently used in the treatment of sewage and degrades several EDCs. There are numerous studies on the biodegradation of BPA, which essentially demonstrate that endocrine-disrupting compounds may be successfully destroyed in both natural water and sewage (Cajthaml et al. 2009).

8.5.1 Bacterial Degradation of Endocrine-Disrupting Chemicals

Since the turn of the century, EDCs have caused some concern about human health due to their ability to alter the way that hormones work naturally and to have either reversible or permanent biological consequences. An organic substance called bisphenol (BPA) poses threats to people's health.

According to Hirano et al. (2000), *Pleurotus ostreatus* was able to effectively break down BPA into phenol, diethylstilbestrol, 4-isopropylphenol, and 4-isopropenylphenol in just 12 days, removing roughly 80% of it. It was discovered that *Coriolopsis polyzona* laccase may concurrently get rid of EDCs such BPA, NP, and TCS. Using high-performance liquid chromatography, two thermostable laccases from *Pycnoporus sanguineus* were tested to degrade NP and TCS with more than 95% clearance after 8 hours of treatment with 100 U/L at pH 5 (Cabana et al. 2007; Ramírez-Cavazos et al. 2014).

The ability of ligninolytic bacteria (Hirano et al. 2000) to produce some lignin-modifying enzymes, such as laccase and manganese peroxidase, which are primarily responsible for the EDC degradation as well as their skill in the degradation of recalcitrant compounds, make them promising candidates for the degradation of EDC. In light of their remarkable adaptability to a variety of settings and their genome's flexibility, bacteria appear to have a greater potential for EDC breakdown. Furthermore, due to its low cost and minimal disruption of the contaminated environment when compared to other cleanup techniques, bacterial degradation has been recommended as a simple way to eliminate EDC from wastewater. Several BPA-degrading bacteria and their performance of degradation have been documented by many researchers: *Bacillus megaterium* strain ISO-2, *Sphingomonas bisphenolicum* strain AO1, *Pseudomonas sp.*, and *Achromobacter xylosoxidans* strain B-16, etc. (Larsson et al. 2017; Suyamud et al. 2018; Zhang et al. 2007).

There are some genera that contain EDCs-degrading bacteria: *Bacillus, Virgibacillus, Novosphingobium, Rhodococcus, Acinetobacter, Pseudomonas, Sphingomonas, Enterobacter, Klebsiella, Aeromonas, Comamonas, Thauera, Deinococcus, Stenotrophomonas,* and cyanobacteria (Chen et al. 2017d, 2018; Fernandez et al. 2017; Khattab et al. 2019; Kriszt et al. 2012; Li et al. 2018, 2020; Liu et al. 2016; Menashe et al. 2020; Qiu et al. 2019; Shih et al. 2017; Wang et al. 2019; Xiong et al. 2018, 2020).

8.5.2 Fungal Degradation of Endocrine-Disrupting Chemicals

Ligninolytic fungi (LF) are intriguing microorganisms that can participate in a variety of versatile breakdown processes. Microbes are a useful model group because they can actively be used for cleanup via LF or their enzymes. The possibility for decontamination process optimization is a major rationale for the requirement of understanding EDC transformation pathways. This chapter describes the most recent data on the LF biodegradation pathways and discusses the most prominent micropollutants (xenoestrogens), including nonylphenols, bisphenol A, 17-ethinylestradiol, and polychlorinated biphenyls as POPs with endocrine-disrupting capabilities. The ability of LF to either polymerize the target pollutants or significantly degrade the original structure using ligninolytic enzymes and cytochrome P-450 is frequently demonstrated. Additionally, most transformation processes are followed by a decline in ecotoxicity or endocrine-disrupting activity (Cajthaml 2015).

Their pervasive presence in the environment makes their clearance from contaminated areas necessary. Utilizing enzyme systems capable of reacting with these compounds is one possible strategy for dealing with this problem. In terms of available enzymes, oxidative enzymes are gaining popularity due to their adaptability, ability to be produced in huge quantities, and ability to have their properties altered. Mediators dramatically improved the effectiveness of the enzymatic process, encouraging the oxidation of substrates resistant to laccase by increasing the breakdown of those materials. The laccase offering the greatest results was picked so that more research could be done on its oxidative powers against micropollutant combinations (Macellaro et al. 2014).

White-rot fungi (WRF), which can produce extracellular lignin-modifying enzymes for the biotransformation of aromatic chemicals, are the focus of several investigations on the fungal breakdown of EDCs, including branched-chain octylphenol (4-t-OP) (Cajthaml et al. 2009; Ming et al. 2017). It was claimed that WRF might eliminate a number of organic pollutants, including textile dyes, pesticides, polychlorinated biphenyls, and polycyclic aromatic hydrocarbons (PAHs) and EDCs. For instance, the ability of the white-rot fungus *Pleurotus ostreatus* HK35 to degrade seven representative EDCs (BPA, E1, 17-estradiol, estriol, 17-ethynylestradiol, TCS, and 4-nnonylphenol) was tested. This fungus is relatively simple to cultivate, fructify, and has great potential for bioremediation (Asgher et al. 2008). *Pleurotus ostreatus* HK35 degraded with an efficiency of more than 90% in 12 days in a model laboratory setting. Trametes versicolor is a multipurpose microbe that can break down a variety of ECDs, including BPA and TCS. Additionally, it has been shown that various microbes in the environment are capable of metabolizing EDCs (Cho & Kim 2000; Harms et al. 2011; Křesinová et al. 2018; Takamiya et al. 2008). The nonligninolytic fungus *Umbelopsis isabellina* decomposed certain EDCs with endocrine action, such as technical nonylphenol (tNP), 4-t-OP, and 4-cumylphenol (4-CP), successfully (Janicki et al. 2016). The clearance rates of tNP, 4-t-OP, and 4-CP were over 90% after 12 hours of incubation. *Fusarium falciforme* RRK20, a filamentous fungal strain from Taiwanese wastewater samples, was used to accelerate the breakdown of 4-t-OP, a compound with stronger estrogenic activity than other compounds (Rajendran et al. 2017). The EDCs used by the strain RRK20 include alkylphenol polyethoxylates, alkylphenols, and both natural and synthetic estrogens. Under aerobic conditions, it has been demonstrated that the yeast *Candida rugopelliculosa* RRKY5 is capable of utilizing a number of chemicals, including 4-methylphenol, BPA, and phenol. Furthermore, both the aromatic ring and the branched alkyl side chain were responsible for the strain RRKY5's (Rajendran et al. 2016) degradation of 4-t-OP. Enzymes can take a role in the regulation and metabolism of organisms as a fundamental biocatalyst. Environmental EDCs that degrade through the action of microorganisms or their enzymes have been used. The enzymes that microorganisms release provide them the ability to break down EDCs. Manganese peroxidase (MnP) and laccase have both been extensively researched for their roles in the biodegradation of EDCs.

Additionally, there are many genera of fungi having the ability to biodegrade or transform hormones: *Trichoderma, Trametes, Phanerochaete, Pleurotus, Aspergillus, Lentinula, Drechslera, Curvularia, Penicillium, Acremonium, Fusarium, Gibberella, Beauvaria, Doaratomyces, Scopulariopsis, Chaetomidium*, and *Thielavia*, etc. (Chatterjee & Abraham 2019; Hosseinabadi et al. 2015; Kollerov et al. 2020; Muszyńska et al. 2018; Pezzella et al. 2017; Savinova et al. 2019; Zoghi et al. 2019).

8.5.3 Algal Degradation of Endocrine-Disrupting Chemicals

Algae have also been shown to be capable of removing EDCs from aquatic habitats, in addition to bacteria and fungi. *Chlorella fusca, Chlamydomonas, Chlorella vulgaris*, and *Cyclotella caspia*, etc., have shown the decomposition EDCs. Two microalgae, *Chlamydomonas reinhardtii* and *Selenastrum capricornutum*, were examined for their potential to degrade the hormones β-estradiol (E2) and 17-ethynylestradiol (EE2) (Hom-Diaz et al. 2015). The results showed that microalgae were an inexpensive photooxidation and/or biodegradation treatment technology for contaminants, since E2 and EE2 were completely removed. *Chlamydomonas mexicana* and *Chlorella vulgaris*, two freshwater microalgae, have demonstrated the ability to biodegrade BPA (Ji et al. 2014).

There are algae that can eliminate EDCs, and these are *Raphidocelis subcapitata, Chlorococcus*, and *Scenedesmus quadricauda* (Fu et al. 2019; Gao et al. 2011; Liu et al. 2018; Shi et al. 2010).

8.6 APPLICATION OF NANOTECHNOLOGY IN BIODEGRADATION OF EDCs

There is no denying that nanotechnology, including nanoparticles and nanomaterials, is now advancing quickly. The prospects for enzymes to link nanoparticles and nanomaterials are quickly expanding as a result of the rising need for nanomaterial design, environmental monitoring, biochemical engineering, and biomedical applications (Chen et al. 2017b, 2017c). Numerous academics now do research on enzymes, nanoparticles, and nanomaterials, and the interactions among these components are receiving a lot of interest (Chen et al. 2016, 2017b). Reports state that, since 2014, one of the main roles or functions of enzymes has been the modification, breakdown, or production of nanoscale particles and nanomaterials. Nanomaterials are used extensively, and as a result, they penetrate the environment. Nanomaterials are said to pose some environmental hazards. Therefore, during the biodegradation of EDCs, these nanomaterials may come into contact with the microbes or enzymes that break down EDCs, which could disrupt the biodegradation process. To better understand how carbon nanotubes and/or graphene interact with manganese peroxidase (MnP), Chen et al. (2017a, 2017b) explored how carbon nanotubes and/or graphene nanomaterials are incorporated into the biodegradation of EDCs and TCS. The findings showed that the addition of nanomaterials altered the binding conformation of BPA, NP, and TCS substrates to MnP, and this alteration also had an impact on the overall biodegradation process (Chen et al. 2016, 2017a, 2017b).

8.7 CONCLUSIONS

This chapter discusses the biodegradation of EDCs using microorganisms. EDCs are chemicals that mimic, block, and interfere with the normal functioning of the endocrine system, and their presence in the environment is a major health concern. They comprise diverse categories of chemicals such as fungicides, industrial chemicals, metals, nonylphenols, pollutants, plasticizers, pesticides, pharmaceutical agents, and phytoestrogens. These EDCs can be found in air, soil, water, food sources, and personal care products. Various adverse effects, such as fertility issues, abnormalities in sex organs, endometriosis, breast development, early puberty, obesity, cancer, problems in the nervous system, immune function, and respiration, are associated with them. A number of physicochemical approaches are being used to remove EDCs from the environment (soil and water sources). Some well-studied approaches are adsorption, membrane-based filtration, ozonation, AOP, photochemical degradation, enzymatic degradation, and biological treatments. These approaches also have some limitations, such as high cost, high energy input, membrane fouling, and production of harmful reaction intermediates, etc., and therefore, pose difficulty to be used for large-scale treatment. Moreover, many studies have suggested that treatment approaches based on chemical and biological approaches are cost-effective and eco-friendly for removing estrogenic pollutants that require minimal input costs, a broad range of applications, and straightforward equipment. There are some specific fungi and bacteria that can efficiently break down EDCs. Some studies showed the efficient biodegradation of BPA in both natural water and sewage. Genera that contain EDCs degrading bacteria are *Bacillus, Virgibacillus, Novosphingobium, Rhodococcus, Acinetobacter, Pseudomonas, Sphingomonas, Enterobacter, Klebsiella, Aeromonas, Comamonas, Thauera, Deinococcus, Stenotrophomonas,* and cyanobacteria. Including this, in order to enhance the removal efficiency, these methods are used in combination.

BIBLIOGRAPHY

Adeleke, O.A., Latiff, A.A.A., Saphira, M.R., Daud, Z., Ismail, N., Ahsan, A., Ab Aziz, N.A., Al-Gheethi, A., Kumar, V., Fadilat, A., Apandi, N. 2019. Principles and Mechanism of Adsorption for the Effective Treatment of Palm Oil Mill Effluent for Water Reuse. In *Nanotechnology in Water and Wastewater Treatment*, ed. A. Ahsan, A.F. Ismail, 1–33. Elsevier: Amsterdam, The Netherlands. ISBN 9780128139028.

Al-Rifai, J.H., Khabbaz, H., Schäfer, A.I. 2011. Removal of pharmaceuticals and endocrine disrupting compounds in a water recycling process using reverse osmosis systems. *Separation and Purification Technology* 77:60–67.

Anway, M.D., Skinner, M.K. 2006. Epigenetic transgenerational actions of endocrine disruptors. *Endocrinology* 147:S43–S49.

Asgher, M., Bhatti, H.N., Ashraf, M., Legge, R.L. 2008. Recent developments in biodegradation of industrial pollutants by white rot fungi and their enzyme system. *Biodegradation* 19:771–783.

Bar-El, D.S., Reifen, R. 2010. Soy as an endocrine disruptor: cause for caution? *Journal of Pediatric Endocrinology and Metabolism* 23(9):855–861. https://doi.org/10.1515/jpem.2010.138

Baud, O., Berkane, N. 2019. Hormonal changes associated with intra-uterine growth restriction: impact on the developing brain and future neurodevelopment. *Frontiers in Endocrinology (Lausanne)* 10:179. https://doi.org/10.3389/fendo.2019.00179

Bell, M.R. 2014. Endocrine-disrupting actions of PCBs on brain development and social and reproductive behaviours. *Current Opinion in Pharmacology* 19:134–144. https://doi.org/10.1016/j.coph.2014.09.020

Borikar, D., Mohseni, M., Jasim, S. 2015. Evaluation and comparison of conventional and advanced oxidation processes for the removal of PPCPs and EDCs and their effect on THM-formation potentials. *Ozone: Science and Engineering* 37:154–169.

Braun, J.M. 2017. Early life exposure to endocrine disrupting chemicals and childhood obesity and neurodevelopment. *Nature Reviews Endocrinology* 13(3):161–173. https://doi.org/10.1038/nrendo.2016.186

Buoso, E., Masi, M., Racchi, M., Corsini, E. 2020. Endocrine-disrupting chemicals' (EDCs) effects on tumour microenvironment and cancer progression: emerging contribution of RACK1. *International Journal of Molecular Sciences* 21(23):9229. https://doi.org/10.3390/ijms21239229

Cabana, H., Jiwan, J.L., Rozenberg, R., Elisashvili, V., Penninckx, M., Agathos, S.N., Jones, J.P. 2007. Elimination of endocrine disrupting chemicals nonylphenol and bisphenol A and personal care product ingredient triclosan using enzyme preparation from the white rot fungus Coriolopsis polyzona. *Chemosphere* 67:770–778.

Cajthaml, T. 2015. Biodegradation of endocrine-disrupting compounds by ligninolytic fungi: mechanisms involved in the degradation. *Environmental Microbiology* 17(12):4822–4834. https://doi.org/10.1111/1462-2920.12460

Cajthaml, T., Křesinová, Z., Svobodová, K., Möder, M. 2009. Biodegradation of endocrine-disrupting compounds and suppression of estrogenic activity by ligninolytic fungi. *Chemosphere* 75:745–750.

Cesaro, A., Belgiorno, V. 2016. Removal of endocrine disruptors from urban wastewater by advanced oxidation processes (AOPs): a review. *The Open Biotechnology Journal* 10:151–172. https://doi.org/10.2174/1874070701610010151

Chatterjee, A., Abraham, J. 2019. Mycoremediation of 17 β-estradiol using Trichoderma citrinoviride strain AJAC3 along with enzyme studies. *Environmental Progress & Sustainable Energy* 38(4). https://doi.org/10.1002/ep.13142

Chen, L., Ma, J., Li, X., Zhang, J., Fang, J., Guan, Y., Xie, P. 2011. Strong enhancement on Fenton oxidation by addition of hydroxylamine to accelerate the ferric and ferrous iron cycles. *Environmental Science & Technology* 45(9):3925–3930.

Chen, M., Qin, X., Li, J., Zeng, G. 2016. Probing molecular basis of single-walled carbon nanotube degradation and nondegradation by enzymes based on manganese peroxidase and lignin peroxidase. *RSC Advances* 6:3592–3599.

Chen, M., Qin, X., Zeng, G. 2017a. Biodiversity change behind wide applications of nanomaterials? *Nano Today* 17:11–13.

Chen, M., Zeng, G., Lai, C., Zhang, C., Xu, P., Yan, M., Xiong, W. 2017b. Interactions of carbon nanotubes and/or graphene with manganese peroxidase during biodegradation of endocrine disruptors and triclosan. *Chemosphere* 184:127–136. https://doi.org/10.1016/j.chemosphere.2017.05.162

Chen, M., Zeng, G., Xu, P., Lai, C., Tang, L. 2017c. How do enzymes 'meet' nanoparticles and nanomaterials? *Trends in Biochemical Sciences* 42:914–930.

Chen, S.C., Liao, T.L., Wei, Y.H., Tzeng, C.R., Kao, S.H. 2010. Endocrine disruptor, dioxin (TCDD)-induced mitochondrial dysfunction and apoptosis in human trophoblast-like JAR cells. *Molecular Human Reproduction* 16(5):361–372. https://doi.org/10.1093/molehr/gaq004

Chen, Y.-L., Fu, H.-Y., Lee, T.-H., Shih, C.-J., Huang, L., Wang, Y.-S., Ismail, W., Chiang, Y.-R. 2018. Estrogen degraders and estrogen degradation pathway identified in an activated sludge. *Applied and Environmental Microbiology* 84(10):e00001–18.

Chen, Y.-L., Yu, C.-P., Lee, T.-H., Goh, K.-S., Chu, K.-H., Wang, P.-H., Ismail, W., Shih, C.-J., Chiang, Y.-R. 2017d. Biochemical mechanisms and catabolic enzymes involved in bacterial estrogen degradation pathways. *Cell Chemical Biology* 24:712–724.

Cho, Y., Kim, H. 2000. Transformation of triclosan by Trametes versicolor and Pycnoporus cinnabarinus. *Applied and Environmental Microbiology* 66:4157–4160.

Combarnous, Y, Nguyen, T.M.D. 2019. Comparative overview of the mechanisms of action of hormones and endocrine disruptor compounds. *Review Toxics* 7(5):1–10.

Delgado, L.F., Charles, P., Glucina, K., Morlay, C. 2012. The removal of endocrine disrupting compounds, pharmaceutically activated compounds and cyanobacterial toxins during drinking water preparation using activated carbon-a review. *Science of the Total Environment* 435–436:509–525. https://doi.org/10.1016/j.scitotenv.2012.07.046

Dharupaneedi, S.P., Nataraj, S.K., Nadagouda, M., Reddy, K.R., Shukla, S.S., Aminabhavi, T.M. 2019. Membrane-based separation of potential emerging pollutants. *Separation and Purification Technology* 210:850–866.

Dhiman, S.K., Dureja, H. 2020. Significance of and challenges in regulating endocrine disruptors—how regulators and industry can conquer? *Endocrine, Metabolic & Immune Disorders—Drug Targets* 20(10):1664–1681. https://doi.org/10.2174/1871530320666200606225104

Diamanti-Kandarakis, E., Bourguignon, J.P., Giudice, L.C., Hauser, R., Prins, G.S., Soto, A.M., Zoeller, R.T., Gore, A.C. 2009. Endocrine-disrupting chemicals: an endocrine society scientific statement. *Endocrine Reviews* 30(4):293–342. https://doi.org/10.1210/er.2009-0002

Dickenson, E.R.V., Drewes, J.E. 2010. Quantitative structure property relationships for the adsorption of pharmaceuticals onto activated carbon. *Water Science and Technology* 62:2270–2276.

Ding, N., Harlow, S.D., Randolph, J.F. Jr., Loch-Caruso, R., Park, S.K. 2020. Perfluoroalkyl and polyfluoroalkyl substances (PFAS) and their effects on the ovary. *Human Reproduction Update* 26(5):724–752. https://doi.org/10.1093/humupd/dmaa018

Egerton, T.A., Purnama, H. 2014. Does hydrogen peroxide really accelerate TiO_2 UV-C photocatalyzed decolouration of azo-dyes such as reactive orange 16? *Dyes and Pigments* 101:280–285.

Endocrine Disrupting Chemicals. www.endocrine.org/patient-engagement/endocrinelibrary/edcs#:~:text=Endocrine%2Ddisrupting%20chemicals%20(EDCs)%20are%20substances%20in%20the%20environment,of%20your%20body's%20endocrine%20system (assessed August 15, 2022).

Fan, X., Tao, Y., Wang, L., Zhang, X., Lei, Y., Wang, Z., Noguchi, H. 2014. Performance of an integrated process combining ozonation with ceramic membrane ultra-filtration for advanced treatment of drinking water. *Desalination* 335:47–54.

Farooq, T., Hameed, A., Raza, A. 2021. Role of Phthalates as EDCs in Metabolic Disorders. In *Endocrine Disrupting Chemicals-Induced Metabolic Disorders and Treatment Strategies*, ed. M.S.H. Akash, K. Rehman, M.Z. Hashmi, 239–250. Springer: Cham. https://doi.org/10.1007/978-3-030-45923-9_15

Fernandez, L., Louvado, A., Esteves, V.I., Gomes, N.C.M., Almeida, A., Cunha, A. 2017. Biodegradation of 17β-estradiol by bacteria isolated from deep sea sediments in aerobic and anaerobic media. *Journal of Hazardous Materials* 323:359–366.

Ferreiro, C., Iker, G., Lombraña, I., De Luis, A., Villota, N., Ros, O., Etxebarria, N. 2020. Contaminants of emerging concern removal in an effluent of wastewater treatment plant under biological and continuous mode ultrafiltration treatment. *Sustainability* 12:1–19.

Frenzilli, G., Martorell-Ribera, J., Bernardeschi, M., Scarcelli, V., Jönsson, E., Diano, N., Moggio, M., Guidi, P., Sturve, J., Asker, N. 2021. Bisphenol A and bisphenol S induce endocrine and chromosomal alterations in brown trout. *Frontiers in Endocrinology* 12:645519. https://doi.org/10.3389/fendo.2021.645519

Fu, M., Deng, B., Lü, H., Yao, W., Su, S., Wang, D. 2019. The bioaccumulation and biodegradation of testosterone by Chlorella vulgaris. *International Journal of Environmental Research and Public Health* 16:1253. https://doi.org/10.3390/ijerph16071253

Gao, Q.T., Wong, Y.S., Tam, N.F.Y. 2011. Removal and biodegradation of nonylphenol by immobilized Chlorella vulgaris. *Bioresource Technology* 102:10230–10238.

Garrido-Cardenas, J.A., Esteban-García, B., Agüera, A., Sánchez-Pérez, J.A., Manzano-Agugliaro, F. 2020. Wastewater treatment by advanced oxidation process and their worldwide research trends. *International Journal of Environmental Research and Public Health* 17(1):170. https://doi.org/10.3390/ijerph17010170

Gernjak, W., Krutzler, T., Glaser, A., Malato, S., Caceres, J., Bauer, R., Fernández-Alba, A.R. 2003. Photo-Fenton treatment of water containing natural phenolic pollutants. *Chemosphere* 50:71–78.

Gmurek, M., Olak-Kucharczyk, M., Ledakowicz, S. 2017. Photochemical decomposition of endocrine disrupting compounds—a review. *Chemical Engineering Journal* 310:437–456.

Gupta, P., Thompson, B.L., Wahlang, B., Jordan, C.T., Zach Hilt, J., Hennig, B., Dziubla, T. 2018. The environmental pollutant, polychlorinated biphenyls, and cardiovascular disease: a potential target for antioxidant nanotherapeutics. *Drug Delivery and Translational Research* 8(3):740–759. https://doi.org/10.1007/s13346-017-0429-9

Harms, H., Schlosser, D., Wick, L.Y. 2011. Untapped potential: exploiting fungi in bioremediation of hazardous chemicals. *Nature Reviews Microbiology* 9:177–192.

Haugland, J.O., Kinney, K.A., Johnson, W.H., Camino, M.M.A., Whitman, C.P., Lawler, D.F. 2019. Laccase removal of 2-chlorophenol and sulfamethoxazole in municipal wastewater. *Water Environment Research* 91(4):281–291.

Hirano, T., Honda, Y., Watanabe, T., Kuwahara, M. 2000. Degradation of bisphenol A by the lignin-degrading enzyme, manganese peroxidase, produced by the white-rot basidiomycete, Pleurotus ostreatus. *Journal of the Agricultural Chemical Society of Japan* 64:1958–1962.

Hites, R.A. 2011. Dioxins: an overview and history. *Environmental Science & Technology* 45(1):16–20. https://doi.org/10.1021/es1013664

Hom-Diaz, A., Llorca, M., Rodríguez-Mozaz, S., Vicent, T., Barceló, D., Blánquez, P. 2015. Microalgae cultivation on wastewater digestate: β-estradiol and 17α-ethynylestradiol degradation and transformation products identification. *Journal of Environmental Management* 155:106–113.

Hosseinabadi, T., Vahidi, H., Nickavar, B., Kobarfard, F. 2015. Biotransformation of progesterone by whole cells of filamentous fungi Aspergillus brasiliensis. *Iranian Journal of Pharmaceutical Research* 14:919–924.

Hu, Z., Si, X., Zhang, Z., Wen, X. 2014. Enhanced EDCs removal by membrane fouling during the UF process. *Desalination* 336:18–23.

Husain, Q., Qayyum, S. 2013. Biological and enzymatic treatment of bisphenol A and other endocrine disrupting compounds: a review. *Critical Reviews in Biotechnology* 33(3):260–292.

James, C.P., Germain, E., Judd, S. 2014. Micropollutant removal by advanced oxidation of microfiltered secondary effluent for water reuse. *Separation and Purification Technology* 127:77–83.

Janicki, T., Krupinski, M., Długonski, J. 2016. Degradation and toxicity reduction of the endocrine disruptors nonylphenol, 4-tert-octylphenol and 4-cumylphenol by the non-ligninolytic fungus Umbelopsis isabellina. *Bioresource Technology* 200:223–229.

Ji, M.K., Kabra, A.N., Choi, J., Hwang, J.H., Kim, J.R., Abou-Shanab, R.A.I., Oh, Y.K., Jeon, B.H. 2014. Biodegradation of bisphenol A by the freshwater microalgae Chlamydomonas mexicana and Chlorella vulgaris. *Ecological Engineering* 73:260–269.

Kang, G.-D., Cao, Y-M. 2014. Application and modification of poly(vinylidene fluoride) (PVDF) membranes—a review. *Journal of Membrane Science* 463:145–165.

Khattab, R.A., Elnwishy, N., Hannora, A., Mattiasson, B., Omran, H., Alharbi, O.M.L., Ali, I. 2019. Biodegradation of 17-β-estradiol in water. *International Journal of Environmental Science and Technology* 16:4935–4944.

Kim, S., Cho, H., Joo, H., Her, N., Han, J., Yi, K., Kim, J.O., Yoon, J. 2017. Evaluation of performance with small and scale-up rotating and flat reactors; photocatalytic degradation of bisphenol A, 17B–estradiol, and 17A–ethynyl estradiol under solar irradiation. *Journal of Hazardous Materials* 336:21–32.

Kollerov, V., Shutov, A., Kazantsev, A., Donova, M. 2020. Biotransformation of androstenedione and androstadienedione by selected Ascomycota and Zygomycota fungal strains. *Phytochemistry* 169:112160. https://doi.org/10.1016/j.phytochem.2019.112160

Křesinová, Z., Linhartová, L., Filipová, A., Ezechiáš, M., Mašín, P., Cajthaml, T. 2018. Biodegradation of endocrine disruptors in urban wastewater using Pleurotus ostreatus bioreactor. *New Biotechnology* 43:53–61.

Kriszt, R., Krifaton, C., Szoboszlay, S., Cserhati, M., Kriszt, B., Kukolya, J., Czeh, A., Feher-Toth, S., Torok, L., Szoke, Z., Kovács, K.J., Barna, T., Ferenczi, S. 2012. A new zearalenone biodegradation strategy using non-pathogenic Rhodococcus pyridinivorans K408 strain. *PLoS One* 7:e43608. https://doi.org/10.1371/journal.pone.0043608

Kumar, M., Sarma, D.K., Shubham, S., Kumawat, M., Verma, V., Prakash, A., Tiwari, R. 2020. Environmental endocrine-disrupting chemical exposure: role in non-communicable diseases. *Frontiers in Public Health* 8:553850. https://doi.org/10.3389/fpubh.2020.553850

Larsson, K., Lindh, C.H., Jönsson, B.A.G., Giovanoulis, G., Bibi, M., Bottai, M., Bergström, A., Berglund, M. 2017. Phthalates, non-phthalate plasticizers and bisphenols in Swedish preschool dust in relation to children's exposure. *Environment International* 102:114–124.

Li, C., Kong, X., Lan, L., Tadda, M.A., Liu, D. 2020. Effects of carbon sources on 17 beta-estradiol degradation by Sphingomonas sp. and the involved intracellular metabolomics analysis. *Environmental Science: Processes & Impacts* 22:197–206.

Li, M., Zhao, X., Zhang, X., Wu, D., Leng, S. 2018. Biodegradation of 17β-estradiol by bacterial co-culture isolated from manure. *Scientific Reports* 8:3787. https://doi.org/10.1038/s41598-018-22169-0

Li, Q., Ma, X., Gao, N., Chu, W. 2017. Removal of 17α-ethynylestradiol from aqueous solutions by a hybrid PAC/UF process. *Water SA* 43:116–121.

Lin, H., Peng, W., Zhang, M., Chen, J., Hong, H., Zhang, Y. 2013. A review on anaerobic membrane bioreactors: applications, membrane fouling and future perspectives. *Desalination* 314:169–188.

Liu, C., Liu, K., Zhao, C., Gong, P., Yu, Y. 2020. The characterization of a short chain dehydrogenase/reductase (SDRx) in Comamonas testosterone. *Toxicology Reports* 7:460–467.

Liu, J., Li, S., Li, X., Gao, Y., Ling, W. 2018. Removal of estrone, 17β-estradiol, and estriol from sewage and cow dung by immobilized Novosphingobium sp. ARI-1. *Environmental Technology* 39:2423–2433.

Liu, J., Liu, J., Xu, D., Ling, W., Li, S., Chen, M. 2016. Isolation, immobilization, and degradation performance of the 17β-estradiol-degrading bacterium Rhodococcus sp. JX-2. *Water, Air, & Soil Pollution* 227:422. https://doi.org/10.1007/s11270-016-3122-6

Liu, W., Chen, Q., He, N., Sun, K., Sun, D., Wu, X., Duan, S. 2018. Removal and biodegradation of 17β-estradiol and diethylstilbestrol by freshwater microalgae Raphidocelis subcapitata. *International Journal of Environmental Research and Public Health* 15:452–466.

Liu, Y., Wang, H., Chai, L., Li, X., Wu, M., Wang, H. 2022. Effects of perchlorate and exogenous T4 exposures on development, metamorphosis and endochondral ossification in Bufo gargarizans larvae. *Aquatic Toxicology* 242:106036. https://doi.org/10.1016/j.aquatox.2021.106036

Luo, Y., Guo, W., Hao, H., Duc, L., Ibney, F., Zhang, J., Liang, S., Wang, X.C. 2014. Science of the total environment a review on the occurrence of micropollutants in the aquatic environment and their fate and removal during wastewater treatment. *Science of the Total Environment* 473–474:619–641.

Macellaro, G., Pezzella, C., Cicatiello, P., Sannia, G., Piscitelli, A. 2014. Fungal laccases degradation of endocrine disrupting compounds. *BioMed Research International* 2014:614038. https://doi.org/10.1155/2014/614038

Mailler, R., Gasperi, J., Coquet, Y., Derome, C., Buleté, A., Vulliet, E., Bressy, A., Varrault, G., Chebbo, G., Rocher, V. 2016. Removal of emerging micropollutants from wastewater by activated carbon adsorption: experimental study of different activated carbons and factors influencing the adsorption of micropollutants in wastewater. *Journal of Environmental Chemical Engineering* 4:1102–1109.

Malhotra, M., Suresh, S., Garg, A. 2018. Tea waste derived activated carbon for the adsorption of sodium diclofenac from wastewater: adsorbent characteristics, adsorption isotherms, kinetics, and thermodynamics. *Environmental Science and Pollution Research* 25:32210–32220.

Menashe, O., Raizner, Y., Kuc, M.E., Cohen-Yaniv, V., Kaplan, A., Mamane, H., Avisar, D., Kurzbaum, E. 2020. Biodegradation of the endocrine-disrupting chemical 17α-ethynylestradiol (EE2) by Rhodococcus zopfii and Pseudomonas putida encapsulated in small bioreactor platform (SBP) capsules. *Applied Sciences* 10:336. https://doi.org/10.3390/app10010336

Michalska, J., Pinski, A., Zur, J., Mrozik, A. 2020. Selecting bacteria candidates for the bioaugmentation of activated sludge to improve the aerobic treatment of landfill leachate. *Water* 12:140. https://doi.org/10.3390/w12010140

Milanović, M., Đurić, L., Milošević, N., Milic, N. 2021. Comprehensive insight into triclosan—from widespread occurrence to health outcomes. *Environmental Science and Pollution Research International* 1–22. https://doi.org/10.1007/s11356-021-17273-0

Ming, C., Qin, X., Zeng, G. 2017. Biodegradation of carbon nanotubes, graphene, and their derivatives. *Trends in Biotechnology* 35:836–846.

Muszyńska, B., Zmudzki, P., Lazur, J., Kała, K., Sułkowska-Ziaja, K., Opoka, W. 2018. Analysis of the biodegradation of synthetic testosterone and 17α-ethynylestradiol using the edible mushroom Lentinula edodes. *3 Biotech* 8:424. https://doi.org/10.1007/s13205-018-1458-x

Nasseri, S., Ebrahimi, S., Abtahi, M., Saeedi, R. 2018. Synthesis and characterization of polysulfone/graphene oxide nano-composite membranes for removal of bisphenol A from water. *Journal of Environmental Management* 205:174–182.

Neale, P.A., Schäfer, A.I. 2012. Quantification of solute-solute interactions in steroidal hormone removal by ultrafiltration membranes. *Separation and Purification Technology* 90:31–38.

Nguyen, P.D., Le, T.M.T., Vo, T.K.Q., Nguyen, P.T., Vo, T.D.H., Dang, B.T., Son, N.T., Nguyen, D.D., Bui, X.T. 2021. Submerged membrane filtration process coupled with powdered activated carbon for nonylphenol ethoxylates removal. *Water Science and Technology* 84(7):1793–1803.

Oliveira, A.C., Rebelo, A.R., Homem, C.C.F. 2021. Integrating animal development: how hormones and metabolism regulate developmental transitions and brain formation. *Developmental Biology* 475:256–264.

Papalou, O., Kandaraki, E.A., Papadakis, G., Diamanti-Kandarakis, E. 2019. Endocrine disrupting chemicals: an occult mediator of metabolic disease. *Frontiers in Endocrinology (Lausanne)* 10:112. https://doi.org/10.3389/fendo.2019.00112

Pezzella, C., Macellaro, G., Sannia, G., Raganati, F., Olivieri, G., Marzocchella, A., Schlosser, D., Piscitelli, A. 2017. Exploitation of Trametes versicolor for bioremediation of endocrine disrupting chemicals in bioreactors. *PLoS One* 12:e0178758. https://doi.org/10.1371/journal.pone.0178758

Qiu, Q., Wang, P., Kang, H., Wang, Y., Tian, K., Huo, H. 2019. Genomic analysis of new estrogen-degrading bacterial strain, Acinetobacter sp. DSSKY-A-001. *International Journal of Genomics* 2019:2804134. https://doi.org/10.1155/2019/2804134

Rajendran, R.K., Huang, S.L., Lin, C.C., Kirschner, R. 2016. Biodegradation of the endocrine disrupter 4-tert-octylphenol by the yeast strain Candida rugopelliculosa RRKY5 via phenolic ring hydroxylation and alkyl chain oxidation pathways. *Bioresource Technology* 226:55–64.

Rajendran, R.K., Lin, C.C., Huang, S.L., Kirschner, R. 2017. Enrichment, isolation, and biodegradation potential of long-branched chain alkylphenol degrading non-ligninolytic fungi from wastewater. *Marine Pollution Bulletin* 125:416–425.

Rajesha, B.J., Vishaka, V.H., Balakrishna, G.R., Padaki, M., Nazri, N.A.M. 2019. Effective composite membranes of cellulose acetate for removal of benzophenone-3. *Journal of Water Process Engineering* 30:100419. https://doi.org/10.1016/j.jwpe.2017.06.003

Ramírez-Cavazos, L.I., Junghanns, C., Ornelas-Soto, N., Cárdenas-Chávez, D.L., Hernández-Luna, C., Demarche, P., Enaud, E., García-Morales, R., Agathos, S.N., Parra, R. 2014. Purification and characterization of two thermostable laccases from Pycnoporus sanguineus and potential role in degradation of endocrine disrupting chemicals. *Journal of Molecular Catalysis B: Enzymatic* 108:32–42.

Rana, D., Narbaitz, R.M., Garand-Sheridan, A.M., Westgate, A., Matsuura, T., Tabe, S., Jasim, S.Y. 2014. Development of novel charged surface modifying macromolecule blended PES membranes to remove EDCs and PPCPs from drinking water sources. *Journal of Materials Chemistry A* 2:10059–10072.

Sahar, E., David, I., Gelman, Y., Chikurel, H., Aharoni, A., Messalem, R., Brenner, A. 2011. The use of RO to remove emerging micropollutants following CAS/UF or MBR treatment of municipal wastewater. *Desalination* 273:142–147.

Savinova, O.S., Solyev, P.N., Vasina, D.V., Tyazhelova, T.V., Fedorova, T.V., Savinova, T.S. 2019. Biotransformation of progesterone by Aspergillus nidulans VKPM F-1069 (wild type). *Steroids* 149:108421. https://doi.org/10.1016/j.steroids.2019.05.013

Shi, W., Wang, L., Rousseau, D.P.L., Lens, P.N.L. 2010. Removal of estrone, 17α-ethinylestradiol, and 17ß-estradiol in algae and duckweed-based wastewater treatment systems. *Environmental Science and Pollution Research* 17:824–833.

Shih, C.J., Chen, Y.L., Wang, C.H.H., Wei, S.T.S., Lin, I.T., Ismail, W.A., Chiang, Y.R. 2017. Biochemical mechanisms and microorganisms involved in anaerobic testosterone metabolism in estuarine sediments. *Frontiers in Microbiology* 8:1520. https://doi.org/10.3389/fmicb.2017.01520

Si, X., Hu, Z., Ding, D., Fu, X. 2019. Effects of effluent organic matters on endocrine disrupting chemical removal by ultrafiltration and ozonation in synthetic secondary effluent. *Journal of Environmental Sciences* 76:57–64.

Si, X., Hu, Z., Huang, S. 2018. Applied sciences combined process of ozone oxidation and ultrafiltration as an effective treatment technology for the removal of endocrine-disrupting chemicals. *Applied Sciences* 8:1240. https://doi.org/10.3390/app8081240

Sin, J.C., Lam, S.M., Lee, K.T., Mohamed, A.R. 2013. Degrading two endocrine-disrupting chemicals from water by UV irradiation with the presence of nanophotocatalysts. *Desalination and Water Treatment* 51:3505–3520.

Su-Hua, W., Bing-Zhi, D., Yu, H. 2010. Adsorption of bisphenol A by polysulphone membrane. *Desalination* 253:22–29.

Suyamud, B., Inthorn, D., Panyapinyopol, B. 2018. Biodegradation of Bisphenol a by a newly isolated *Bacillus megaterium* strain ISO-2 from a polycarbonate industrial wastewater. *Water, Air, & Soil Pollution* 229:348. https://doi.org/10.1007/s11270-018-3983-y

Syahida, N., Anan, M., Jaafar, J., Hafiz, M., Othman, D., Rahman, M.A., Aziz, F., Shahrodin, N.S. 2019. Titanium dioxide incorporated thin film composite membrane for bisphenol A removal. *Malaysian Journal of Fundamental and Applied Sciences* 15:755–760.

Takamiya, M., Magan, N., Warner, P.J. 2008. Impact assessment of bisphenol A on lignin-modifying enzymes by basidiomycete Trametes versicolor. *Journal of Hazardous Materials* 154:33–37.

Wang, Q., Yang, C., Zhang, G., Hu, L., Wang, P. 2017. Photocatalytic Fe-doped TiO_2/PSF composite UF membranes: characterization and performance on BPA removal under visible-light irradiation. *Chemical Engineering Journal* 319:39–47.

Wang, R., Ma, X., Liu, T., Li, Y., Song, L., Tjong, S.C., Cao, L., Wang, W., Yu, Q., Wang, Z. 2020. Degradation aspects of endocrine disrupting chemicals: a review on photocatalytic processes and photocatalysts. *Applied Catalysis A: General* 597:117547. https://doi.org/10.1016/j.apcata.2020.117547

Wang, Y., Shao, H., Zhu, S., Tian, K., Qiu, Q., Huo, H. 2019. Degradation of 17 ß-estradiol and products by a mixed culture of Rhodococcus equi DSSKP-R-001 and Comamonas testosteroni QYY20150409. *Biotechnology & Biotechnological Equipment* 33:268–277.

Werkneh, A.A., Gebru, S.B., Redae, G.H., Tsige, A.G. 2022. Removal of endocrine disrupters from the contaminated environment: public health concerns, treatment strategies and future perspectives-A review. *Heliyon* 8(4):e09206. https://doi.org/10.1016/j.heliyon.2022.e09206

Xiong, W., Peng, W., Liang, R. 2018. Identification and genome analysis of Deinococcus actinosclerus SJTR1, a novel 17β-estradiol degradation bacterium. *3 Biotech* 8(10):433. https://doi.org/10.1007/s13205-018-1466-x

Xiong, W., Yin, C., Peng, W., Deng, Z., Lin, S., Liang, R. 2020. Characterization of an 17β-estradiol-degrading bacterium Stenotrophomonas maltophilia SJTL3 tolerant to adverse environmental factors. *Applied Microbiology and Biotechnology* 104:1291–1305.

Yilmaz, B., Terekeci, H., Sandal, S., Kelestimur, F. 2020. Endocrine disrupting chemicals: exposure, effects on human health, mechanism of action, models for testing and strategies for prevention. *Reviews in Endocrine and Metabolic Disorders* 21(1):127–147. https://doi.org/10.1007/s11154-019-09521-z

Yoon, Y., Westerhoff, P., Snyder, S.A., Wert, E.C. 2006. Nanofiltration and ultrafiltration of endocrine disrupting compounds, pharmaceuticals and personal care products. *Journal of Membrane Science* 270:88–100.

Yuan, H., He, Z. 2015. Integrating membrane filtration into bioelectrochemical systems as next generation energy-efficient wastewater treatment technologies for water reclamation: a review. *Bioresource Technology* 195:202–209.

Zahari, A.M., Shuo, C.W., Sathishkumar, P., Yusoff, A.R.M., Gu, F.L., Buang, N.A., Lau, W.J., Gohari, R.J., Yusop, Z. 2018. A reusable electrospun PVDF-PVP-MnO_2 nanocomposite membrane for bisphenol A removal from drinking water. *Journal of Environmental Chemical Engineering* 6:5801–5811.

Zhang, C., Zeng, G., Yuan, L., Yu, J., Li, J., Huang, G., Xi, B., Liu, H. 2007. Aerobic degradation of bisphenol a by Achromobacter xylosoxidans strain B-16 isolated from compost leachate of municipal solid waste. *Chemosphere* 68:181–190.

Zhang, J., Chua, H.C., Zhou, J., Fane, A.G. 2006. Factors affecting the membrane performance in submerged membrane bioreactors. *Journal of Membrane Science* 284:54–66.

Zhao, F.B., Tang, C.C., Liu, X.Y., Shi, F.J., Song, X.R., Tian, Y., Li, Z.S. 2015. Transportation characteristics of bisphenol A on ultrafiltration membrane with low molecule weight cut-off. *Desalination* 362:18–25.

Zhao, Y., Cho, C.W., Cui, L. et al. 2019. Adsorptive removal of endocrine-disrupting compounds and a pharmaceutical using activated charcoal from aqueous solution: kinetics, equilibrium, and mechanism studies. *Environmental Science and Pollution Research* 26:33897–33905. https://doi.org/10.1007/s11356-018-2617-7

Zhu, Y., Wei, J., Zhang, H., Liu, K., Kong, Z., Dong, Y., Jin, G., Tian, J., Qin, Z. 2018. Fabrication of composite membrane with adsorption property and its application to the removal of endocrine disrupting compounds during filtration process. *Chemical Engineering Journal* 352: 53–63.

Zoghi, M., Gandomkar, S., Habibi, Z. 2019. Biotransformation of progesterone and testosterone enanthate by Circinella muscae. *Steroids* 151:108446. https://doi.org/10.1016/j.steroids.2019.108446

9 Bioremediation of Heavy Metals Mediated by Cyanobacteria

Nikita, Uma Singh, G. L. Tiwari, Richa Tandon and Ravi Tandon

9.1 INTRODUCTION

A global environmental crisis has emerged in the past few decades due to the rapid urbanization and industrialization required to satisfy human requirements. These developments have led to issues like exploiting existing resources and growing pollution in the land, air, and water environments. Heavy metals are often found in the crust of the earth; however, they are released into the environment due to anthropogenic activities such as mining, smelting, and agricultural etc. (Figure 9.1). Aqueous discharges are generated by industries including mining, steel mills, and galvanizing that contain a significant number of heavy metals, including cobalt, nickel, chromium, copper, cadmium, mercury, zinc, and lead (Briffa, Sinagra, and Blundell 2020). Therefore, it has become clear that pollution from heavy metals poses a significant threat to living things in the environment such as cancer, nausea, vomiting, mental disorder, etc., as shown in Figure 9.2 (Igiri et al. 2018; Okolo et al. 2016).

These are non-biodegradable with a high density of about 5 g/cm^3; even their small concentrations in the water can be toxic to the aquatic organism (Nowicka 2022). Once the aquatic organisms consume them and reach their threshold limit (1–2000 ppm based on the type of metal ion), they tend to accumulate in the biosystem and damage the biological pathways by inhibiting enzymes and acting as a catalyst in redox reaction, resulting in the production of free radicals (Pandi, Shashirekha, and Swamy 2009). Therefore, the heavy metal ions might have a role in DNA damage, ageing, and serious illnesses like cancer (Briffa, Sinagra, and Blundell 2020). Recent studies have shown that bacteria, fungi, and algae are capable of bioremediating heavy metals. Several microorganisms can degrade, detoxify, and even accumulate organic and inorganic compounds. Cyanobacteria are a renewable, natural biome with different affinities for different metals; thus, these can be considered an ideal candidate as a biosorbent material. Cyanobacteria remove heavy metals through two major mechanisms: the adsorption process – i.e., intracellular – and metal uptake by absorption – i.e., extracellular process (Al-Amin et al. 2021).

9.2 CYANOBACTERIA IN BIOREMEDIATION

Several strains of cyanobacteria have been identified as useful in bioremediating heavy metals, including removing radioactive metals also referred as "radionuclides" such as ^{238}U and ^{226}Ra through biosorption in *Chroococcidiopsis thermalis* S.M/S9 and the bioremediation of ^{226}Ra and ^{90}Sr with 99% efficacy in *Gleomargarita lithophora* (Heidari et al. 2018; Mehta et al. 2019). According to previous reports, toxic heavy metals like Pb, Cd, and Cr can effectively detoxify by the *Nostoc* and *Phormidium* with more than 90% efficiency through biosorption (Chakdar et al. 2022). Recently, it has been recorded the heavy-metal-contaminated sewage water was treated with various cyanobacterial strains such as *Anabaena oryzae*, *Nostoc muscorum*, *Cyanosarcina fontana*, and *Stigonema ocellatum* (Fawzy and Mohamed 2017). Out of them, *Anabaena oryzae*, had been shown to remove the heavy

DOI: 10.1201/9781003391487-9

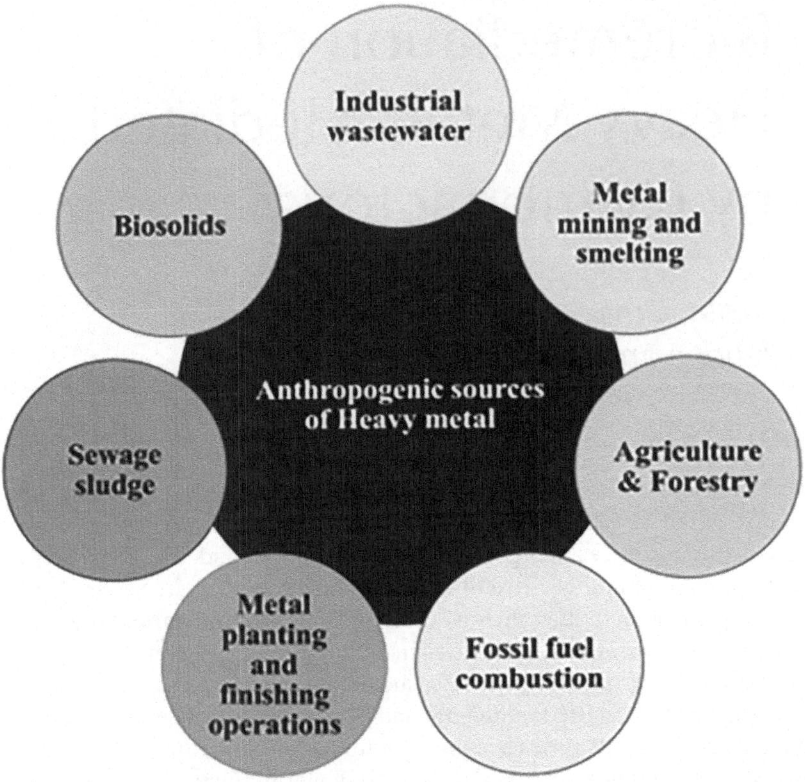

FIGURE 9.1 Various sources of heavy metal in the environment.

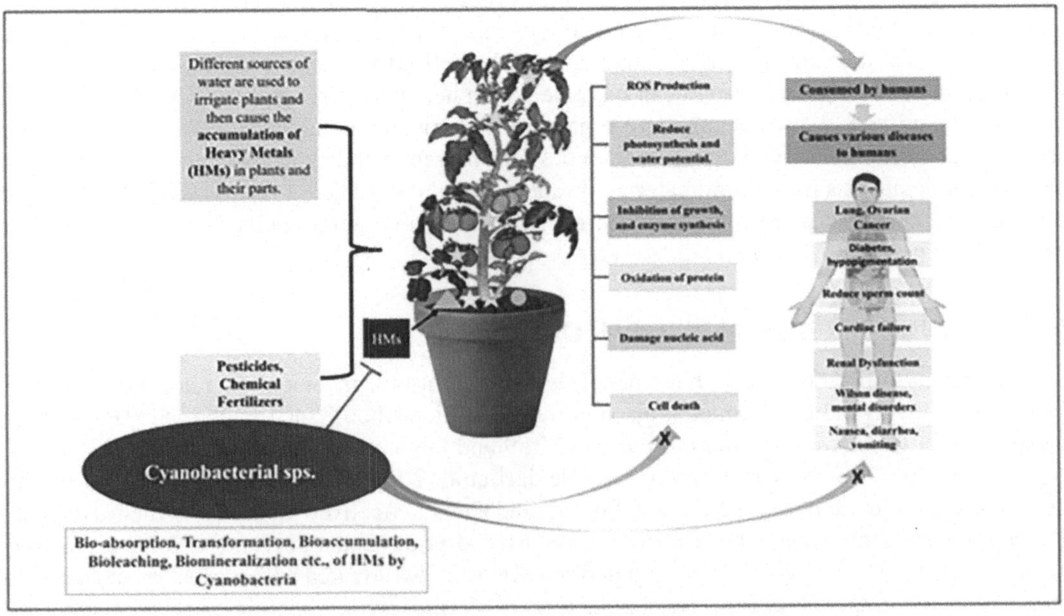

FIGURE 9.2 Effect of heavy metals (HMs) in plants, humans, and bioremediation of HMs by cyanobacteria.

TABLE 9.1
Table Showing the Cyanobacterial Species in Bioremediation of Heavy Metal Ions

Cyanobacteria	Heavy metal	References
Anabaena doliolum	Cd	(Goswami et al. 2015)
Cyanospira capsulate and *Nostoc*	Cu	(Colica, Mecarozzi, and De Philippis 2010; De Philippis et al. 2003)
Cyanothece	Cu, Cr, Ni	(Micheletti et al. 2008)
Gloeothece magna	Cd, Mn	(Mohamed 2001)
Lyngbya putealis	Cr	(Kiran and Kaushik 2008)
Lyngbya wollei	Cu	(Bishop and Rodgers 2012)
Lyngbya sp., *Aulosira* sp., and *Scytonema* sp.	Cu, Cd, Pb	(Kumar et al. 2010)
Microcystis sp.	Cd, Cu, Cr	(Rai and Tripathi 2007)
Nostoc sp.	Pb, Cd, Cr	(Warjri and Syiem 2018)
Oscillatoria quadripunctulata *Oscillatoria tenius*	Cd, Co, Pb, Ni, Zn	(Azizi, Hosseinzadeh Colagar, and Hafeziyan 2012; Ajayan, Selvaraju, and Thirugnanamoorthy 2011)
Phormidium sp., *P. valderium, P. bohner, P.ambigunum, P. corium*	Cd, Cr, Hg, Cd, Zn, Pb, Ni, Cu	(Shanab, Essa, and Shalaby 2012; Gabriel et al. 1994; Dwivedi 2012)
Synechococcus sp.	Cu, Ni. Pb, Cd. Cr	(Gardea-Torresdey et al. 2001)
Scenedesmus acutus *Scenedesmus quadricauda*	Cd, Zn, Cr, Hg, Pb	(Cañizares-Villanueva, González-Moreno, and Domínguez-Bocanegra 2001; Shanab, Essa, and Shalaby 2012; Travieso et al. 1999)
Scytonema	As	(Prasad, Banerjee, and Lakshmi 2006)
Spirulina maxima	Cd	(Costa and de França 2003)
Arthrospira platensis	Pb	(Wajda et al. 2017)
Tolypothrix Tolypothrix ceytonica ceytonica	Cu, Zn	(El-Bestawy 2008)
Nostoc minutum and *Anabaena spiroides*	Pb, Cd, Ni	(Al-Sherif et al. 2015)
Synechococcus sp. *Aphaanpcapsa* sp.	Cr, Pb	(de Souza et al. 2022)
Arthrospira (Spirulina) platensis	Zn	(Zinicovscaia et al. 2018)
Nostoc carneum, N. insulare, Oscillatoria geminata and Spirulina laxissima	^{134}Cs, ^{85}Sr, ^{226}Ra, ^{241}Am	(Pohl and Schimmack 2006)
Chroococcidiopsis thermalis S.M/S9	^{238}U and ^{226}Ra	(Heidari et al. 2018; Mehta et al. 2019)
Gleomargarita lithophora	^{226}Ra and ^{90}Sr	(Heidari et al. 2018; Mehta et al. 2019)
Synechocystis sp. BASO671	Cr(VI), Cd(II) and Cr(VI) + Cd(II)	(Ozturk et al. 2014)
Synechocystis sp. PCC6803	Cd^{2+}	(Shen et al. 2021)
Nostoc muscorum	Cd(II)	(Raghavan et al. 2020)

metal (such as Cd, Pb, Fe, and Cu) more efficiently than other strains. We have summarized the involvement of different strains of cyanobacteria in the bioremediation of heavy metals in Table 9.1.

9.3 METAL REMOVAL BY BIOSORPTION AND TRANSPORTATION (ADSORPTION) IN CYANOBACTERIA

Biosorption is a passive and physio-chemical process that comprises two components: solid phase biosorbent and liquid phase biosorbent. Plant biomass/microorganisms (also known as bio sorbent) are referred to as solid phase biosorbents that carry biosorption by the slow uptake of the metal from the surface reaction, and that adsorbs the liquid phase containing heavy metal, also known as sorbate (Ali Redha 2020). The leading site exists on the cell wall of cyanobacteria and acts as a biosorbent due to the presence of the possible functional groups that can interact with heavy metals such as hydroxyl group (a part of carbohydrates), carboxyl group (a part of sugar, fatty acids, and nucleic acid), ester, sulfhydryl, carbonyl, and phosphate groups that are found on various biomolecules in the cells (Cruz et al. 2020).

Metals can be removed from the environment by microorganism using a several mechanisms, including metabolically driven activities and passive metal adsorption on the charged macromolecules of the cell envelope. Exopolysaccharides (EPSs) are long chain polymers of carbohydrates secreted by various microorganisms, including cyanobacteria (Ali Redha 2020). EPS-producing cyanobacteria carry a negative charge on the external cell layer acting as chelating agents for removing charged heavy-metal ions from the aqueous solutions (Philippis, Colica, and Micheletti 2011). The negative charge on the external surface of the cyanobacteria is due to the presence of heterogeneous macromolecules such as proteins, carbohydrates, nucleic acid, and lipids that show high affinity to positively charged ions. EPS is present on the surface of the cell; thus, it prevents the metal ions from going inside the cells of microorganisms (Shen et al. 2021). Cyanobacteria have resistance of heavy-metal toxicity by increasing the Glutathione Reductase (GSH), the precursor of phyto-chelation that can sequester heavy metals such as Cd^{2+} (Shen et al. 2021). Thus, GSH plays a significant role inside the cyanobacteria in reducing the heavy metal toxicity. In addition, significant upregulation of the various genes is associated with heavy metals removal, for example, MYO_14270 gene that encodes cadmium-translocating P-type ATPase (Cad A), plays important role in preventing Cd^{2+}-induced toxicity inside the cell (Shen et al. 2021).

The heavy-metal concentration has a huge impact on the gene's expression, such as high concentration of cadmium, and causes high expression of genes such as MYO_05725 gene (responsible for the synthesis of exopolysaccharides synthesis) (Shen et al. 2021). Besides, heavy metals such as Hg^{2+} biosorption also have been reported by the two strains of the cyanobacteria such as *Spirulina platensis* and *Aphanothece flocculus* (Cain, Vannela, and Woo 2008). Therefore, biosorption has many advantages for heavy metal bioremediation, although there are very few reports of heavy metal bioremediation reducing the toxicity at a contaminated place through cyanobacteria at exceptionally large scale in real life situations. Thus, it is a big issue that should be resolved by exploiting the mechanisms of the metal removal by the cyanobacteria in a sustainable way. A report revealed that uronic acid content and monosaccharides composition such as glucuronic acid-galacturonic acid of EPS has a role in the removal of the Cr(VI) and Cd(II) mixture in *Synechocystis sp.* In addition, it was found that C-H and C-O group are available binding sites for the divalent cations (Ozturk et al. 2014). Thus, there is need to exploit bioinformatics-based approach to identify the putative genes for the biosynthesis of exopolysaccharides along with the biophysical approach for the identification and characterization of the biomolecules having a role in heavy metal biosorption such as EPS.

9.4 METAL REMOVAL BY BIOACCUMULATION IN CYANOBACTERIA

Metal ions are permeable through the cell envelope of the cyanobacteria, like other cells. In addition, cyanobacteria contain metallothioneins (MTs) and phytochelatins (PCs) (Chakdar et al. 2022). It involves the accumulation of metal ions by enzymatic sequestration. In bioaccumulation, the metal gradient inside and outside the cell initiates the process of the metal ion, where it should be higher intracellularly compared to the extracellular space (Arunakumara and Zhang 2007). There are specific channels through which these metals enter different parts of the cells, such as porins, letting the heavy metal ions translocation from extracellular to periplasmic space, and three different kinds of proteins are present for the transportation of the ions from periplasmic space to the cytoplasmic space such as channels, secondary carriers, and primary active transporter. ATP hydrolysis is required in transporting the ion; in the case of the primary active transporter, secondary carriers mediate ion transportation by protein motive force (Al-Amin et al. 2021). Metal ions transport through the channels depending on the gradient of metal ions and energy-intensive passive transport. The cytoplasm metal ion is sequestered by metal binding entities such as metallothionine, phytochelatins, and polyphosphates (Balzano et al. 2020). Then, these metal ions convert into less toxic types by the metabolic activity of the cells. Therefore, the cyanobacterial gene of the transporter can be genetically recombined to enhance the mechanism of the metal accumulation. Thus, genetic modification enhances the expression of the metal binding protein (such as cysteine-rich metallothionine [MTs], polyphosphate, and phytochelatin) that leads to an increase the storage system of the metal ion.

It has been reported that, in algal bloom HD-103, the metal uptake is achieved due to the participation of two moieties – i.e., phosphate and peptide – and, in the case of HD-104, biosorption of metal ions such as Cr $^{(+3)}$ is mediated by the carboxylate and phosphate, respectively. The main constituents of these two different algal blooms including a green algae *Cladophora sp.* in HD-103 and a blue green algae *Spirulina sp.* in HD-104. HD 103 showed the efficient adsorption of Ni (II) and Cr (III) (Doshi et al. 2008). As a result, using a consortium of two or more types of algae can increase the efficiency of metal absorption.

9.5 EFFECT OF HEAVY METAL ON CYANOBACTERIA

Several heavy metals, when present in optimum concentrations, are beneficial for the growth of cyanobacteria due to their essential roles in the physiological and metabolic process. For example, Fe, Cu, and Zn play a crucial role in various metabolic process and support the overall health and growth of cyanobacteria. On the other hand, heavy metals such as As, Cd, Pb, etc., are detrimental to the growth of cyanobacteria, even in the presence of a lower concentration (Tiwari et al. 2018). Cyanobacteria can grow in different environments, ranging from halophilic to cryophilic conditions (Gaysina, Saraf, and Singh 2019). Hence, they are ubiquitous in nature. It has been shown from the previous report that *Arthrospira platensis* can grow for three cycles of chromium and copper-containing effluents (Cepoi et al. 2020). As well as in one study, *Spirulina* is found to withstand and accumulate the polymetallic system at the recurrent event of pollution (Yadav, Dwivedi, et al. 2021). ROS generation can have both positive and negative effects on the development process of cyanobacteria; for instance, the low concentration of reactive oxygen species (ROS) aid in the development process, and a high concentration has a toxic effect (Yadav, Dwivedi, et al. 2021).

Heavy metals also cause a toxic effect in some cases such as 1 mg/L Pb concentration showing the repression of the growth of the *Cyanothece* and *Anabaena* (Choudhary et al. 2007). Heavy metal accumulation also decreases nitrogen metabolism and photosynthesis. Besides, it displays a harmful effect on the light-harvesting complex (LHC) (such as carotenoids, chlorophyll, and phycocyanin), and this is due to the surge in the level of the ROS because of the metal ions (Kaňa, Prášil, and Mullineaux 2009). For example, there is a negative effect of Cr on the photosynthetic activity in *Nostoc* (Tiwari, Patel, and Prasad 2018). Therefore, cyanobacteria evolved through various mechanisms to survive under high stress due to the heavy metal in their surroundings, such as biosorption and bioaccumulation. Moreover, cyanobacteria regulate heavy metal detoxification at the molecular level. Thus, modification of the cyanobacterial genes to enhance the survival of the cyanobacteria against heavy metals through advanced genetic engineering techniques like CRISPR will be the best option to help the cyanobacteria escape from the harmful effect of heavy metals. In the coming section, we have briefly mentioned the cyanobacterial strains' genetic engineering aspects.

9.6 ADVANTAGES OF BIOREMEDIATION BY CYANOBACTERIA

It is less expensive and highly effective when compared to conventional physical and chemical procedures. It has a much higher rate of CO_2 fixation than plants and has better advantages to fix the problems from CO_2 emission (Chisti 2007). Cyanobacteria such as thermophilic cyanobacteria such as *Synechococcus lividus, Mastigocladus laminosus, Chlorogloeopsis sp.*, and *Synechococcus aquatilis* can tolerate high temperatures (Jansson and Northen 2010). Due to their thermophilic nature, these thermophilic cyanobacteria can be used to sequester CO_2 from flue gas. Besides, this is an eco-friendly way to degrade the pollutants and convert them into simple, non-toxic components, such as CO_2 and H_2O, by microbial metabolism. It can re-enter the biogeochemical as carbon, phosphorus, and nitrogen (Pandey et al. 2005; Yadav, Gupta, et al. 2021) In addition, it has a leading advantage in sewage treatment plants (STPs) and effluent treatment plants (ETPs) that can produce biofuels and biofertilizers. It has been observed that STPs and ETPs are unaffordable in low-income countries where land is limited. Therefore, bioremediation of heavy metal in sewage system by cyanobacteria can be economical and eco-friendly (Chi et al. 2011).

9.7 CHALLENGES ASSOCIATED WITH BIOREMEDIATION VIA CYANOBACTERIA

In the case of aquatic bioremediation, one of the significant challenges is the trouble in the biomass separation from the effluent before discharge using immobilized cyanobacteria that can reduce the harvesting problem (Vijayakumar 2012). Cyanobacteria can withstand heavy metals up to a certain point, but beyond that point, heavy metals become toxic for their growth and development (Patel, Tiwari, and Prasad 2018; Tóth et al. 2012). In addition, there are several difficulties associated with isolating and cultivating of the cyanobacterial strains for bioremediation, which we will briefly describe in the following paragraphs.

Only some of the strains have the same growth rate. For better bioremediation, cyanobacteria should have a short doubling time, stable growth in the polluted areas, and the good sequestering power of heavy metal. Therefore, it is one of the challenges to find the cyanobacterial strain with all these characteristics required for the remediation (Włodarczyk et al. 2020). Recently, certain strains of *Synechococcus* were isolated, including *Synechococcus sp.* PC11901 with short doubling time, high light, and salt tolerance, along with high bioaccumulation of the biomass; the other two strains, *S. elongatus* PCC 11801 and *S. elongatus* 2973, were found to show the promising characteristics for the bioremediation (Włodarczyk et al. 2020).

The concentration of heavy-metal detoxification in lab studies and polluted sites differ drastically. The actual potential of heavy metal would cause a different impact cyanobacteria. Therefore, bioremediation may vary in natural conditions, such as contaminated industrial areas (Chakdar et al. 2022). For example, the concentration of Cr remediation used in the lab was in a lower (0.1–100 mg/L) range compared to the higher concentration of the Cr in industrial areas, which is about 240 mg/L (Cui et al. 2021). Another challenge faced for the bioremediation of HMs, in realistic scenarios, is that HMs exist together. Therefore, the selection of the cyanobacterial strain should consider the removal of the multiple heavy metals adequately from the contaminated sites (Chakdar et al. 2022). However, it is the biggest challenge to grow the highly potential algae for considerable heavy toxic-metal removal, as the affinity of the metal varies based on the variable affinity of the different methods with the metal-sequestering cyanobacteria. For example, Pb biosorption in *Gloethece sp.* is higher as compared to Cu when both these metals co-exist together, and the combination of the two metals exerts a detrimental impact on the efficiency of the biosorption of the *Gloethece* (Pereira et al. 2011).

Although cyanobacteria use for bioremediation is cost-effective, the process is much slower than conventional chemical methods. Besides, they are toxic to other aquatic lifeforms such as *Cylindrospermopsis raciborskii, Anabaena circinalis, Nodularia spumigena* CCY 9414, *Microcystis aeruginosa*, and *Planktothrix sp.* (Falconer 2004). It has been reported that the best way to use them is in dead form, as it can reduce toxicity (Doshi et al. 2008). *Spirulina platensis* is one cyanobacterium whose dead parts may reduce toxicity towards the heavy metal. According to research, the dead biomass of this cyanobacterium is quite effective at removing heavy metals such as lead, cadmium, and mercury from polluted water. The cell wall of *Spirulina platensis* is rich in negatively charged functional groups that can attach to positively charged heavy-metal ions. The biosorption process includes dead-biomass-adsorbing heavy-metal ions onto its surface, reducing their concentration in the water. As a result, *Spirulina platensis* is an efficient and environmentally friendly solution for heavy metal pollution (Vijayaraghavan and Yun 2008).

9.8 GENETIC ENGINEERING ASPECTS OF THE CYANOBACTERIA

We have discussed the challenges associated with bioremediation in the previous section – the possible ways to improve the bioremediation of heavy metals. Editing of the cyanobacteria genome in all the aspects, inducing biosorption, bioaccumulation, and a two-component system, can be done to improve the metal tolerance. Here in preceding section, we summarized the few reports related to the genetic engineering of the cyanobacterial strains to enhance heavy metal bioremediation.

9.8.1 Enhancement of Biosorption via Genome Manipulation of the EPS

Microorganisms have the potential to degrade pollutants; the major enzymes which play a vital role in pollutant degradation are hydrolases, oxidoreductases, laccase, and lyases, although this naturally occurring bioremediation has some drawbacks, including less yield and low efficiency. As we have discussed, the presence of the exopolysaccharides (EPS) on the surface of the cyanobacteria aid them in removing heavy metals (Mota et al. 2016). Thereby, EPS is found to be amenable to better tolerance against toxicity to Cd and Zn by accumulations of higher dry weight possessing approximately 40% EPS, as compared to the lower dry weight of the immature biofilms having EPS (Ivorra et al. 2000). Therefore, to enhance the efficacy of heavy-metal bioremediation, targeting the EPS through the genetic engineering is one of the strategies. Thus, to increase the metal interaction with the EPS, there is a need to identify the genes through bioinformatics and biophysical approaches. Targeting the transcriptional activators and metabolic enzymes involved in EPS synthesis serves as the potential targets to enhance these natural detoxifying agents (Chakdar et al. 2022). Therefore, modulations of the transcriptional regulators can alleviate the transcription of the target genes. The sigma factor in the cyanobacteria is a crucial master regulator to produce the EPS. Therefore, from recent reports, it has been seen that accumulation of the EPS in *Anabaena sp.* PCC7120 increases by 3.2-fold through the overexpression of the Sig J (Yoshimura et al. 2007; Srivastava et al. 2017). Another recent report suggested that the secreted EPS, also known as released polysaccharides (RPS) secretion, had augmented after deletion mutation of sigma factor (SigF) in *Synechocystis sp.* PCC 6803 (Flores et al. 2019). In addition, sigma factor plays an important role as regulator in the metabolic pathway. Thus, genome shuffling and genome editing of the cyanobacterial strain can result into the improved production of the EPS.

9.8.2 Alleviation of the Bioaccumulation of Heavy Metal by Genetic Engineering

Metal binding proteins are produced by the cyanobacteria that play a huge role in the accumulation and storage of the toxic, heavy metal (Blindauer 2011; Tsuji et al. 2004). Metallothioneins (MTs) are the cysteine-rich metal binding proteins (MBPs) that bind to the heavy metal via the thiol group (Cui et al. 2021). MT was discovered for the first time in *Synechocoystis elongates* PCC 7942 that involves in the binding of heavy metal (Blindauer 2011). Genetic mutation of the gene-encoding MTs can improve the certain metal uptake ability of the cyanobacteria; for example, the mutation in K45C of the SmtA (MT protein) can augment the capacity to uptake the Cd (II) in Syn 7942 (Saffar et al. 2015). Recently, compared to the wild-type strain, higher cadmium tolerance was found in the genetically modified *Anabaena sp.* PCC7120 in the presence of the cadmium that was associated with increased expression of MT – NmtA (Divya, Chandwadkar, and Acharya 2018). In addition, overexpression of the *alr*0975 enhances the growth of engineered *Anabaena sp.* PCC7120 – approximately 22.3% – in the presence of the $CdCl_2$ (Chaurasia et al. 2017). Protein engineering techniques such as phage-assisted continuous evolution (PACE) and error-prone, PCR-based mutation can be utilized to enhance the efficiency and effectiveness of the metal binding proteins (Morrison, Podracky, and Liu 2020). Cyanobacteria contain another metal-binding molecule known as phytochelatins (PCs), which are oligomers of glutathione and cysteine-rich peptides synthesized by PC synthase. To sequester toxic heavy metals, inhibiting PC synthase can reduce the production of PCs. Therefore, the use of advanced bioinformatics and biotechnology can aid in the efficient removal of heavy metals.

9.9 CONCLUSIONS

Rapid urbanization and industrialization have increased land, air, and water pollution. Due to anthropogenic activities such as mining and smelting, heavy metals that are found in the earth's crust have been released into the environment and pose a significant threat to living things. Bioremediation is a very effective and desirable strategy for repairing polluted settings because it uses microbial activity to clean and recycle the region. In a recent study, the efficacy of various cyanobacterial species in eliminating heavy metals from contaminated locations was comprehensively compared. Due to the numerous advantages of bioremediation, it has become a more desirable alternative to conventional methods for ecological management and contaminant control. For instance, cyanobacteria may efficiently remove heavy metals from contaminated locations through biosorption and bioaccumulation methods. Cyanobacteria are a diverse and remarkable group of microorganisms found in both aquatic and terrestrial environments. However, identifying cyanobacterial strains with high metal-sorption capabilities remains challenging, despite their widespread presence. According to our study, engineering the EPS synthesis pathway may enhance cyanobacteria's metal binding abilities. Furthermore, it has been reported that indigenous cyanobacterial strains – especially *Limnococcus sp., Nostoc muscorum,* and *Synechococcus sp.* PCC 7942 – have remarkable heavy-metal-removal efficiencies via both biosorption and bioaccumulation over a brief period; additionally, *Synechococcus sp.* PCC 7942 has been investigated for its ability to sequester the broadest possible range of heavy metal ions from water. More research is needed to boost cyanobacteria's energy and carbon availability, analyze recombinant proteins, and bring together the genetically modified strains with the circadian cycle. Such developments could enhance cyanobacteria's resilience to adverse conditions. Utilizing waste biomass for environmental cleanup can support environmentally friendly and renewable products, fostering a more robust and sustainable economy.

ACKNOWLEDGMENTS

The authors are thankful to Prof. G. Kumar, Head, Botany of Department, University of Allahabad, Prayagraj, India for providing laboratory facilities. Ravi Tandon is grateful to UGC-Faculty Recharge Programme (FRP).

BIBLIOGRAPHY

Ajayan, K.V., M. Selvaraju, and K. Thirugnanamoorthy. 2011. "Growth and Heavy Metals Accumulation Potential of Microalgae Grown in Sewage Wastewater and Petrochemical Effluents." *Pakistan Journal of Biological Sciences: PJBS* 14 (16): 805–811. https://doi.org/10.3923/pjbs.2011.805.811.

Al-Amin, Abdullah, Fahmida Parvin, Joydeep Chakraborty, and Yong Ick Kim. 2021. "Cyanobacteria Mediated Heavy Metal Removal: A Review on Mechanism, Biosynthesis, and Removal Capability." *Environmental Technology Reviews* 10 (1): 44–57. https://doi.org/10.1080/21622515.2020.1869323.

Ali Redha, Ali. 2020. "Removal of Heavy Metals from Aqueous Media by Biosorption." *Arab Journal of Basic and Applied Sciences* 27 (1): 183–193. https://doi.org/10.1080/25765299.2020.1756177.

Al-Sherif, Emad A., Mohammad S. Abd El-Hameed, Mervat A. Mahmoud, and Heba S. Ahmed. 2015. "Use of Cyanobacteria and Organic Fertilizer Mixture as Soil Bioremediation." *American-Eurasian Journal of Agricultural & Environmental Sciences* 15 (5): 794–799. https://doi.org/10.5829/idosi.aejaes.2015.15.5.93245.

Arunakumara, K.K.I.U., and Xuecheng Zhang. 2007. "Heavy Metal Bioaccumulation and Toxicity with Special Reference to Microalgae." *Journal of Ocean University of China* 7 (1): 60–64. https://doi.org/10.1007/s11802-008-0060-y.

Azizi, Seyed Naser, Abasalt Hosseinzadeh Colagar, and Seyede Maryam Hafeziyan. 2012. "Removal of Cd(II) from Aquatic System Using Oscillatoria Sp. Biosorbent." *The Scientific World Journal* 2012. https://doi.org/10.1100/2012/347053.

Balzano, Sergio, Angela Sardo, Martina Blasio, Tamara Bou Chahine, Filippo Dell'Anno, Clementina Sansone, and Christophe Brunet. 2020. "Microalgal Metallothioneins and Phytochelatins and Their Potential Use in Bioremediation." *Frontiers in Microbiology* 11 (April): 1–16. https://doi.org/10.3389/fmicb.2020.00517.

Bishop, W.M., and J.H. Rodgers. 2012. "Responses of Lyngbya Wollei to Exposures of Copper-Based Algaecides: The Critical Burden Concept." *Archives of Environmental Contamination and Toxicology* 62 (3): 403–410. https://doi.org/10.1007/s00244-011-9711-x.

Blindauer, Claudia A. 2011. "Bacterial Metallothioneins: Past, Present, and Questions for the Future." *Journal of Biological Inorganic Chemistry* 16 (7): 1011–1024. https://doi.org/10.1007/s00775-011-0790-y.

Briffa, Jessica, Emmanuel Sinagra, and Renald Blundell. 2020. "Heavy Metal Pollution in the Environment and Their Toxicological Effects on Humans." *Heliyon* 6 (9): e04691. https://doi.org/10.1016/j.heliyon.2020.e04691.

Cain, Amber, Raveender Vannela, and L. Keith Woo. 2008. "Cyanobacteria as a Biosorbent for Mercuric Ion." *Bioresource Technology* 99 (14): 6578–6586. https://doi.org/10.1016/J.BIORTECH.2007.11.034.

Cañizares-Villanueva, R.O., S. González-Moreno, and A.R. Domínguez-Bocanegra. 2001. "Growth, Nutrient Assimilation and Cadmium Removal by Suspended and Immobilized Scenedesmus Acutus Cultures: Influence of Immobilization Matrix." F. Chen, and Y. Jiang (Eds) *Algae and Their Biotechnological Potential: Proceedings of the 4th Asia-Pacific Conference on Algal Biotec*, pp. 147–161. Springer Netherlands.

Cepoi, Liliana, Inga Zinicovscaia, Ludmila Rudi, Tatiana Chiriac, Vera Miscu, Svetlana Djur, Ludmila Strelkova, and Dmitrii Grozdov. 2020. "Spirulina Platensis as Renewable Accumulator for Heavy Metals Accumulation from Multi-Element Synthetic Effluents." *Environmental Science and Pollution Research* 27 (25): 31793–31811. https://doi.org/10.1007/s11356-020-09447-z.

Chakdar, Hillol, Shobit Thapa, Amit Srivastava, and Pratyoosh Shukla. 2022. "Genomic and Proteomic Insights into the Heavy Metal Bioremediation by Cyanobacteria." *Journal of Hazardous Materials* 424 (PC): 127609. https://doi.org/10.1016/j.jhazmat.2021.127609.

Chaurasia, Neha, Yogesh Mishra, Antra Chatterjee, Ruchi Rai, Shivam Yadav, and L.C. Rai. 2017. "Overexpression of Phytochelatin Synthase (Pcs) Enhances Abiotic Stress Tolerance by Altering the Proteome of Transformed Anabaena Sp. PCC 7120." *Protoplasma* 254 (4): 1715–1724. https://doi.org/10.1007/s00709-016-1059-7.

Chi, Zhanyou, Yubin Zheng, Anping Jiang, and Shulin Chen. 2011. "Lipid Production by Culturing Oleaginous Yeast and Algae with Food Waste and Municipal Wastewater in an Integrated Process." *Applied Biochemistry and Biotechnology* 165 (2): 442–453. https://doi.org/10.1007/s12010-011-9263-6.

Chisti, Yusuf. 2007. "Biodiesel from Microalgae." *Biotechnology Advances* 25 (3): 294–306. https://doi.org/10.1016/J.BIOTECHADV.2007.02.001.

Choudhary, Meenakshi, Umesh Kumar Jetley, Mohammed Abash Khan, Sunaina Zutshi, and Tasneem Fatma. 2007. "Effect of Heavy Metal Stress on Proline, Malondialdehyde, and Superoxide Dismutase Activity in the Cyanobacterium Spirulina Platensis-S5." *Ecotoxicology and Environmental Safety* 66 (2): 204–209. https://doi.org/10.1016/J.ECOENV.2006.02.002.

Colica, Giovanni, Pier Cesare Mecarozzi, and Roberto De Philippis. 2010. "Treatment of Cr(VI)-Containing Wastewaters with Exopolysaccharide-Producing Cyanobacteria in Pilot Flow through and Batch Systems." *Applied Microbiology and Biotechnology* 87 (5): 1953–1961. https://doi.org/10.1007/s00253-010-2665-5.

Costa, Antonio Carlos Augusto da, and Francisca Pessôa de França. 2003. "Cadmium Interaction with Microalgal Cells, Cyanobacterial Cells, and Seaweeds; Toxicology and Biotechnological Potential for Wastewater Treatment." *Marine Biotechnology (New York, N.Y.)* 5 (2): 149–156. https://doi.org/10.1007/s10126-002-0109-7.

Cruz, Diogo, Vitor Vasconcelos, Guillaume Pierre, Philippe Michaud, and Cédric Delattre. 2020. "Exopolysaccharides from Cyanobacteria: Strategies for Bioprocess Development." *Applied Sciences (Switzerland)* 10 (11). https://doi.org/10.3390/app10113763.

Cui, Jinyu, Yaru Xie, Tao Sun, Lei Chen, and Weiwen Zhang. 2021. "Deciphering and Engineering Photosynthetic Cyanobacteria for Heavy Metal Bioremediation." *Science of the Total Environment* 761: 144111. https://doi.org/10.1016/j.scitotenv.2020.144111.

Divya, T.V., Pallavi Chandwadkar, and Celin Acharya. 2018. "NmtA, a Novel Metallothionein of Anabaena Sp. Strain PCC 7120 Imparts Protection against Cadmium Stress but Not Oxidative Stress." *Aquatic Toxicology* 199: 152–161. https://doi.org/10.1016/j.aquatox.2018.03.035.

Doshi, Hiren, Chetan Seth, Arabinda Ray, and I.L. Kothari. 2008. "Bioaccumulation of Heavy Metals by Green Algae." *Current Microbiology* 56 (3): 246–255. https://doi.org/10.1007/s00284-007-9070-z.

Dwivedi, Seema. 2012. "Bioremediation of Heavy Metal by Algae : Current and Future Perspective." *Journal of Advanced Laboratory Research in Biology* 3 (3): 3–7.

El-Bestawy, Ebtesam. 2008. "Treatment of Mixed Domestic-Industrial Wastewater Using Cyanobacteria." *Journal of Industrial Microbiology and Biotechnology* 35 (11): 1503–1516. https://doi.org/10.1007/s10295-008-0452-4.

Falconer, I.R. (2004). *Cyanobacterial Toxins of Drinking Water Supplies* (1st ed.). CRC Press. https://doi.org/10.1201/9780203022870.

Fawzy, Mustafa A., and Abdel Kareem S.H. Mohamed. 2017. "Bioremediation of Heavy Metals from Municipal Sewage by Cyanobacteria and Its Effects on Growth and Some Metabolites of Beta Vulgaris." *Journal of Plant Nutrition* 40 (18): 2550–2561. https://doi.org/10.1080/01904167.2017.1380822.

Flores, Carlos, Marina Santos, Sara B. Pereira, Rita Mota, Federico Rossi, Roberto De Philippis, Narciso Couto, et al. 2019. "The Alternative Sigma Factor SigF is a Key Player in the Control of Secretion Mechanisms in Synechocystis Sp. PCC 6803." *Environmental Microbiology* 21 (1): 343–359. https://doi.org/10.1111/1462-2920.14465.

Gabriel, J., M. Mokrejš, J. Bílý, and P. Rychlovský. 1994. "Accumulation of Heavy Metals by Some Wood-Rotting Fungi." *Folia Microbiologica* 39 (2): 115–118. https://doi.org/10.1007/BF02906805.

Gardea-Torresdey, Jorge Luis, Socorro Arteaga, Kirk J. Tiemann, Russell Chianelli, Nicholas Pingitore, and William Mackay. 2001. "Absorption of Copper(II) by Creosote Bush (Larrea Tridentata): Use of Atomic and x-Ray Absorption Spectroscopy." *Environmental Toxicology and Chemistry* 20 (11): 2572–2579. https://doi.org/10.1002/etc.5620201124.

Gaysina, Lira A., Aniket Saraf, and Prashant Singh. 2019. "Chapter 1—Cyanobacteria in Diverse Habitats." A.K. Mishra, D.N. Tiwari, and A.N. Rai (Eds) *Cyanobacteria*, 1–28. Academic Press. https://doi.org/10.1016/B978-0-12-814667-5.00001-5.

Goswami, Smita, Omega L. Diengdoh, Mayashree B. Syiem, Kannan Pakshirajan, and Mothe Gopi Kiran. 2015. "Zn(Ii) and Cu(Ii) Removal by Nostoc Muscorum: A Cyanobacterium Isolated from a Coal Mining Pit in Chiehruphi, Meghalaya, India." *Canadian Journal of Microbiology* 61 (3): 209–215. https://doi.org/10.1139/cjm-2014-0599.

Heidari, Fatemeh, Hossein Riahi, Mahmoud Reza Aghamiri, Farideh Zakeri, Zeinab Shariatmadari, and Tomáš Hauer. 2018. "226Ra, 238U and Cd Adsorption Kinetics and Binding Capacity of Two Cyanobacterial Strains Isolated from Highly Radioactive Springs and Optimal Conditions for Maximal Removal Effects in Contaminated Water." *International Journal of Phytoremediation* 20 (4): 369–377. https://doi.org/10.1080/15226514.2017.1393392.

Igiri, Bernard E., Stanley I.R. Okoduwa, Grace O. Idoko, Ebere P. Akabuogu, Abraham O. Adeyi, and Ibe K. Ejiogu. 2018. "Toxicity and Bioremediation of Heavy Metals Contaminated Ecosystem from Tannery Wastewater: A Review." 2018. https://doi.org/10.1155/2018/2568038.

Ivorra, Núria, Simone Bremer, Helena Guasch, Michiel H.S. Kraak, and Wim Admiraal. 2000. "Differences in the Sensitivity of Benthic Microalgae to Zn and Cd Regarding Biofilm Development and Exposure History." *Environmental Toxicology and Chemistry* 19 (5): 1332. https://doi.org/10.1897/1551-5028(2000)019<1332:ditsob>2.3.co;2.

Jansson, Christer, and Trent Northen. 2010. "Calcifying Cyanobacteria—The Potential of Biomineralization for Carbon Capture and Storage." *Current Opinion in Biotechnology* 21 (3): 365–371. https://doi.org/10.1016/J.COPBIO.2010.03.017.

Kaňa, Radek, Ondřej Prášil, and Conrad W. Mullineaux. 2009. "Immobility of Phycobilins in the Thylakoid Lumen of a Cryptophyte Suggests That Protein Diffusion in the Lumen is Very Restricted." *FEBS Letters* 583 (4): 670–674. https://doi.org/10.1016/j.febslet.2009.01.016.

Kiran, Bala, and Anubha Kaushik. 2008. "Cyanobacterial Biosorption of Cr(VI): Application of Two Parameter and Bohart Adams Models for Batch and Column Studies." *Chemical Engineering Journal* 144 (3): 391–399. https://doi.org/10.1016/j.cej.2008.02.003.

Kumar, Dhananjay, Bhanu Prakash, Lalit K. Pandey, and J.P. Gaur. 2010. "Sorption of Paraquat and 2,4-D by an Oscillatoria Sp.-Dominated Cyanobacterial Mat." *Applied Biochemistry and Biotechnology* 160 (8): 2475–2485. https://doi.org/10.1007/s12010-009-8710-0.

Mehta, Neha, Karim Benzerara, Benjamin D. Kocar, and Virginie Chapon. 2019. "Sequestration of Radionuclides Radium-226 and Strontium-90 by Cyanobacteria Forming Intracellular Calcium Carbonates." *Environmental Science & Technology* 53 (21): 12639–12647. https://doi.org/10.1021/acs.est.9b03982.

Micheletti, E., G. Colica, C. Viti, P. Tamagnini, and R. De Philippis. 2008. "Selectivity in the Heavy Metal Removal by Exopolysaccharide-Producing Cyanobacteria." *Journal of Applied Microbiology* 105 (1): 88–94. https://doi.org/10.1111/j.1365-2672.2008.03728.x.

Mohamed, Zakaria A. 2001. "Removal of Cadmium and Manganese by a Non-Toxic Strain of the Freshwater Cyanobacterium Gloeothece Magna." *Water Research* 35 (18): 4405–4409. doi: 10.1016/S0043-1354(01)00160-9.

Morrison, Mary S., Christopher J. Podracky, and David R. Liu. 2020. "The Developing Toolkit of Continuous Directed Evolution." *Nature Chemical Biology* 16 (6): 610–619. https://doi.org/10.1038/s41589-020-0532-y.

Mota, Rita, Federico Rossi, Luisa Andrenelli, Sara Bernardes Pereira, Roberto De Philippis, and Paula Tamagnini. 2016. "Released Polysaccharides (RPS) from Cyanothece Sp. CCY 0110 as Biosorbent for Heavy Metals Bioremediation: Interactions between Metals and RPS Binding Sites." *Applied Microbiology and Biotechnology* 100 (17): 7765–7775. https://doi.org/10.1007/s00253-016-7602-9.

Nowicka, Beatrycze. 2022. "Heavy Metal–Induced Stress in Eukaryotic Algae—Mechanisms of Heavy Metal Toxicity and Tolerance with Particular Emphasis on Oxidative Stress in Exposed Cells and the Role of Antioxidant Response." *Environmental Science and Pollution Research* 29 (12): 16860–16911. https://doi.org/10.1007/s11356-021-18419-w.

Okolo, N.V, E.A. Olowolafe, I. Akawu, and S.I.R. Okoduwa. 2016. "Effects of Industrial Effluents on Soil Resources in Challawa Industrial Area, Kano, Nigeria." *Journal of Global Ecology and Environment* 5 (1): 1–10.

Ozturk, Sahlan, Belma Aslim, Zekiye Suludere, and Sema Tan. 2014. "Metal Removal of Cyanobacterial Exopolysaccharides by Uronic Acid Content and Monosaccharide Composition." *Carbohydrate Polymers* 101 (1): 265–271. https://doi.org/10.1016/J.CARBPOL.2013.09.040.

Pandey, Kapil D., Prabhu N. Shukla, Deen D. Giri, and Ajai K. Kashyap. 2005. "Cyanobacteria in Alkaline Soil and the Effect of Cyanobacteria Inoculation with Pyrite Amendments on Their Reclamation." *Biology and Fertility of Soils* 41 (6): 451–457. https://doi.org/10.1007/s00374-005-0846-7.

Pandi, M., V. Shashirekha, and Mahadeswara Swamy. 2009. "Bioabsorption of Chromium from Retan Chrome Liquor by Cyanobacteria." *Microbiological Research* 164 (4): 420–428. https://doi.org/10.1016/j.micres.2007.02.009.

Patel, Anuradha, Sanjesh Tiwari, and Sheo Mohan Prasad. 2018. "Toxicity Assessment of Arsenate and Arsenite on Growth, Chlorophyll a Fluorescence and Antioxidant Machinery in Nostoc Muscorum." *Ecotoxicology and Environmental Safety* 157: 369–379. https://doi.org/10.1016/j.ecoenv.2018.03.056.

Pereira, Sara, Ernesto Micheletti, Andrea Zille, Arlete Santos, Pedro Moradas-Ferreira, Paula Tamagnini, and Roberto De Philippis. 2011. "Using Extracellular Polymeric Substances (EPS)-Producing Cyanobacteria for the Bioremediation of Heavy Metals: Do Cations Compete for the EPS Functional Groups and Also Accumulate inside the Cell?" *Microbiology* 157 (2): 451–458. https://doi.org/10.1099/mic.0.041038-0.

Philippis, Roberto De, Giovanni Colica, and Ernesto Micheletti. 2011. "Exopolysaccharide-Producing Cyanobacteria in Heavy Metal Removal from Water: Molecular Basis and Practical Applicability of the Biosorption Process." *Applied Microbiology and Biotechnology* 92 (4): 697–708. https://doi.org/10.1007/s00253-011-3601-z.

Philippis, Roberto De, Raffaella Paperi, Claudio Sili, and Massimo Vincenzini. 2003. "Assessment of the Metal Removal Capability of Two Capsulated Cyanobacteria, Cyanospira Capsulata and Nostoc PCC7936." *Journal of Applied Phycology* 15 (2–3): 155–161. https://doi.org/10.1023/A:1023889410912.

Pohl, Peter, and Wolfgang Schimmack. 2006. "Adsorption of Radionuclides (134Cs, 85Sr, 226Ra, 241Am) by Extracted Biomasses of Cyanobacteria (Nostoc Carneum, N. Insulare, Oscillatoria Geminata and Spirulina Laxis-Sima) and Phaeophyceae (Laminaria Digitata and L. Japonica; Waste Products from Alginat." *Journal of Applied Phycology* 18 (2): 135–143. https://doi.org/10.1007/s10811-006-9084-0.

Prasad, Bhim Bali, Souri Banerjee, and Dhana Lakshmi. 2006. "An AlgaSORB Column for the Quantitative Sorption of Arsenic(III) from Water Samples." *Water Quality Research Journal of Canada* 41 (2): 190–197. https://doi.org/10.2166/wqrj.2006.022.

Raghavan, Prashanth S., Akhilesh A. Potnis, Kaustava Bhattacharyya, Darshana A. Salaskar, and Hema Rajaram. 2020. "Axenic Cyanobacterial (Nostoc Muscorum) Biofilm as a Platform for Cd(II) Sequestration from Aqueous Solutions." *Algal Research* 46: 101778. https://doi.org/10.1016/j.algal.2019.101778.

Rai, Prabhat Kumar, and B.D. Tripathi. 2007. "Removal of Heavy Metals by the Nuisance Cyanobacteria Microcystis in Continuous Cultures: An Eco-Sustainable Technology." *Environmental Sciences* 4 (1): 53–59. https://doi.org/10.1080/15693430601164956.

Saffar, B., A. Mehri Ghahfarrokhi, K. Mahnam, and M. Mobini-Dehkordi. 2015. "Improvement of Cd2+ Uptake Ability of SmtA Protein by Lys/Cys Mutation; Experimental and Theoretical Studies." *Journal of Biomolecular Structure and Dynamics* 33 (11): 2347–2359. https://doi.org/10.1080/07391102.2015.1054431.

Shanab, Sanaa, Ashraf Essa, and Emad Shalaby. 2012. "Bioremoval Capacity of Three Heavy Metals by Some Microalgae Species (Egyptian Isolates)." *Plant Signaling and Behavior* 7 (3): 392–399. https://doi.org/10.4161/psb.19173.

Shen, Li, Ran Chen, Junjun Wang, Ling Fan, Linlin Cui, Yejuan Zhang, Jinju Cheng, Xueling Wu, Jiaokun Li, and Weimin Zeng. 2021. "Biosorption Behavior and Mechanism of Cadmium from Aqueous Solutions by: Synechocystis Sp. PCC6803." *RSC Advances* 11 (30): 18637–18650. https://doi.org/10.1039/d1ra02366g.

Souza, Priscila Oliveira de, Vanderleia Sinhor, Marcelo Guerreiro Crizel, Natanael Pires, Pedro José Sanches Filho, Rochele Sogari Picoloto, Fabio Andrei Duarte, Claudio Martin Pereira Pereira, and Marcia Foster Mesko. 2022. "Bioremediation of Chromium and Lead in Wastewater from Chemistry Laboratories Promotes by Cyanobacteria." *Bioresource Technology Reports* 19 (September): 101161. https://doi.org/10.1016/J.BITEB.2022.101161.

Srivastava, Amit, Klaus Brilisauer, Ashutosh K. Rai, Anand Ballal, Karl Forchhammer, and Anil K. Tripathi. 2017. "Down-Regulation of the Alternative Sigma Factor SIGJ Confers a Photoprotective Phenotype to Anabaena PCC 7120." *Plant and Cell Physiology* 58 (2): 287–297. https://doi.org/10.1093/pcp/pcw188.

Tiwari, Sanjesh, Parul Parihar, Anuradha Patel, Rachana Singh, and Sheo Mohan Prasad. 2018. *Metals in Cyanobacteria: Physiological and Molecular Regulation. Cyanobacteria: From Basic Science to Applications*. Elsevier Inc. https://doi.org/10.1016/B978-0-12-814667-5.00013-1.

Tiwari, Sanjesh, Anuradha Patel, and Sheo Mohan Prasad. 2018. "Kinetin Alleviates Chromium Toxicity on Growth and PS II Photochemistry in Nostoc Muscorum by Regulating Antioxidant System." *Ecotoxicology and Environmental Safety* 161 (October): 296–304. https://doi.org/10.1016/J.ECOENV.2018.05.052.

Tóth, Tünde, Ottó Zsiros, Mihály Kis, Gyozo A.A. Garab, and László Kovács. 2012. "Cadmium Exerts Its Toxic Effects on Photosynthesis via a Cascade Mechanism in the Cyanobacterium, Synechocystis PCC 6803." *Plant, Cell and Environment* 35 (12): 2075–2086. https://doi.org/10.1111/j.1365-3040.2012.02537.x.

Travieso, L., R.O. Cañizares, R. Borja, F. Benítez, A.R. Domínguez, R. Dupeyrón, and Y.V. Valiente. 1999. "Heavy Metal Removal by Microalgae." *Bulletin of Environmental Contamination and Toxicology* 62 (2): 144–151. https://doi.org/10.1007/s001289900853.

Tsuji, Naoki, Shingo Nishikori, Osamu Iwabe, Kentaro Shiraki, Hitoshi Miyasaka, Masahiro Takagi, Kazumasa Hirata, and Kazuhisa Miyamoto. 2004. "Characterization of Phytochelatin Synthase-like Protein Encoded by Alr0975 from a Prokaryote, Nostoc Sp. PCC 7120." *Biochemical and Biophysical Research Communications* 315 (3): 751–755. https://doi.org/10.1016/j.bbrc.2004.01.122.

Vijayakumar, Subramaniyan. 2012. "Potential Applications of Cyanobacteria in Industrial Effluents—A Review." *Journal of Bioremediation and Biodegradation* 3. https://doi.org/10.4172/2155-6199.1000154.

Vijayaraghavan, K., and Yeoung-Sang Yun. 2008. "Bacterial Biosorbents and Biosorption." *Biotechnology Advances* 26 (3): 266–291. https://doi.org/10.1016/j.biotechadv.2008.02.002.

Wajda, Łukasz, Aleksandra Duda-Chodak, Tomasz Tarko, and Paweł Kamiński. 2017. "Application of Principal Component Analysis for the Optimisation of Lead(II) Biosorption." *World Journal of Microbiology and Biotechnology* 33 (10): 1–10. https://doi.org/10.1007/s11274-017-2358-7.

Warjri, Susan Moore, and Mayashree B. Syiem. 2018. "Analysis of Biosorption Parameters, Equilibrium Isotherms and Thermodynamic Studies of Chromium (VI) Uptake by a Nostoc Sp. Isolated from a Coal Mining Site in Meghalaya, India." *Mine Water and the Environment* 37 (4): 713–723. https://doi.org/10.1007/s10230-018-0523-3.

Włodarczyk, Artur, Tiago Toscano Selão, Birgitta Norling, and Peter J. Nixon. 2020. "Newly Discovered Synechococcus Sp. PCC 11901 Is a Robust Cyanobacterial Strain for High Biomass Production." *Communications Biology* 3 (1). https://doi.org/10.1038/s42003-020-0910-8.

Yadav, Ajit Pratap Singh, Vinay Dwivedi, Satyendra Kumar, Anamika Kushwaha, Lalit Goswami, and Bezawada Sridhar Reddy. 2021. "Cyanobacterial Extracellular Polymeric Substances for Heavy Metal Removal: A Mini Review." *Journal of Composites Science* 5 (1): 1–18. https://doi.org/10.3390/jcs5010001.

Yadav, Priya, Rajan Kumar Gupta, Rahul Prasad Singh, Praveen Kumar Yadav, Jalaluddin, Arun Kumar Patel, and Kapil Deo Pandey. 2021. "Role of Cyanobacteria in Green Remediation." *Sustainable Environmental Clean-Up: Green Remediation* (August): 187–210. https://doi.org/10.1016/B978-0-12-823828-8.00009-8.

Yoshimura, Hidehisa, Shinobu Okamoto, Yoichi Tsumuraya, and Masayuki Ohmori. 2007. "Group 3 Sigma Factor Gene, SigJ, a Key Regulator of Desiccation Tolerance, Regulates the Synthesis of Extracellular Polysaccharide in Cyanobacterium Anabaena Sp. Strain PCC 7120." *DNA Research* 14 (1): 13–24. https://doi.org/10.1093/dnares/dsm003.

Zinicovscaia, Inga, Nikita Yushin, Margarita Shvetsova, and Marina Frontasyeva. 2018. "Zinc Removal from Model Solution and Wastewater by Arthrospira (Spirulina) Platensis Biomass." *International Journal of Phytoremediation* 20 (9): 901–908. https://doi.org/10.1080/15226514.2018.1448358.

10 Metal Recovery from Lithium-Ion Batteries Using Bioleaching

A Greener Approach

Prashant Ram Jadhao, Shally Gupta, K. K. Pant and K. D. P. Nigam

10.1 INTRODUCTION

Burgeoning global energy needs have motivated the quest for alternative sources of sustainable energy and storage technologies [1]. Nowadays, rechargeable batteries are being considered as an alternative energy storage technology and are experiencing unparalleled, rapid development due to the integration of renewable resources and electric transportation [2]. The global battery market demand breakdown in terms of applications between 2018 and 2030 is presented in Figure 10.1. The market is expected to increase by 25% per year to 2.6 GWh in 2030. The application of batteries in electric transportation and electricity grids is the major driver for this rapid growth in global battery demand. The demand for electric transport is expected to see a massive energy growth from 142 to 2,333 GWh, which is mainly attributed to EV passenger cars [3].

Among the rechargeable batteries, lithium-ion batteries (LIBs) are extensively employed for various equipment and electric vehicles (EVs), owing to superior properties – i.e., high voltage window, energy density, large operating range of temperature, and long cycle-life [4]. The production of LIBs reached 5.86 billion in 2012 from 2.05 billion in 2005, and the global LIBs demand is estimated to touch $95 billion by 2025, growing to 16.5% annually [5, 6]. Apart from the demand for traditional consumer electronics, other evolving fields such as energy storage and electric transportation also require huge LIBs volumes [7]. The demand for LIB packs only for the EV market is projected to touch 1 million in 2030 [8]. LIBs application in EVs and continuous upgradation of electronic devices will certainly result in a significant number of end-of-life (EoL) LIBs, owing to the shorter life-periods. The weight of EoL LIBs is expected to reach 464,000 tons in 2025 from 10,700 tons in 2012, with an increment of 59% [9]. Therefore, it is crucial to consider the recycling of EoL LIBs to mitigate the future demand for resource material and to reduce the harmful impact of the unregulated discharge of EoL batteries to the environment.

The cathode of LIBs usually comprises of an Al foil coated with a layer of organic binder and cathode-active material – i.e., $LiCoO_2$ (LCO), $LiMn_2O4$ (LMO), $LiNiO_2$ (LNO), and $LiNixCoyMnzO_2$ (LNCM) [10]. The heavy metals and electrolytes present in the EoL LIBs are harmful to the environment and mankind [11]. The unregulated disposal of spent LIBs via landfilling can cause severe environmental threats, such as soil and groundwater pollution due to the leaching of toxic materials. Moreover, EoL LIBs also contain substantial amounts of valuable metals. In general, EoL LIBs include 5 – 20 wt% cobalt (Co), 5 – 7 wt% lithium (Li), 5 – 7 wt% nickel (Ni), etc. [12]. Thus, the recycling of EoL LIBs is a necessity for environmental remediation and also to mitigate future metals supply risk.

Pyrometallurgy, hydrometallurgy, and biohydrometallurgy are the primarily employed techniques for the recycling of LIBs. In this context, pyrometallurgical processes require heating of the

 DOI: 10.1201/9781003391487-10

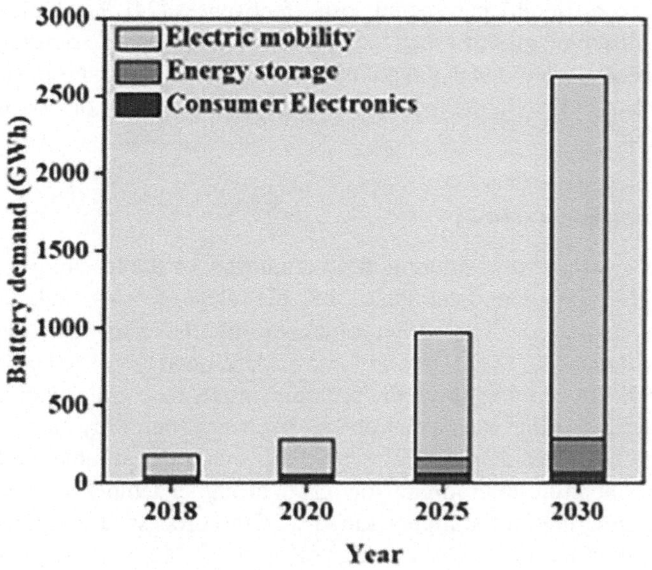

FIGURE 10.1 Trends of global battery demand in terms of the application over the years (2018 to 2030).

battery waste at elevated temperatures (1200–1400 °C) to recover the metals in the form of immiscible molten layers resulting in huge energy consumption and emission of obnoxious gases [13]. Although the pyrometallurgy process has already been scaled up, its wide application at the industrial scale remains rather limited due to some inherent drawbacks. However, CO_2 generation and the loss of Al, Mn, and Li in the slag are the major downsides of pyrometallurgy [14]. Hydrometallurgy processes provide an efficient alternative to recover precious metals [15]. However, the generation of acidic effluents during the hydrometallurgical process creates secondary pollution; thus, the use of acidic leachants (HCl, H_2SO_4, and HNO_3) is not considered environment-friendly.

The biohydrometallurgy, or bioleaching, process follows a mechanism similar to that of the hydrometallurgy process, except it involves microorganisms that produce acids through metabolism to extract metals [16]. Nowadays, bioleaching is considered an emerging technology for treating the spent LIBs waste through environmentally friendly routes. In this microbe-assisted process, the lixiviant, or leaching agent, is biologically produced by microorganisms, which are then further utilized to dissolve the metals via bio-oxidation [17]. Bioleaching is mediated by different types of microbes showing the ability to secrete organic or inorganic acids, which enhance oxidation/reduction or complex formation [18]. Chemolithoautotrophic bacteria and heterotrophic fungi are two major groups of microorganisms used for metal recovery from LIBs [19]. Among various categories of microbes, acidophilic S/Fe oxidizing microbes are primarily employed during bioleaching [20]. This chapter focuses on the recent advancements in the bioleaching process to recycle valuable metals from EoL LIBs. Further, bioleaching mechanisms – i.e., contact and non-contact mechanisms – have been explicitly discussed.

10.2 BIOLEACHING

Bioleaching is a greener and promising approach that uses different microorganisms such as bacteria, fungi, and archaea for metal recovery from different metal sources [21–23]. These microorganisms produce different leaching agents like organic acid, H_2SO_4, etc., via their metabolism, which then reacts with metals and causes their extraction. Recently, bioleaching has received great attention owing to its eco-friendly nature and cost-effectiveness [24]. The bioleaching process uses microorganisms of different groups, such as chemolithoautotrophic bacteria and heterotrophic fungi [25, 26]. Microorganism-aided metal recovery from EoL LIBs is a promising environmentally friendly approach that will help in the resource recovery, management of EoL LIBs, and reduction in ecological pollution.

10.2.1 Bacteria for Bioleaching

The S/Fe oxidizing bacteria have commonly been employed for the leaching of metals. The bacterial leaching takes place in acidic conditions – i.e., pH range of 1–3 – and, at this condition, the metal ions remain in the solution. These microorganisms use the energy generated by the oxidation of Fe^{2+} and elemental/reduced S to Fe^{3+} and sulfuric acid, respectively. The Fe^{3+} and H_2SO_4 promote the extraction of metals. In addition, these microorganisms also use atmospheric CO_2 as a source of carbon. The microorganisms that use Fe and/or S as an energy source are categorized as mesophilic and thermophilic [27, 28]. The mesophilic microorganisms grow at temperature less than 40 °C [20, 25, 29, 30]. Thermophilic microorganisms grow at higher temperatures (40–80 °C) and are beneficial for bioleaching due to the higher kinetics of the process at elevated temperatures. The moderately thermophilic microorganisms can grow up to 50 °C, while the extreme thermophilic microorganisms can grow till 75–80 °C [31–33].

Naseri et al. investigated metal recovery from spent coin cells (SCCs) using *Acidithiobacillus thiooxidans* (*A. thiooxidans*) [34]. The authors reported that a logarithmic phase of growth for *A. thiooxidans* happened after incubation for 10 days, and the maximum cell count was observed to be around 5.5 x 10^8 cells mL^{-1}. Also, pH was reduced to 1.4 due to bacterial activity and acid production in the logarithmic phase – i.e., 10 days after the incubation. The conversion of elemental sulfur to sulfate ions as shown in equation (10.1) provides the energy to *A. thiooxidans* [35]. The two-step bioleaching was employed as the single-step bioleaching of SCCs didn't show any cell growth. It has been reported that the growth of microbes is more supported in the absence of waste [36]. The metal extraction was performed in the second step, and the maximum recovery of metals was measured after 11 days. In addition, the authors reported that the enhanced pulp density resulted in the lower extraction of metals due to insufficient concentration of bioleaching agents in the solution. The authors considered 30 g/L pulp density as optimum which resulted in 99% Li, 60% Co, and 20% Mn recovery after 11 days of bioleaching.

$$S^0(s)+1.5\,O_2(aq)+H_2O(aq)\rightarrow 2H^+(aq)+SO_4^{2-}(aq) \tag{10.1}$$

Similarly, Naseri et al. examined Li, Co, and Mn recovery from SCCs using *Acidithiobacillus ferrooxidans* (*A. ferrooxidans*) and reported that 40 g/L of pulp density is optimum for metal recovery [37]. At this pulp density, the maximum recovery of Li, Co, and Mn was around 100%, 88%, and 20%, respectively. The authors further investigated the kinetics of metal extraction and reported that the diffusion of ferric ions is a key factor for metal extraction. Roy et al. also investigated metal recovery using *A. ferrooxidans* from EoL LIBs [38]. The *A. ferrooxidans* were cultured in a media having a composition as shown in Table 10.1. The higher concentration of $FeSO_4$ was used for greater production of sulfuric acid and also to maintain a higher concentration of Fe^{3+} in culture media. Authors reported Ni, Mn, Co, and Li recovery with efficiencies of 90%, 92%, 82%, and 89%, respectively, from spent LIBs at 72 h and a pulp density of 100 g/L. In this study, the higher extraction of metals can be due to the reuse of the bacterial culture for three cycles and the higher concentration of H_2SO_4 and Fe^{3+} in *A. ferrooxidans* culture.

In another study, Roy et al. examined Co and Li recovery from $LiCoO_2$-based EoL LIBs using *A. ferrooxidans* [39]. The pH of the culture media was maintained at 2 using the 5 N H_2SO_4. The bioleaching was performed at 160 rpm and 30 °C. The replenishing study was also carried out and till three cycles bioleaching experiments were performed. The authors majorly examined the effect of

TABLE 10.1
Composition of the Modified Culture Media

S.N.	Compound	Amount (g/L)
1	$(NH_4)_2SO_4$	0.50
2	KCl	0.05
3	K_2HPO_4	0.50
4	$MgSO_4$	0.50
5	$Ca(NO_3)_2$	0.01
6	$FeSO_4$	150.00
	pH—2.0	

TABLE 10.2
Extraction of Metals Using *A. ferrooxidans* (Reproduced with Permission from [39])

		Metal extraction (%)			
$FeSO_4$	Metal	1st extraction 24 h	2nd extraction 24 h	3rd extraction 24 h	Total extraction (%) in 72 h
45 g/L	Co	22.76 ± 0.19	10.32 ± 0.19	7.53 ± 0.13	40.61
	Li	19.33 ± 0.06	13.94 ± 0.02	9.65 ± 0.02	42.92
90 g/L	Co	35.75 ± 0.42	14.53 ± 0.23	4.32 ± 0.04	54.59
	Li	29.11 ± 0.11	17.56 ± 0.10	7.94 ± 0.01	54.61
150 g/L	Co	45.24 ± 0.31	33.06 ± 0.40	15.72 ± 0.04	94.02
	Li	28.48 ± 0.40	21.70 ± 0.04	10.12 ± 0.01	60.30

pulp density on metal extraction and reported that metal extraction decreases with increasing pulp density. This is owing to the high toxicity of metals, lesser dissolved O_2, and insufficient distribution of air at higher pulp density [37]. The higher pulp density increases the viscosity and concentration of the culture medium that limits the O_2 penetration into the culture medium. This negatively affects the mechanism of the metal leaching, which reduces the leaching efficiency [40]. The biogenic concentration of H_2SO_4 is an important parameter for metal extraction from LIBs. Therefore, it is important to have a higher concentration of $FeSO_4$ in the culture medium to increase the H_2SO_4 concentration. Therefore, the authors increased the concentration of $FeSO_4$ from 45 to 90 and 150 g/L. Further, the authors reported that the extraction of Co increased from 40.61% to 94.02%, and for Li, extraction increased from 42.92% to 60.30% at 100 g/L pulp density and a leaching time of 72 h. Table 10.2 gives the details about the extraction of Co and Li using *A. ferrooxidans* using three cycles of replenishing bacteria culture medium.

Heydarian et al. used a mixed culture of *A. ferrooxidans* and *A. thiooxidans* for metal recovery [41]. Initially, a separate culture medium was used for the growth of both bacteria. Table 10.3 shows the composition of the culture media used for the growth of the bacteria. The authors further investigated the effect of the inoculum ratio of *A. ferrooxidans to A. thiooxidans (A.f/A.t)* and also determined the best process conditions for metal extraction. In the mixed bacteria culture, Fe^{2+} ions provide energy for the growth of bacteria and oxidize to Fe^{3+} ions. Similarly, S is also a source of energy that oxidizes to H_2SO_4, and O_2 acts as an electron acceptor in a process of sulfur bioxidation. In the mixed culture, *A. thiooxidans* assists in creating the required acidic conditions for the growth of *A. ferrooxidans*. The bacteria help in balancing the Fe^{2+}/Fe^{3+} cycle to maintain high ORP, generate H_2SO_4 for providing proton, and maintain a high concentration of Fe^{3+} ions. *A. ferrooxidans* help to increase the ORP, and *A. thiooxidans* increase H_2SO_4 production. The

TABLE 10.3
The Composition of Culture Media Used for the Growth of Microbes

A. ferrooxidans		A. thiooxidans		Mixed bacteria	
Compound	**Amount (g/L)**	**Compound**	**Amount (g/L)**	**Compound**	**Amount (g/L)**
$(NH_4)_2SO_4$	3	$(NH_4)_2SO_4$	2	$(NH_4)_2SO_4$	4
KCl	0.1	KCl	0.1	KCl	0.1
K_2HPO_4	0.5	$K_2HPO_4{\cdot}3H_2O$	0.25	K_2HPO_4	0.5
$MgSO_4.7H_2O$	0.5	$MgSO_4.7H_2O$	0.25	$MgSO_4.7H_2O$	0.5
$Ca(NO_3)_2$	0.01	S^0	10	$Ca(NO_3)_2$	0.01
$FeSO_4.7H_2O$	44.2			$FeSO_4.7H_2O$	22.1
				$CaCl_2.2H_2O$	0.13
				S^0	5

authors reported 99.2% Li, 50.4% Co, and 89.4% Ni recovery at the optimized process conditions – i.e., pH = 1.5, $FeSO_4$ = 36.7 g/L, sulfur = 5 g/L, pulp density = 40 g/L, and *A.f/A.t* ratio = 3.2. In LIBs, the lithium is in the form of $LiCoO_2$ and $LiNiO_2$, and the leaching of Li can be expressed using the following equations

$$4LiCoO_{2(s)} + 6H_2SO_{4(aq)} \rightarrow 2Li_2SO_{4(aq)} + 4CoSO_{4(aq)} + 6H_2O_{(aq)} + O_{2(g)} \tag{10.2}$$

$$2FeSO_4 + 2LiCoO_2 + 4H_2SO_4 \rightarrow Fe_2(SO_4)_3 + 2CoSO_4 + Li_2SO_4 + 4H_2O \tag{10.3}$$

$$4LiNiO_{2(s)} + 6H_2SO_{4(aq)} \rightarrow 2Li_2SO_{4(aq)} + 4NiSO_{4(aq)} + 6H_2O_{(aq)} + O_{2(g)} \tag{10.4}$$

10.2.2 Heterotrophic Fungi

It is a type of bioleaching where organic carbon is the source of energy for microorganisms. The metabolic process of organic carbon generates organic acids as by-products. Organic acids react with metals and are responsible for metal extraction. Moreover, there is also a generation of non-acidic complexing agents from protein catabolism. These complexing agents can be used in the alkaline system for metal extraction. Heterotrophic bioleaching can be employed in the wider pH range and moderate alkaline system. Metal recovery from electronic waste is mainly carried out using cyanide- and organic-acid-producing microorganisms. The microorganisms producing cyanide are majorly used to recover precious metals, whereas organic-acid-producing microorganisms are used to recover base and heavy metals.

Bahaloo-Horeh and Mousavi have examined metal recovery using *Aspergillus niger (A. niger)* [42]. The authors optimized the generation of organic acids by investigating the effect and interactions between various process parameters using response surface methodology. The pathway for the synthesis of organic acids by *A. niger* is shown in Figure 10.2. The synthesis of citric acid and malic acid takes place in the Krebs cycle. In addition, the malic acid can also be formed via an anaplerotic reaction from pyruvate. Oxalic acid is formed by the hydrolysis of oxaloacetate. The synthesis of oxaloacetate occurs in the Krebs cycle as well as from pyruvate. The oxalic acid hampers the efficiency of metal extraction, specifically of Co and Ni from LIBs; thus, authors tried to minimize the production of oxalic acid. The excretion of gluconic acid takes place at a high concentration of glucose, which depends upon the pH. Gluconic acid production happens at higher pH than the value of pH apt to produce citric acid and lower than the value suitable for the excretion of oxalic acid.

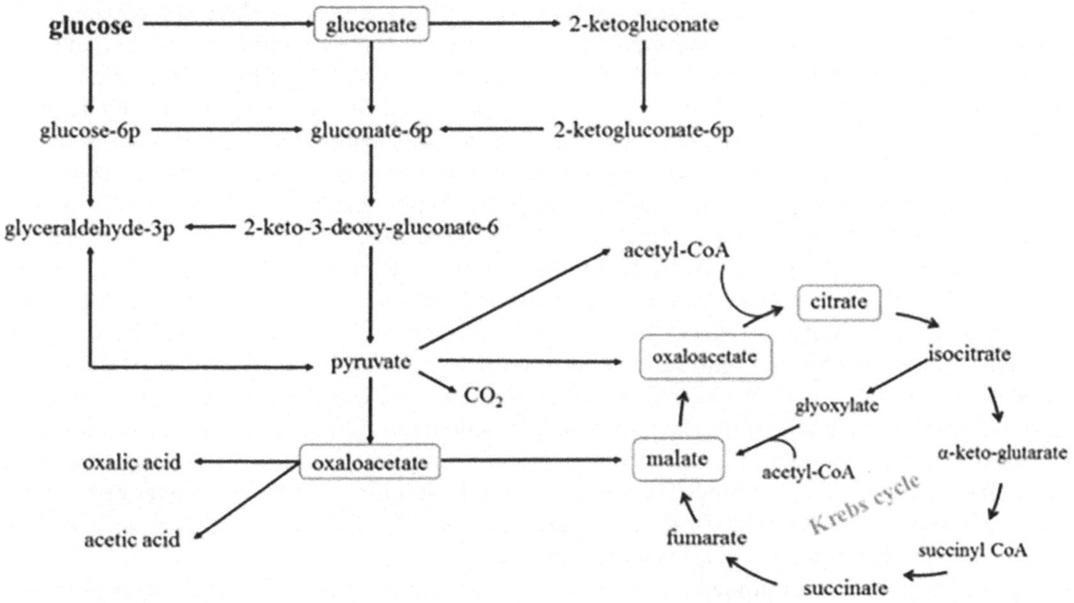

FIGURE 10.2 Biogenic synthesis of organic acids (reproduced with permission from [42]).

TABLE 10.4
Composition of the Culture Media Used for the Growth of the Fungi

Compound	Sucrose	Yeast extract	$NaNO_3$	KH_2PO_4	$MgSO_4.7H_2O$	KCl
Amount (g/L)	100	1.6	1.5	0.5	0.025	0.025

Glucose oxidase inhibits when the value of pH decreases below 3.5. The authors reported that the optimized condition resulted in the maximum production of citric acid (26,478 ppm), malic acid (1832.53 ppm), and gluconic acid (8433.76 ppm). The production of oxalic acid was 305.55 ppm at the optimized conditions. These biogenically synthesized organic acids were further used for the recovery of metals. The maximum recovery of Cu, Li, Mn, and Al was 100%, 100%, 77%, and 75%, respectively, at 2% (w/v) pulp density, whereas the highest recovery of Co and Ni – i.e., 64% and 54%, respectively – was reported at 1% (w/v) pulp density. The authors further stated that citric acid is more important for metal recovery due to the presence of 3 carboxyl groups in it. The more carboxyl groups are beneficial for metal extraction. Thus, gluconic acid is least suitable among produced organic acids for metal leaching due to the presence of only one carboxyl group.

Horeh et al. examined metal recovery using *A. niger* [43]. The composition of the culture media used for the growth of the fungi is shown in Table 10.4. The authors employed different approaches – i.e., one-step, two-step, and spent medium leaching – for the recovery of metals. In the case of a one-step process, fungi with culture medium along with LIB powder were incubated for 30 days. In the two-step process, initially, fungi were cultured in the media without LIB powder. The logarithmic phase ended after 3 days, and afterward, the LIB was added to the culture media. The bioleaching in the two-step process was carried out for 27 days. LIB was added to the culture medium after the start of the stationary phase for spent medium leaching. The stationary phase starts after 14 days, and afterward, the bioleaching was performed for 16 days. The authors reported that the one-step and two-step approach affects the metabolism of *A. niger* and metal recovery. Figure 10.3(a) depicts metal recovery using different approaches as reported by the authors. Metal recovery was highest by the spent medium bioleaching, which resulted in the 100% Cu, 95% Li, 70% Mn, 65% Al, 45% Co, and 38% Ni extraction. The leaching of Li, Al, Cu, and Mn was 100%, 58%, 11%, and 8%, respectively, by one-step leaching. In the two-step process, the leaching of Li, Al, Mn, and Cu was 100%, 61%, 10%, and 6%, respectively. The recovery of Co and Ni was negligible in both one-step and two-step approaches. The higher recovery of metals in spent medium bioleaching is due to the higher amount of citric acid. Citric acid forms a more stable metal-ligand complex compared to oxalic, malic, and gluconic acid, which results in the higher leaching of metals. Oxalic acid is the major organic acid that is formed biogenically during one-step and two-step processes, which hamper the metal leaching. The oxalic acid forms the insoluble complex with metals – i.e., nickel oxalate and cobalt oxalate – which were precipitated and remained with the residue of LIB powder. This justifies the negligible recovery of Co and Ni in one-step and two-step approaches. The spent medium bioleaching provides some advantages such as higher metal recovery, lesser process time, no contamination of fungi, etc. Therefore, the authors reported that spent medium bioleaching is more efficient for the leaching of heavy metals. In addition, the authors also reported metal recovery using chemical leaching. Figure 10.3(b) shows the comparative results for the spent medium bioleaching and chemical leaching. The spent medium bioleaching results in a higher recovery of metals compared to chemical leaching, except that of Li and Mn. This shows that bioleaching using biogenically produced organic acids is more efficient.

Bahaloo-Horeh et al. examined metal recovery using adapted metal-tolerant *A. niger* [44]. The composition of the culture media used for the growth of the fungi is shown in Table 10.4. In this study, the fungi were adapted to increase the resistance to metal toxicity. The adaptation was carried out by a series of sub-culturing processes. Initially, the unadapted *A. niger* was mixed with sucrose medium containing LIB powder (0.3% w/v), and it was incubated at 130 rpm and 30 °C. Further, 10% (v/v) of inoculation of the last step was mixed with the sucrose medium containing LIB (0.4% w/v). The series of adaptions was carried out till 1% (w/v) LIB powder and adapted fungi were further used for bioleaching. The variation in the different parameters for adapted and unadapted strains is shown in Figure 10.4. The authors reported that both adapted and unadapted *A. niger* produces malic, citric, oxalic, and gluconic acid, as shown in Figure 10.2. The authors reported that an approximately similar amount of oxalic acid was produced by both adapted and unadapted *A. niger*. The unadapted *A. niger* produced a higher amount of malic and citric acids, whereas the adapted *A. niger* produced a higher amount of gluconic acid, as shown in Figure 10.4. This can be attributed to the change in the metabolism of the fungi due to the adaption process. The authors further stated that metabolic adaption can result in the synthesis of new enzymes or pathways to substitute not-working ones. Moreover, the adapted *A. niger* requires a shorter period for entering the logarithmic phase, and also, the production of acid is faster compared to unadapted *A. niger*. The authors reported that the unadapted *A. niger* resulted in 100% Li, 58% Al, 11% Cu, and 8% Mn, whereas the leaching of Co and Ni was negligible. In the case of adapted *A. niger*, the leaching of Li, Cu, Mn, Al, Ni, and Co was around 100%, 94%, 72%, 62%, 45%, and 38%, respectively.

10.3 BIOLEACHING MECHANISMS

10.3.1 Autotrophic Bacteria Leaching

Autotrophic bacteria leaching takes place via two mechanisms – i.e., contact and non-contact leaching [45, 46].

10.3.1.1 Contact Leaching

In this type of mechanism, the bacteria and metal substrate are in direct contact, and the metal substrate directly receives the electron from the reduced metals [47]. In the case of contact leaching, surface properties and electrostatic interactions govern the attachment of bacteria to the metal substrate [45]. The contact between the bacteria and its attachment with the metal substrate is an

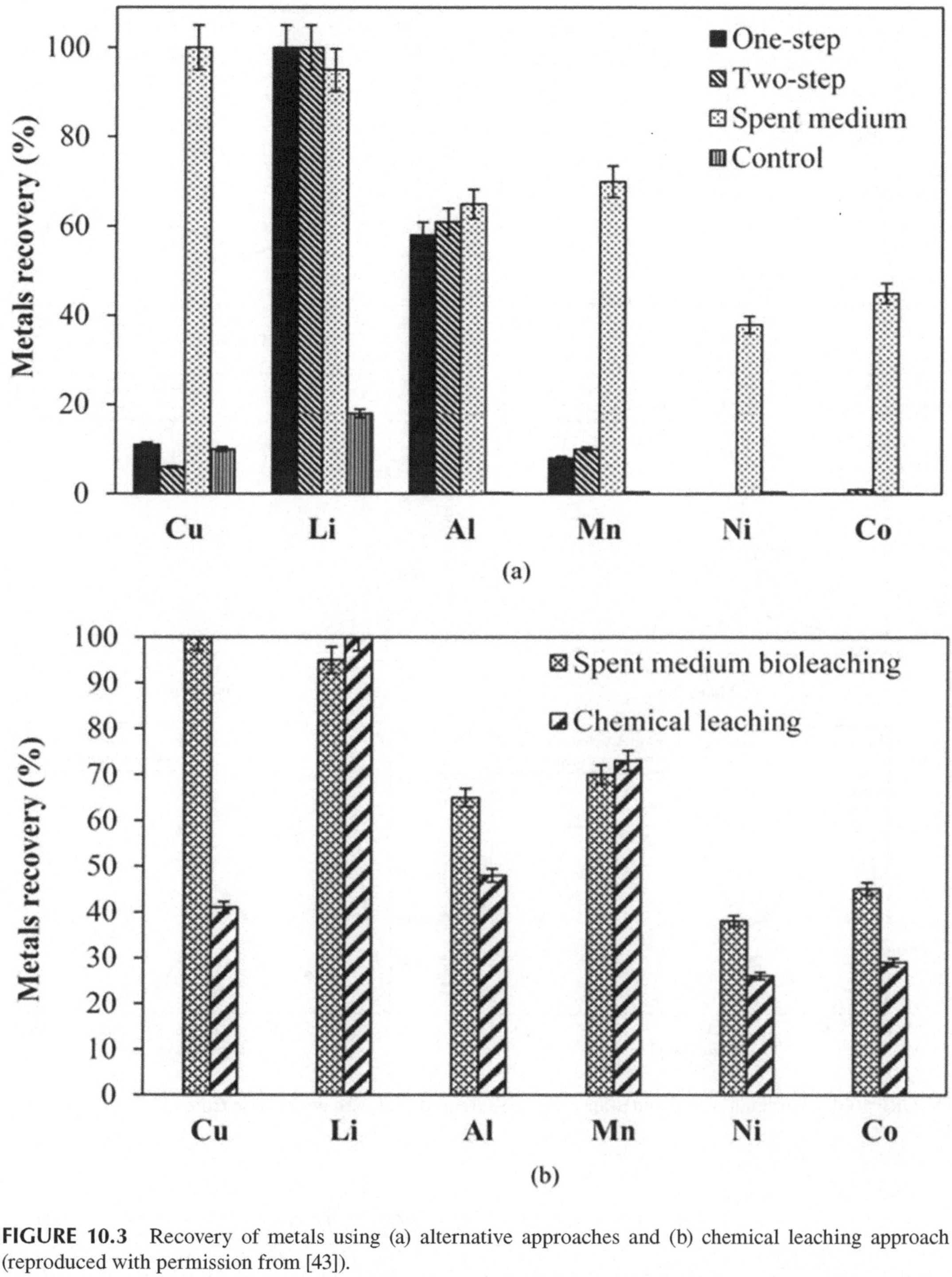

FIGURE 10.3 Recovery of metals using (a) alternative approaches and (b) chemical leaching approach (reproduced with permission from [43]).

important factor and defines the efficiency of the metal leaching. In contact leaching, S^0 and Fe^{2+} are the primary energy source for autotrophic bacteria and also act as electron donors [17]. The oxidation of S^0 to H_2SO_4 and Fe^{2+} to Fe^{3+} takes place during the process, which helps for the leaching of metals. The following equations show the reaction responsible for the biogenic production of H_2SO_4 and Fe^{3+} and the leaching of metals [17].

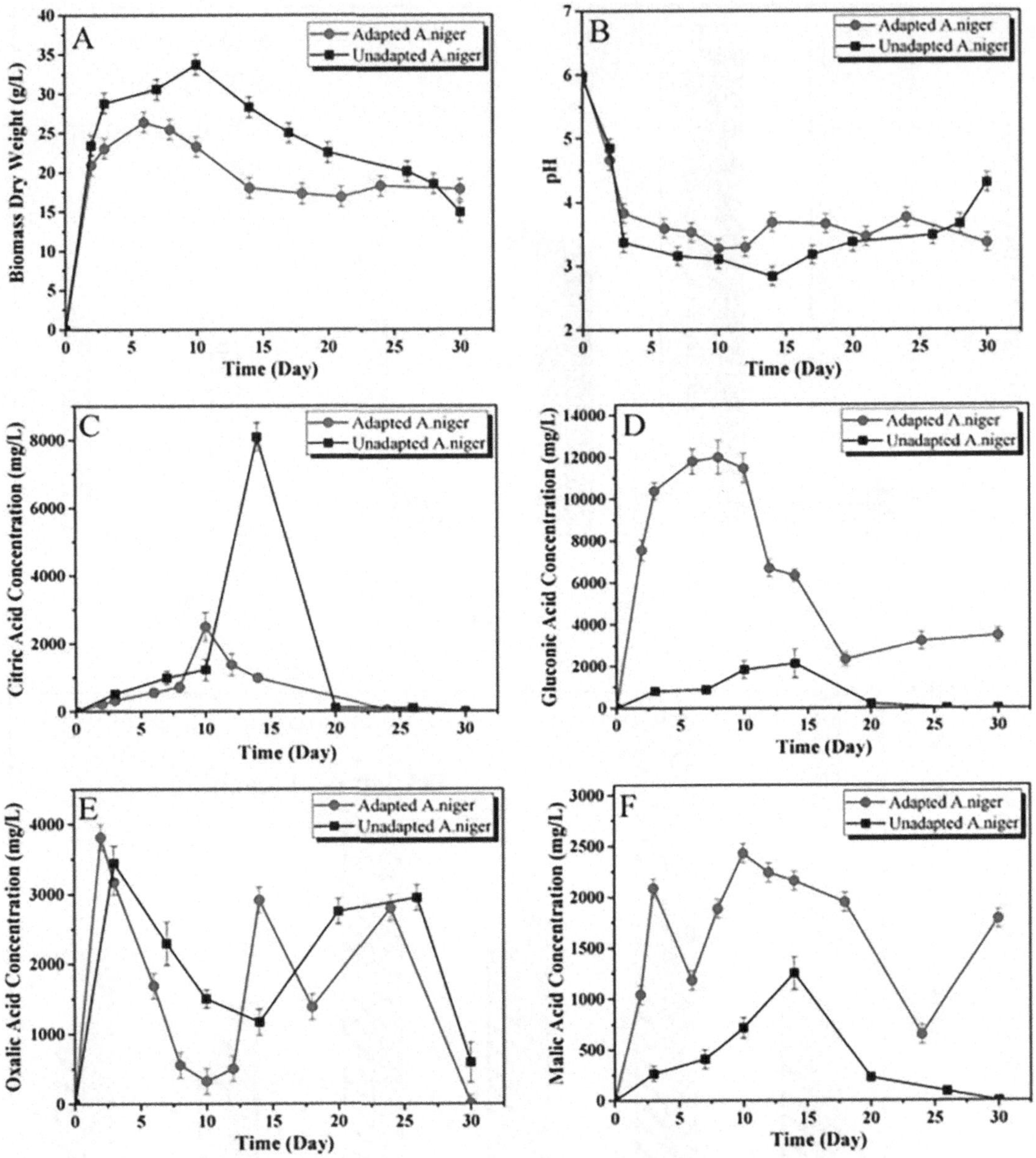

FIGURE 10.4 Variation before and after adaptation (reproduced with permission from [44]).

$$S^0 + 1.5O_2 + H_2O \rightarrow 2H^+ + SO_4^{2-} \rightarrow H_2SO_4 \tag{10.5}$$

$$H_2SO_4 + M \rightarrow MSO_4 + 2H^+ \ (M = Metal) \tag{10.6}$$

$$4LiCoO_2 + 6H_2SO_4 \rightarrow 2Li_2SO_4 + 4CoSO_4 + 6H_2O + O_2 \tag{10.7}$$

$$2Fe^{2+} + \frac{1}{2}O_2 + 2H^+ \rightarrow 2Fe^{3+} + H_2O \tag{10.8}$$

Xin et al. have examined the recovery of metals from NMC-based LIBs and proposed mechanism reactions involved in metal leaching as follows [48]:

$$4Fe^{2+} + O_2 + 4H^+ \rightarrow 4Fe^{3+} + 2H_2O \tag{10.9}$$

$$Li_2O + 2H^+ \rightarrow 2Li^+ + H_2O \tag{10.10}$$

$$M^{4+} / M^{3+} (M = Mn, Co, Ni) + Fe^{2+} \rightarrow M^{2+} (MO) + Fe^{3+} \tag{10.11}$$

$$MO + 2H^+ \rightarrow M^{2+} + H_2O \tag{10.12}$$

$$\begin{aligned} 6LiNi_{1/3}Mn_{1/3}Co_{1/3}O_2 + 12H_2SO_4 + 6FeSO_4 &\rightarrow 3Fe(SO_4)_3 + 2MnSO_4 + 2NiSO_4 \\ &+ 2CoSO_4 + 3Li_2SO_4 + 12H_2O \end{aligned} \tag{10.13}$$

Silva et al. studied the effect of bacteria attachment to the metal substrate and reported that the leaching efficiency of the contact mode is 25% higher than that of the non-contact mode bioleaching [49]. Xin et al. have reported that, in non-contact mode bioleaching, the metal leaching decreased from 43.5% to 25.0% for Co, 38.3% to 15% for Ni, and 93% to 42% for Mn [48].

10.3.1.2 Non-Contact Mechanism

It is the mechanism of bioleaching of metals without contact between the metal substrate and the microorganisms [50, 51]. In this type of mechanism, the bacteria only act as a catalyst and enhance the reoxidation of Fe^{2+} ion [30, 52]. Similar to the contact mechanism, the S^0 and Fe^{2+} oxidize to form H_2SO_4 and Fe^{3+}, respectively. In a non-contact mechanism, the H_2SO_4 is responsible for the leaching of metals via acidolysis and redoxolysis, whereas the Fe^{3+} ions play the role of oxidizing agent and enhance the leaching rate as shown in Eq. (10.14)-(10.16) [17].

$$S^0 + 3O_2 + 2H_2O \rightarrow 2H_2SO_4 \tag{10.14}$$

$$Fe^{2+} \rightarrow Fe^{3+} + e^- \tag{10.15}$$

$$2Fe^{3+} + Cu^0 \rightarrow Cu^{2+} + 2Fe^{2+} \tag{10.16}$$

10.3.2 Mechanism of Metal Leaching Using Heterotrophic Fungi

In heterotrophic fungi bioleaching, the leaching of metals takes place via organic acids formed biogenically during the metabolism of the fungi [17]. The metals are oxidized by O_2/H_2O redox couple, and subsequently, the organic acids form a complex with protonated metals [29, 44]. The protonation of metals also occurs, and it increases the mobility of free metal cations, which assists in the leaching of metals at lower pH. Metal cations form the complex with the biogenically formed organic acids as shown in Eq. (10.17)–(10.24), and hence, complexolysis is a major part of fungal bioleaching [42, 43].

$$C_{12}H_{22}O_{11} + H_2O \rightarrow C_6H_{12}O_6 + C_6H_{12}O_6 \tag{10.17}$$

$$C_6H_{12}O_6 + 1.5O_2 \rightarrow C_6H_8O_7 + H_2O \tag{10.18}$$

$$C_6H_8O_7 \rightarrow (C_6H_7O_7)^- + H^+ \tag{10.19}$$

$$(C_6H_7O_7)^- + M^+ \rightarrow \textit{metal citrate complex} \tag{10.20}$$

$$C_6H_{12}O_6 + 4.5O_2 \rightarrow 3C_2H_2O_4 + 3H_2O \tag{10.21}$$

$$C_2H_2O_4 \rightarrow (C_2HO_4)^- + H^+ \tag{10.22}$$

$$(C_2HO_4)^- + M^+ \rightarrow metal\ oxalate\ complex \tag{10.23}$$

$$7C_2H_2O_4 + 2LiCoO_2 \rightarrow 2LiHC_2O_4 + 2Co(HC_2O_4)_2 + 4H_2O + 2CO_2 \tag{10.24}$$

10.4 CONCLUSIONS

The global battery market is expected to grow by 25% per annum to 2.6 GWh in 2030, and the global LIBs demand is estimated to touch $95 billion by 2025. LIBs application and continuous upgradation of electronic devices will certainly result in a significant number of spent LIBs, owing to shorter life periods. In general, spent LIBs include 5–20 wt % cobalt (Co), 5–7 wt % lithium (Li), 5–7 wt % nickel (Ni), etc. Therefore, it is essential to consider spent LIBs recycling to mitigate the future demand for resource material and to reduce the harmful impact of the unregulated discharge of spent batteries to the environment. Bioleaching is a greener and promising approach that uses different microorganisms – i.e., bacteria and fungi – for metal recovery from different types of metal sources. These microorganisms produce different leaching agents like organic acid, H_2SO_4, etc., via their metabolism, which then reacts with metals and causes their extraction. The S/Fe-oxidizing bacteria have commonly been employed for the leaching of metals. These microorganisms use the energy generated by the oxidation of Fe^{2+} and elemental/reduced S to Fe^{3+} and H_2SO_4, respectively. Fe^{3+} and H_2SO_4 promote the extraction of metals. The most commonly used bacteria for bioleaching are *A. ferrooxidans* and *A. thiooxidans*. Bioleaching using heterotrophic fungi uses organic carbon as a source of energy. The metabolic process of organic carbon generates various organic acids as by-products. These organic acids react with metals and are responsible for metal extraction. *A. niger* is the most widely used fungi for metal recovery from LIBs. Bioleaching mechanism for bacterial bioleaching is of two types – i.e., contact and non-contact leaching. The leaching efficiency of the contact mode is higher than that of the non-contact mode bioleaching. In the mechanism of heterotrophic fungi, bioleaching involves the leaching of metals via organic acids formed biogenically during the metabolism of the fungi. Overall, the bioleaching of LIBs for the recovery of metals is a promising, greener approach and should be further explored at the pilot scale to investigate the industrial application of biohydrometallurgy.

BIBLIOGRAPHY

1. S. Quereshi, P.R. Jadhao, A. Pandey, E. Ahmad, K.K. Pant, Overview of sustainable fuel and energy technologies 1, in: S. Dutta, M.H. Chaudhery (Eds.), *Sustainable Fuel Technologies Handbook*, Elsevier, 2021: pp. 3–25.
2. X. Zhang, L. Li, E. Fan, Q. Xue, Y. Bian, F. Wu, R. Chen, Toward sustainable and systematic recycling of spent rechargeable batteries, *Chemical Society Reviews* 47 (2018) 7239–7302. https://doi.org/10.1039/c8cs00297e.
3. Global Battery Alliance—World Economic Forum, A vision for a sustainable battery value chain in 2030: Unlocking the full potential to power sustainable development and climate change mitigation, 2019. https://www3.weforum.org/docs/WEF_A_Vision_for_a_Sustainable_Battery_Value_Chain_in_2030_Report.pdf.
4. X. Zhang, Y. Xie, X. Lin, H. Li, H. Cao, An overview on the processes and technologies for recycling cathodic active materials from spent lithium-ion batteries, *Journal of Material Cycles and Waste Management* 15 (2013) 420–430. https://doi.org/10.1007/s10163-013-0140-y.
5. J.E.C. Sabisch, A. Anapolsky, G. Liu, A.M. Minor, Evaluation of using pre-lithiated graphite from recycled Li-ion batteries for new LiB anodes, *Resources, Conservation and Recycling* 129 (2018) 129–134. https://doi.org/10.1016/j.resconrec.2017.10.029.

6. B. Swain, Recovery and recycling of lithium: A review, *Separation and Purification Technology* 172 (2017) 388–403. https://doi.org/10.1016/j.seppur.2016.08.031.
7. C. Zhang, J. Jiang, W. Zhang, Y. Wang, S.M. Sharkh, R. Xiong, A novel data-driven fast capacity estimation of spent electric vehicle lithium-ion batteries, *Energies* 7 (2014) 8076–8094. https://doi.org/10.3390/en7128076.
8. M. Chen, X. Ma, B. Chen, R. Arsenault, P. Karlson, N. Simon, Y. Wang, Recycling end-of-life electric vehicle lithium-ion batteries, *Joule* 3 (2019) 2622–2646. https://doi.org/10.1016/j.joule.2019.09.014.
9. S. Zhao, W. He, G. Li, Recycling technology and principle of spent lithium-ion battery, in: L. An (Ed.), *Recycling of Spent Lithium-Ion Batteries: Processing Methods and Environmental Impacts*, Springer, Cham, 2019: pp. 1–26. https://doi.org/10.11817/j.ysxb.1004.0609.2021-42256.
10. E. Gratz, Q. Sa, D. Apelian, Y. Wang, A closed loop process for recycling spent lithium ion batteries, *Journal of Power Sources* 262 (2014) 255–262. https://doi.org/10.1016/j.jpowsour.2014.03.126.
11. D.H.P. Kang, M. Chen, O.A. Ogunseitan, Potential environmental and human health impacts of rechargeable lithium batteries in electronic waste, *Environmental Science & Technology* 47 (2013) 5495–5503. https://doi.org/10.1021/es400614y.
12. S.M. Shin, N.H. Kim, J.S. Sohn, D.H. Yang, Y.H. Kim, Development of a metal recovery process from Li-ion battery wastes, *Hydrometallurgy* 79 (2005) 172–181. https://doi.org/10.1016/j.hydromet.2005.06.004.
13. P.G. Schiavi, P. Altimari, M. Branchi, R. Zanoni, G. Simonetti, M.A. Navarra, F. Pagnanelli, Selective recovery of cobalt from mixed lithium ion battery wastes using deep eutectic solvent, *Chemical Engineering Journal* 417 (2021) 129249. https://doi.org/10.1016/j.cej.2021.129249.
14. B. Makuza, Q. Tian, X. Guo, K. Chattopadhyay, D. Yu, Pyrometallurgical options for recycling spent lithium-ion batteries: A comprehensive review, *Journal of Power Sources* 491 (2021) 229622. https://doi.org/10.1016/j.jpowsour.2021.229622.
15. D.L. Thompson, J.M. Hartley, S.M. Lambert, M. Shiref, G.D.J. Harper, E. Kendrick, P. Anderson, K.S. Ryder, L. Gaines, A.P. Abbott, The importance of design in lithium ion battery recycling—A critical review, *Green Chemistry* 22 (2020) 7585–7603. https://doi.org/10.1039/d0gc02745f.
16. J. Chun Lee, B.D. Pandey, Bio-processing of solid wastes and secondary resources for metal extraction—A review, *Waste Management* 32 (2012) 3–18. https://doi.org/10.1016/j.wasman.2011.08.010.
17. J.J. Roy, B. Cao, S. Madhavi, A review on the recycling of spent lithium-ion batteries (LIBs) by the bioleaching approach, *Chemosphere* 282 (2021) 130944. https://doi.org/10.1016/j.chemosphere.2021.130944.
18. F. Vakilchap, S.M. Mousavi, S.A. Shojaosadati, Role of Aspergillus niger in recovery enhancement of valuable metals from produced red mud in Bayer process, *Bioresource Technology* 218 (2016) 991–998. https://doi.org/10.1016/j.biortech.2016.07.059.
19. M. Baniasadi, F. Vakilchap, N. Bahaloo-Horeh, S.M. Mousavi, S. Farnaud, Advances in bioleaching as a sustainable method for metal recovery from e-waste: A review, *Journal of Industrial and Engineering Chemistry* 76 (2019) 75–90. https://doi.org/10.1016/j.jiec.2019.03.047.
20. B. Xin, D. Zhang, X. Zhang, Y. Xia, F. Wu, S. Chen, L. Li, Bioleaching mechanism of Co and Li from spent lithium-ion battery by the mixed culture of acidophilic sulfur-oxidizing and iron-oxidizing bacteria, *Bioresource Technology* 100 (2009) 6163–6169. https://doi.org/10.1016/j.biortech.2009.06.086.
21. J.J. Roy, S. Rarotra, V. Krikstolaityte, K.W. Zhuoran, Y.D. Cindy, X.Y. Tan, M. Carboni, D. Meyer, Q. Yan, M. Srinivasan, Green recycling methods to treat lithium-ion batteries e-waste: A circular approach to sustainability, *Advanced Materials* 34 (2022) 2103346.
22. P.R. Jadhao, S. Mishra, A. Pandey, K.K. Pant, K.D.P. Nigam, Biohydrometallurgy: A sustainable approach for urban mining of metals and metal refining, in: K.K. Pant, S.K. Gupta, E. Ahmad (Eds.), *Catalysis for Clean Energy and Environmental Sustainability*, Springer International Publishing, 2021: pp. 865–892. https://doi.org/10.1007/978-3-030-65017-9_27.
23. A.H. Kaksonen, N.J. Boxall, Y. Gumulya, H.N. Khaleque, C. Morris, T. Bohu, K.Y. Cheng, K.M. Usher, A.M. Lakaniemi, Recent progress in biohydrometallurgy and microbial characterisation, *Hydrometallurgy* 180 (2018) 7–25. https://doi.org/10.1016/j.hydromet.2018.06.018.
24. Z. Yu, H. Han, P. Feng, S. Zhao, T. Zhou, A. Kakade, S. Kulshrestha, S. Majeed, X. Li, Recent advances in the recovery of metals from waste through biological processes, *Bioresource Technology* 297 (2020) 122416. https://doi.org/10.1016/j.biortech.2019.122416.
25. K. Pollmann, S. Kutschke, S. Matys, J. Raff, G. Hlawacek, F.L. Lederer, Bio-recycling of metals: Recycling of technical products using biological applications, *Biotechnology Advances* 36 (2018) 1048–1062. https://doi.org/10.1016/j.biotechadv.2018.03.006.

26. M. Jafari, H. Abdollahi, S.Z. Shafaei, M. Gharabaghi, H. Jafari, A. Akcil, S. Panda, Acidophilic bioleaching: A review on the process and effect of organic——Inorganic reagents and materials on its efficiency, *Mineral Processing and Extractive Metallurgy Review* 40 (2019) 87–107. https://doi.org/10.1080/08827508.2018.1481063.
27. W. Krebs, C. Brombacher, P.P. Bosshard, R. Bachofen, H. Brandl, Microbial recovery of metals from solids, *FEMS Microbiology Reviews* 20 (1997) 605–617.
28. G. Chauhan, P. Kaur, K.K. Pant, K.D.P. Nigam, *Sustainable Metal Extraction from Waste Streams*, WILEY-VCH Verlag GmbH, 2020.
29. B.K. Biswal, U.U. Jadhav, M. Madhaiyan, L. Ji, E. Yang, B. Cao, Biological leaching and chemical precipitation methods for recovery of Co and Li from spent lithium-ion batteries, *ACS Sustainable Chemistry & Engineering* 6 (2018) 12343–12352. https://doi.org/10.1021/acssuschemeng.8b02810.
30. W. Wu, X. Liu, X. Zhang, X. Li, Y. Qiu, M. Zhu, W. Tan, Mechanism underlying the bioleaching process of LiCoO2 by sulfur-oxidizing and iron-oxidizing bacteria, *Journal of Bioscience and Bioengineering* 128 (2019) 344–354. https://doi.org/10.1016/j.jbiosc.2019.03.007.
31. S. Ilyas, M.A. Anwar, S.B. Niazi, M.A. Ghauri, Bioleaching of metals from electronic scrap by moderately thermophilic acidophilic bacteria, *Hydrometallurgy* 88 (2007) 180–188. https://doi.org/10.1016/j.hydromet.2007.04.007.
32. S. Ilyas, M. Kim, J. Lee, Integration of microbial and chemical processing for a sustainable metallurgy, *Journal of Chemical Technology & Biotechnology* 93 (2018) 320–332.
33. C.A. Plessis, J.D. Batty, D.W. Dew, Commercial applications of thermophile bioleaching, in: D.E. Rawlings, D.B. Johnson (Eds.), *Biomining*, Springer, Berlin, Heidelberg, 2007: pp. 57–80.
34. T. Naseri, N. Bahaloo-Horeh, S.M. Mousavi, Environmentally friendly recovery of valuable metals from spent coin cells through two-step bioleaching using Acidithiobacillus thiooxidans, *Journal of Environmental Management* 235 (2019) 357–367. https://doi.org/10.1016/j.jenvman.2019.01.086.
35. H.L. Liu, B.Y. Chen, Y.W. Lan, Y.C. Cheng, Biosorption of Zn(II) and Cu(II) by the indigenous Thiobacillus thiooxidans, *Chemical Engineering Journal* 97 (2004) 195–201. https://doi.org/10.1016/S1385-8947(03)00210-9.
36. Y. Qu, B. Lian, B. Mo, C. Liu, Bioleaching of heavy metals from red mud using Aspergillus niger, *Hydrometallurgy* 136 (2013) 71–77. https://doi.org/10.1016/j.hydromet.2013.03.006.
37. T. Naseri, N. Bahaloo-Horeh, S.M. Mousavi, Bacterial leaching as a green approach for typical metals recovery from end-of-life coin cells batteries, *Journal of Cleaner Production* 220 (2019) 483–492. https://doi.org/10.1016/j.jclepro.2019.02.177.
38. J. Jegan Roy, M. Srinivasan, B. Cao, Bioleaching as an eco-friendly approach for metal recovery from spent NMC-based lithium-ion batteries at a high pulp density, *ACS Sustainable Chemistry & Engineering* 9 (2021) 3060–3069. https://doi.org/10.1021/acssuschemeng.0c06573.
39. J.J. Roy, S. Madhavi, B. Cao, Metal extraction from spent lithium-ion batteries (LIBs) at high pulp density by environmentally friendly bioleaching process, *Journal of Cleaner Production* 280 (2021) 124242. https://doi.org/10.1016/j.jclepro.2020.124242.
40. G.A. Somerville, R.A. Proctor, Cultivation conditions and the diffusion of oxygen into culture media: The rationale for the flask-to-medium ratio in microbiology, *BMC Microbiology* 13 (2013) 9. https://doi.org/10.1186/1471-2180-13-9.
41. A. Heydarian, S.M. Mousavi, F. Vakilchap, M. Baniasadi, Application of a mixed culture of adapted acidophilic bacteria in two-step bioleaching of spent lithium-ion laptop batteries, *Journal of Power Sources* 378 (2018) 19–30. https://doi.org/10.1016/j.jpowsour.2017.12.009.
42. N. Bahaloo-Horeh, S.M. Mousavi, Enhanced recovery of valuable metals from spent lithium-ion batteries through optimization of organic acids produced by Aspergillus niger, *Waste Management* 60 (2017) 666–679. https://doi.org/10.1016/j.wasman.2016.10.034.
43. N.B. Horeh, S.M. Mousavi, S.A. Shojaosadati, Bioleaching of valuable metals from spent lithium-ion mobile phone batteries using Aspergillus Niger, *Journal of Power Sources* 320 (2016) 257–266. https://doi.org/10.1016/j.jpowsour.2016.04.104.
44. N. Bahaloo-Horeh, S.M. Mousavi, M. Baniasadi, Use of adapted metal tolerant Aspergillus niger to enhance bioleaching efficiency of valuable metals from spent lithium-ion mobile phone batteries, *Journal of Cleaner Production* 197 (2018) 1546–1557. https://doi.org/10.1016/j.jclepro.2018.06.299.
45. M. Sethurajan, S. Gaydardzhiev, Bioprocessing of spent lithium ion batteries for critical metals recovery—A review, *Resources, Conservation and Recycling* 165 (2021) 105225. https://doi.org/10.1016/j.resconrec.2020.105225.
46. D. Mishra, Y.H. Rhee, Microbial leaching of metals from solid industrial wastes, *Journal of Microbiology* 52 (2014) 1–7. https://doi.org/10.1007/s12275-014-3532-3.

47. T. Rohwerder, T. Gehrke, K. Kinzler, W. Sand, Bioleaching review part A: Progress in bioleaching: Fundamentals and mechanisms of bacterial metal sulfide oxidation, *Applied Microbiology and Biotechnology* 63 (2003) 239–248. https://doi.org/10.1007/s00253-003-1448-7.
48. Y. Xin, X. Guo, S. Chen, J. Wang, F. Wu, B. Xin, Bioleaching of valuable metals Li, Co, Ni and Mn from spent electric vehicle Li-ion batteries for the purpose of recovery, *Journal of Cleaner Production* 116 (2016) 249–258. https://doi.org/10.1016/j.jclepro.2016.01.001.
49. R.A. Silva, J. Park, E. Lee, J. Park, S.Q. Choi, H. Kim, Influence of bacterial adhesion on copper extraction from printed circuit boards, *Separation and Purification Technology* 143 (2015) 169–176. https://doi.org/10.1016/j.seppur.2015.01.038.
50. S. Dev, A. Sachan, F. Dehghani, T. Ghosh, B.R. Briggs, S. Aggarwal, Mechanisms of biological recovery of rare-earth elements from industrial and electronic wastes: A review, *Chemical Engineering Journal* 397 (2020) 124596. https://doi.org/10.1016/j.cej.2020.124596.
51. M. Sethurajan, E.D. Hullebusch, D. Fontana, A. Akcil, H. Deveci, B. Batinic, J.P. Leal, T.A. Gasche, M. Ali, K. Kuchta, I.F.F. Neto, H.M.V.M. Soares, A. Chmielarz, Recent advances on hydrometallurgical recovery of critical and precious elements from end of life electronic wastes—A review, *Critical Reviews in Environmental Science and Technology* 49 (2019) 212–275. https://doi.org/10.1080/10643389.2018.1540760.
52. T. Huang, L. Liu, S. Zhang, Recovery of cobalt, lithium, and manganese from the cathode active materials of spent lithium-ion batteries in a bio-electro-hydrometallurgical process, *Hydrometallurgy* 188 (2019) 101–111. https://doi.org/10.1016/j.hydromet.2019.06.011.

11 Microplastics Pollution
A Perspective on the Source, Fate, Impact, Identification and Extraction from the Environment

Pradeep Kumar, Pinky, Naseeb, Ankush Yadav and Kashyap Kumar Dubey

11.1 INTRODUCTION

Plastics have many attractive properties such as being durable, cheap, water-tight and lightweight. Because of these interesting properties, the manufacturing and consumption of plastics are increasing continuously. Now, plastic is present in all ecosystems of Earth, but only a small part are distinguishable as environmental issues (Yu et al., 2018). Through different processes, plastics in the environment break down into smaller particles and form microplastics (Hahladakis et al., 2018; Wiesinger et al., 2021). The small, water-insoluble particles of plastic debris having a size less than 5 mm are called microplastics. These microplastics affect aquatic life, soil and plants (Nicholson et al., 2022). It is a global issue because it can enter the food chain through plants and aquatic species. Due to their large surface area and water-insoluble nature, microplastics are susceptible to many substances, including heavy metals (Yuan et al., 2022; Ziani et al., 2023).

Due to their widespread distribution, microplastics constitute a global problem (Alimba and Faggio, 2019). Over 359 million tons (Mt) of plastics were produced globally in 2018, up from 348 Mt in 2017 (Plastics Europe, 2019). Global plastic manufacture has significantly increased since the 1950s, when plastic products were widely available, from 0.5 Mt/year in 1960 to 348 Mt/year in 2017. China is the world's top manufacturer of plastics, accounting for 30% of global plastic output in 2018 with 107.7 Mt (Plastics, 2019). By 2030, plastic emissions to the aquatic environment were forecasted for 173 nations by Borrelle et al. (2020), who predicted that these emissions would range from 20 to 53 Mt/year. Because microplastics are multiple disruptors with a variety of physical-chemical characteristics, understanding their impact is very complicated. On the one hand, microplastics act as vectors of transport for toxic chemicals in ecosystems, but on the other, they are a mixture of dangerous chemicals that are voluntarily added during their production as additives to improve polymer properties and lengthen their life. The key additives of concern utilized in the plastics industry, what happens once microplastics are released into the environment, and their ensuing impacts on human health when linked to micro- and nanoplastics are all still mostly unknown at this time.

11.1.1 Sources of Microplastics

Microplastics originated from primary and secondary sources (Figure 11.1). Primary sources of microplastics include the particles in cosmetic and medical products. Secondary microplastics are produced due to breakdown of plastics in tiny particles through different physical, chemical and biological processes (Rogers, 2022). Primary microplastics can enter the water through sewage sludge and spillage of plastic resin powders, and it can also enter soil when sewage sludge containing synthetic fibers from household products is applied to land (Kallenbach et al., 2022). Secondary

 DOI: 10.1201/9781003391487-11

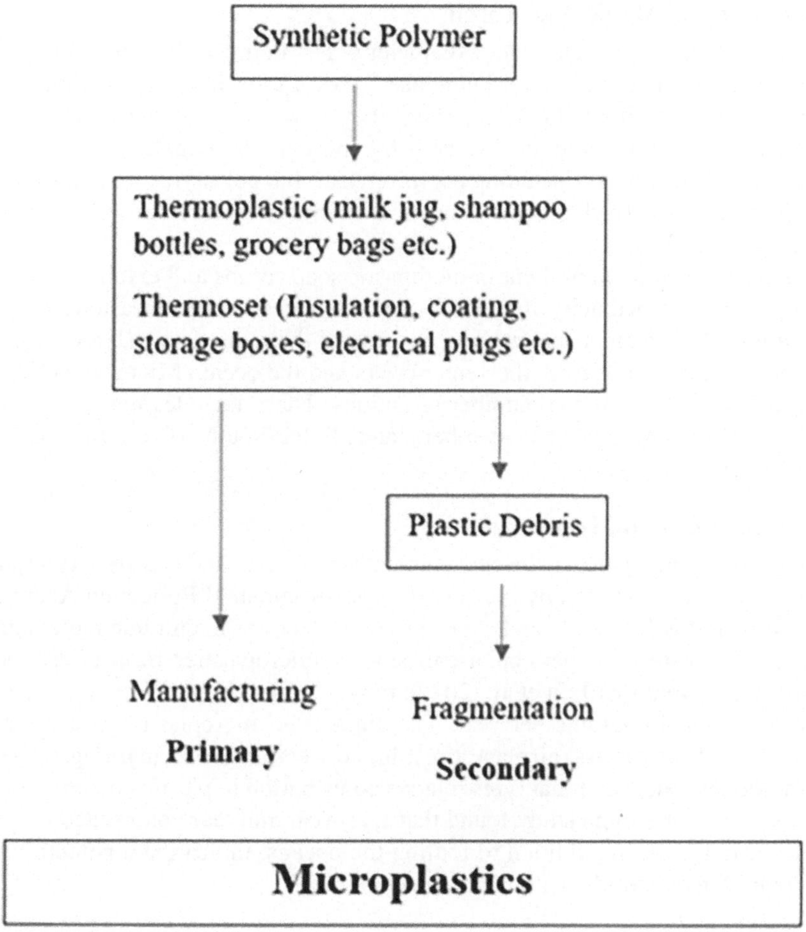

FIGURE 11.1 Represents various sources of microplastics.

microplastic can enter the aquatic ecosystem through wind dispersal, soil erosion or surface runoff. Microplastic can spread in the earth surface by the wind movement (Haque and Fan, 2023). Some major sources of microplastic pollution are following:

11.1.1.1 Clothing and Fabric

The various textiles used for various types of clothing items are composed of tiny, synthetic fibers known as microfibers. These clothing items include nylon or polyester, the heat-resistant synthetic fiber Kevlar – i.e., Aramid (poly-para-phenylene terephthalamidean) fibers are known by the trade name Kevlar – used in the manufacture of military-grade gear (McCann, 2022). Microfibers are also being used by manufacturers to create a variety of products in place of cotton and leather. These include bed coverings and blankets for the bedroom as well as microfiber towels or cloths for cleaning cars and glass surfaces. Thermal insulation materials are also used in sleeping bags and car covers.

The environment becomes more and more polluted with microplastics as a result of both poor product disposal and frequent use. Cai et al. (2020) investigated a study and found that, in the wastewater from residential washing machines, a single cloth produced more than 1,900 fibers per wash. This demonstrates that one of the primary sources of microplastics that build up in natural bodies of water is wastewater from houses and businesses.

11.1.1.2 Cosmetic and Medical Research

Microbeads are solid plastic particles with a diameter of less than 5 millimeters that have been used in cosmetic and medicinal studies. After being used, they are flushed into the drain and deposited in wastewater treatment plants (WWTPs), where they have the potential to spill into the water (Anagnosti et al., 2021). Although polyethylene is the most common material used in their production, other petrochemical plastics, including polypropylene and polystyrene, can also be used. These materials are frequently used in biomedical and health science research as well as in personal care products (Britannica, 2023).

Microbeads are a common ingredient in toothpaste, soap scrubs and exfoliating facial cleansers in the cosmetics sector. Immediately after use, these tiny pieces end up in wastewater. In wastewater treatment facilities, filters cannot screen them because of their small size (Jan-Georg et al., 2022). As a result, when effluent is released, they enter rivers and the ocean. Microbeads have been disallowed in personal care products in a number of nations. These include, among others, the United States, the United Kingdom, a few EU member states, India, South Africa, Taiwan, Australia and New Zealand (Ziani et al., 2023).

11.1.1.3 Vehicles Tyres and Footwears

One of the causes of microplastics in the environment is also wear and tear from tyres and footwear. According to a Denmark study led by the country's Environmental Protection Agency, secondary microplastics from goods like tyres and shoes are two orders of magnitude more significant than primary plastics. The estimated per-capita emission of microplastics from tyres, according to a different study investigated by Mian et al. (2022), ranges from 0.23 to 4.7 kilograms per year, with a global average of 0.81 kilograms per year. The amount of microplastics in automobile tyres is far higher than that of fake grass, airplane tires, brake wear and road markings. The researchers came to the additional conclusion that tyres relative contribution to plastic contamination in oceans ranged from 5 to 10%. The same study found that tyre wear and tear contributed to 3 to 7% of the particulate matter in the air, in addition to fouling the oceans, indicating a potential source of the burden of air pollution on world health (Rødland et al., 2022).

11.1.1.4 Industries

Granules and tiny resin pellets are used as raw materials in the production of end-use plastics. Plastic manufacturing facilities may contribute to the unintentionally leakage of microplastics during land and water transport, the improper use of packing materials and direct outflow from processing facilities (Reddy and Nair, 2022). Diggle and Walker (2022) investigated a study and found microplastic concentrations in range from 150 to 2,400 particles per cubic meter in a seaport close to a plastic manufacturing facility. The actions of fishing vessels and the entire fishing industries, including commercial and recreational fishing, have all contributed to the unjustified disposal of plastic items in various bodies of water. These polymers would degrade over time and produce secondary microplastics as a result. These materials include, among others, nylon netting, plastic monofilament line and fishing equipment. Another significant source of marine pollution has been the shipping sector (Alberghini et al., 2022).

11.2 ADVERSE EFFECTS OF MICROPLASTICS IN THE FOOD CHAIN

Nowadays, we are facing different types of pollution, such as air, soil and water pollution, thermal pollution and are trying to reduce these pollutants by different measures. But a large manufacturing and consumption of plastics products may result in one more significant type of pollution, which is microplastic pollution. Microplastics have been found in the atmosphere, hydrosphere and in lithosphere (Kibria et al., 2023). Microplastics can affect humans and the environment in many ways. Research studies found that microplastics have detrimental effects on marine biota. However, scientific studies have shown less attention towards the presence of microplastics in land ecosystems, but a large amount of plastic is disposed of on land that breaks down slowly and forms microplastics. So, the concentration of microplastics may be higher in terrestrial ecosystems than aquatic (mainly coastal and marine) ecosystems (Zolotova et al., 2022). Microplastic may stay on the surface of water or go down below the surface of water and accumulate on the ocean floor. Microplastics which have density less than seawater float on sea surface and have higher density, accumulating in the deep sea. Other chemical substances are also used during manufacturing of plastics to improve quality, which make plastics more durable, heat resistant, waterproof, etc. After the disposal of plastic products as waste, these chemicals leach into the seawater and cause various adverse effects in marine organisms (Dissanayake et al., 2022). Microplastics are inhaled by marine organisms from different trophic levels. Plastics in seawater not only include chemical additives but some other organic substances like polychlorinated biphenyls, dioxins, etc., which are adsorbed on sea surface by other anthropogenic activities. These organic compounds have more affinity towards plastics than seawater because of their hydrophobic nature (Yang J. et al., 2023). These pollutants have a long lifespan. These are bio accumulative and toxic compounds. Microplastics act as a porter of these toxic substances to wildlife and also enhance their persistence. Recently, studies have shown that microplastic is found in the digestive systems of many marine organisms. These are further converted into nanoplastics (Hassan et al., 2023).

11.2.1 Microplastics in the Food Chain

A biological phenomenon known as a food chain connects organisms either vertically or horizontally, depending on how they feed. This is arranged according to several trophic levels. The food web, in contrast to them, consists of relationships between species that are not straight-lined (Britannica, 2023). One trophic-level organism on the food web may be linked to other organisms in a different trophic level (Reichle, 2023). However, feeding chains can be either terrestrial or aquatic, grazing or detritus-based. According to recent studies, 10% of the energy in a food chain will move from a lower trophic stage to a higher one (Pavluk and Bij, 2008).

Even though there are hundreds of thousands of plastic compounds in use worldwide, only six are widely used: polyethylene (high and low density), polypropylene, polyvinyl chloride, polystyrene, polyurethane and polyethylene terephthalate (Nawab et al., 2023). Primary and secondary microplastics must be distinguished when evaluating the effect of microplastics on the food supply. Primary microplastics are chemicals designed specifically for that size range. On the other hand, secondary microplastics are the breakdown results of plastic parts from larger things (Seyyedi et al., 2023). This difference will help us find the sources of contamination, create mitigation plans for them and reduce their impact on the food-supply chain. Among the main sources are plastic microbeads used in cosmetic formulations, plastic powders used in molding and plastic nanoparticles utilized in a variety of industrial processes (Mohanan et al., 2020).

Additionally, the production of plastics makes extensive use of virgin resin pellets. As bigger plastic things break up and deteriorate, secondary microplastics are created. This could occur at any point in the lifespan of a plastic product, including tyres, clothing and other items (Chappell et al., 2022). There is widespread microplastic pollution of freshwater and marine life. Due to our position at the top of the food chain, humans could be at risk from microplastics because of their great persistence in the environment. Microplastics may also be a severe hazard to freshwater and marine life (Mubin et al., 2023). For a number of marine and freshwater species, exposure to microplastics mostly comes from drinking water. Recent studies have found that fish, bivalves and crustaceans have all been found to consume microplastic. Aquaculture systems where fish or other farmed species are raised can also lead to contaminated water (Wu et al., 2023). Microplastics have also made it into the food chain. There have been reports of up to 600 microplastic particles per kilogram of salt, 660 microplastic fibers/per kilogram of honey and up to roughly 109 microplastic fragments per liter of beer, according to a number of studies (Aparna, 2019). According to a study, fish, mussels and oysters harvested for human consumption contain 0.36–0.47 particles of microplastic per gramme, meaning that seafood consumers may ingest up to 11,000 pieces of microplastic per year (Madhav et al., 2020). Several aquatic or terrestrial living organisms such as fish ingest microplastics, which accumulate in their gastrointestinal tracts. When the digestive tracts of sheep, goats and other animals are digested, there is a considerable chance that microplastic will be ingested. Bottled water is the most prevalent known source of microplastics that are consumed (Sambolino et al., 2023). Single-use water bottles contain 2 to 44 microplastics per liter, but reusable water bottles (intended for collection under a deposit programme) contain 28 to 241 microplastics per liter (Van Cauwenberghe and Janssen, 2014). Various packing materials can introduce microplastics into the food chain. There is evidence to support the theory that the microplastics discovered in food come from indoor dust. Because of bio-absorption, microplastics are more frequently introduced to the marine ecosystem (Bostan et al., 2023). Microplastics affect marine life due to their physical and chemical properties. Due to the pores of microplastics, contaminants including polychlorinated biphenyls, pesticides and chemical fertilizers can be absorbed by them. This might have disastrous effects on marine life and the food chain that sustains it (Campanale et al., 2020).

According to recent studies, microplastics can enter the marine environment by a variety of methods, such as ingestion, microbial assembly, bioaccumulation, biomagnification, etc., and transit through specific trophic levels in a food chain. Adhesion, entanglement and other processes like these also make microplastic exposure to the food chain easier (Ghosh et al., 2022). Microplastics are accidentally ingested or consumed by animals, where they enter the food chain. At the base of the food chain, tiny organisms known as zooplankton consume small pieces of plastic, which may subsequently wind up in the bodies of fish and other animals that consume them (Kye et al., 2023). When animals eat other animals that have consumed plastic, they also swallow the plastic (Miller et al., 2020).

As evidenced by their presence in the food chain, microplastics are a known environmental contaminant. Zooplankton, chaetognatha, ichtyoplankton, copepods and salps have all been found to have microplastics in the marine environment at lower trophic levels (Figure 11.2). Higher trophic levels of the food chain are also impacted by microplastic contamination in fish, seabirds, mammals and polychaetes, crabs, echinoderms and bivalves (Campillo et al., 2023). Both direct eating and trophic transmission are ways that plastic particles get to them. The fish digestive tracts are typically not ingested, and the bioaccumulation of microplastics there does not pose a significant threat to humans (Khanashyam et al., 2023; Rahman et al., 2023; Rummel et al., 2016). The bioaccumulation of filter-feeding crabs, whose digestive tracts are eaten, is more problematic. It is possible to assess the degree of contamination in organisms and seafood meant for human consumption (Amin et al., 2020). Because they are eaten whole and can be a large source of microplastics, mussels and molluscs are a useful indicator. Fish that are benthic may suggest contaminated sediment. The quantity of microplastics in sardines and anchovies, which are also eaten whole, indicates that the open ocean has become contaminated, and subsequently, human exposure (Hassoun et al., 2023).

Biofouling might be a major factor in organisms mistaking plastic for a food source. According to studies, the development of biofilms alters the vertical distribution of the particle, increasing the likelihood that microplastic will be ingested, and they also serve as an attractant for microorganisms due to the release of the odorant diethyl sulfide during the breakdown of biofilms on plastic surfaces (Wypych, 2023). The same cause that led sea turtles to consume plastics was also discovered, with visual signals playing a significant role. Higher trophic creatures are not the only ones who consume plastic debris. Copepods consumed more old particles with a biofilm when exposed to both clean and polluted particles. Zooplanktons are very important for nutrient cycling or availability in an aquatic ecosystem (Chandra et al., 2023). They form a link between higher trophic creatures and basic producers. They devour the primary producers and transform the organic matter into high-density, fast-sinking faecal pellets, making it available to sedimentary organisms and contributing to the aquatic ecosystem carbon sink. Microplastics consume food as well as faecal pellets, which are afterwards ejected from them. Other coprophagous animals may tropically acquire these plastic particles from the faecal pellet. It is possible that not all of the plastic particles consumed by

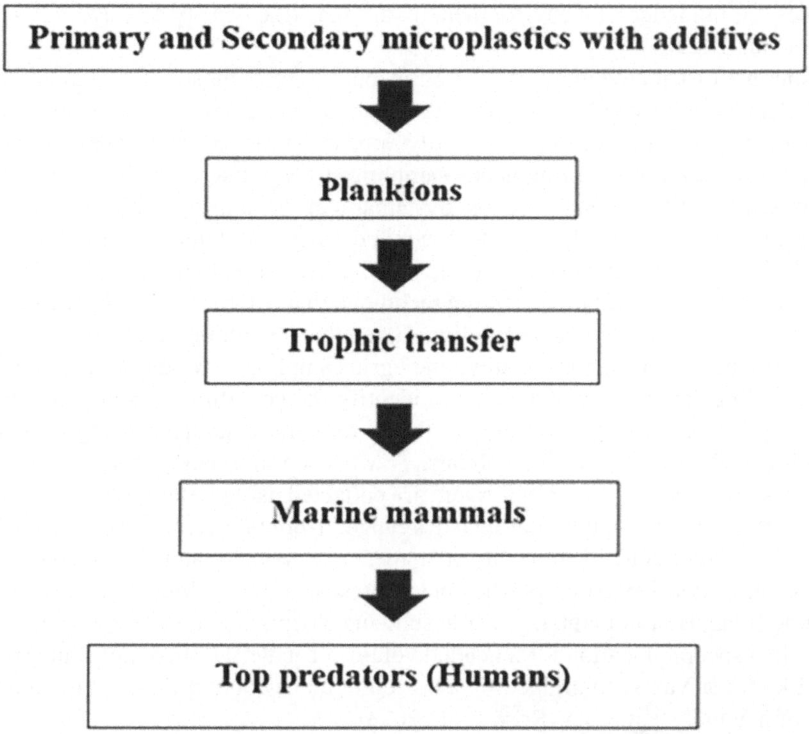

FIGURE 11.2 Steps involves in bioaccumulation of microplastic in food chain.

zooplankton pass through the faeces. Consequently, the leftover plastics in zooplankton's digestive tract directly transferred to the higher trophic levels (Gilfedder et al., 2023). According to a study, 75% of the 166 microplastic particles we found were less than 3 mm. Three porphyrins and nine phthalates were found. The quantity of microplastics and the level of phthalates were shown to be strongly correlated. Both phthalates and porphyrin values were found to be lower in seal tissues than in those of other marine mammals, indicating that they may not yet be having an effect on seals (Hernandez-Milian et al., 2023).

11.3 SAMPLING, EXTRACTION, PURIFICATION AND IDENTICAL APPROACHES FOR MICROPLASTIC

Sampling is one of the influential steps for microplastics analysis. There are various sampling approaches for microplastics, and it depends on the research objectives. Currently, due to the wide variety of options available, research organization may struggle to choose methods for sampling microplastics (Andeobu et al., 2023). We aim to summarize the most important techniques for collecting microplastics from soil and water in this chapter. As microplastics are very small in size, sampling patterns must be well organized for good and authentic results. The research objective, sampling site or area and sampling approach must be well defined.

Microplastics are dispersed throughout the water column, depending on their properties, such as density, form, size, chemical and biofouling adsorption and environmental conditions, such as water density, wind, currents and waves. As a result, the depth and position of the sample have a big influence on how much and what kind of microplastics are recovered. Similar sampling and processing methods can be used to obtain samples from both freshwater and saltwater, which makes it possible to standardize methods in the future. However, each system's microplastic dispersion varies, and these variances are influenced by environmental variables like density and hydrodynamic characteristics (Kye et al., 2023).

Microplastics are distributed unevenly on sediments, which is mostly due to their characteristics and external elements like winds and currents. Since some regions may have larger quantities of microplastics than others, the results will mostly depend on the sampling area and depth (Shahul Hamid et al., 2018). Random sampling is one sampling strategy that could be used to find contamination at a place that is likely homogeneous. Systematic grid sampling, which divides up sampling points into regular patterns, may be used to check the range and contamination hotspots (Thomas et al., 2020). The first point is chosen at random, and the remaining points are placed in a systematic way. A one-dimensional, systematic sampling technique that can be used to determine and confirm the extent of contamination is transect sampling. Given that it combines the benefits of both random sampling and systematic grid sampling, unaligned grid sampling may be a suitable sample strategy. A stratified sampling approach may be used to identify the contamination in a defined subarea of the overall sampling region. Microplastics in the water can be sampled using a dynamic or stationary sampling technique in riverine habitats. Trawls used in dynamic sampling are dragged by boats. Microplastics samples from minor rivers are collected using a stationary sampling approach. According to Zhu et al. (2023), the microplastics ranged from 0.01 items/m^3 to 0.89 items/m^3, with an average of 0.25 0.05 items/m^3. In terms of shape, color, content and size, fibers in the range of 0.33 to 1 mm were extensive. Microplastics being present in the sediments is also noted. In most cases, sampling is carried out manually, while showing a depth and an area for sediments in freshwater bodies. In a specific location for a needed volume of material, there are some specific instruments – i.e., Ekman or Van veen grab corer, a box corer and a corer with a specific diameter can be used instead of manually sampling (Rose et al., 2023).

11.3.1 Extraction

The most popular method for removing microplastics was tissue digestion in oxidative (28% for mussels, 12% for fish) and alkaline (26% for fish) chemicals. Microplastics from other undigested leftovers could be separated by flotation using high-density solutions like sodium chloride (Debraj and Lavanya, 2023). Fourier-transform infrared (FTIR) spectroscopy (58%) and Raman spectroscopy (14% for mussels and 8% for fish) were mostly used for the polymer analysis of microplastics (Dellisanti et al., 2023). Worldwide, mussels and fish have been the main focus of analysis of microplastics in marine biota (Thiele et al., 2019). Through dissection, microplastics can be removed from particular animal organs. The entire soft tissue of mussels was often employed (86%) for the evaluation of microplastics. The two primary fish organs targeted for this aim were the digestive system (56%) and the tract with gills (16%) in order to distinguish between the absorption pathways of microplastics via food exposure and aquatic exposure, respectively (Dellisanti et al., 2023). In physical separation, after the collection of samples, microplastics of different color, size and polymers must be segregated. On-site sieving technique is applied generally for isolating the large microplastics (greater than 1 mm) (Mandal et al., 2023). It is the simplest technique for removing microplastic from water samples utilizing sieves with various types and mesh sizes. Materials that remain on the sieve's upper surface after filtration are gathered and passed. But this technique is not appropriate for microplastics having size less than 1 mm (Reineccius et al., 2021).

11.3.2 Visual Sorting

For microplastic analysis, visual separation and identification serve as essential steps that make it easier to remove non-microplastic material from the sample. Only 20% of the studies on mussels and fish gave precise definitions for suspected microplastics; the majority of them relied on visual screening to separate suspected microplastics before the identification stage. The method that was most frequently employed to describe microplastics is FTIR spectroscopy (Adegoke et al., 2023).

Natural biological remains like shell pieces, seaweed and wood are among the waste material, as are contaminants created by humans like paint, metals and oil residues. In order to limit the number of particles that need to be analyzed and so lower the likelihood of error, it is done either with the naked eye or optical microscopy with the aid of tweezers. However, it takes a lot of time and sometimes results in incorrect identification because it is difficult to distinguish microplastics apart from other materials (Thiele et al., 2021).

11.3.3 Density Separation

In general, the density separation technique, combined with filtering, is employed for the extraction of microplastics smaller than 1 mm. The first step in this method is the introduction of an aqueous solvent with a particular density; the second step involves blending for set amounts of time; the third step involves an equilibration time for settling; and the fourth involves filtering to particular size fractions (Dellisanti et al., 2023).

It is easier to detect polymers when low-density plastics are distinguished from high-density mineral material in environmental matrices using density techniques. Utilizing the different densities of the polymers, density separation is utilized to separate the microplastics. High-density microplastic polymers can float in a solution that has been made by increasing the solvent's density. Typically, high concentration salt solutions like NaCl with varying densities are utilized for separation. This approach also makes use of calcium chloride, sodium iodide, zinc chloride, distilled water and others. Manual or mechanical shaking or stirring (using a centrifuge, shaker or magnetic device) are both possible as well as additional methods, such as aeration and inversion (Tirkey and Upadhyay, 2021).

11.3.4 Filtration

Sieving and vacuum filtration techniques are the most commonly used techniques. Vacuum filtration is generally a wet-sorting technique while sieving is a dry-sorting technique. Dry sorting is done by sieving before density separation to decrease the amount of a large sample and segregate into size fractions. Wet sorting is applied after density separation to separate floating particles by vacuum filtration. Usually, glass fiber filters are used for wet sorting and nitrocellulose filters, zooplankton filters and isopore filters are also used (Widziewicz-Rzońca et al., 2023). Membrane-based separation is one method that effectively removes microplastics from food matrices. The straightforward approach does not need a complex setup because it uses a pressure difference to force the liquid through the membrane (Choudhury et al., 2023). The microplastics can be kept on the surface of different membrane filters with a clear size differentiation by applying pressure. This approach has attracted interest because it has the potential to replace the current ones, which require complex operations and maintenance and consume a lot of energy. The separation procedure can also be split into four groups based on the size of the filtered particles: ultrafiltration, microfiltration, nanofiltration and reverse osmosis (Poerio et al., 2019).

11.3.5 Magnetic Extraction

It has been investigated to use magnetic separation of microplastics in environmental and drinking water samples. On the basis of hydrophobic nature of microplastics, Okoye et al. (2022) introduced the magnetic extraction process; in this technique, iron particles with hydrophobic hydrocarbon tails (hexa-decyl-tri-methoxysilane) are used, iron particles are attached to the magnets and non-iron particles are allowed to cross link with microplastic surface. Although the structural integrity of the sample is not affected by magnetic separation techniques, many gaps are seen when the sample is being separated. Due to this, the bulk of studies have concentrated on employing the extraction approach as a pre-treatment step, rather than as the exclusive method of separation (Kumar and Sharma, 2021).

11.3.6 Matrix Removal

To obtain accurate results and shorten processing times, the matrix removal method is utilized to destroy any organic content in the sample. For the digestion of organic materials, acid and alkaline techniques are typically used. Claessens et al. (2013) discovered that 22.5 M HNO_3 treatment resulted in the most efficient chemical digestion of mussel tissue. The 4:1 acid solution of HNO_3 and $HClO_4$ is recommended by the worldwide council for the exploration of the sea for the digestion of marine animal tissues (De Frond et al., 2023). NaOH can be utilized for digestion of alkaline foods. However, some particular polymers, including polystyrene and polyamide, can also be destroyed by these potent acids and alkalis (Justino et al., 2023). In 2013, Foekema et al. employed 10% KOH to decompose fish tissues at room temperature for 2 to 3 weeks without seeing any polymer breakdown. However, digestion with 10% KOH takes a long time and is not appropriate for plant tissues (Foekema et al., 2013; Langenfeld and Bugbee, 2023).

11.3.6.1 Use of H_2O_2 for Oxidization

According to different studies, 30% H_2O_2 is found to digest the samples having high organic matter. It is used mostly to remove soil organic matter during soil analysis (Yan et al., 2023). Nuelle et al. (2014) found that treatment of sample with 30% H_2O_2 for 7 days dissolves about 50% of organic matter completely, but the polycarbonate, polyamide and polypropylene exhibited the noticeable sign of degradation, like size reduction and discoloration. Zhao et al. (2018) studied about microplastics in marine snow and observed that using 30% H_2O_2 for the samples having high organic matter causes the formation of dense foam, which may result in suspension of a considerable part of the sample above the reagent. So, they used 15% H_2O_2 for 24 hrs. at 75 °C and found the results effective just the same as 30% H_2O_2.

Hurley et al. (2018) observed that, for organic rich samples, oxidization by 30% H_2O_2 with Fenton's reagent at room temperature shows highest organic matter digestion rate for soil samples and sewage sludge. They did not observe any effect on tested microplastic particles (Zhou et al., 2022).

11.3.6.2 Enzymatic Digestion

The problem of undesirable organic waste and contaminants on microplastics, which is frequently encountered in the dissection approach, can be solved using a variety of chemical treatments. Acidic, oxidative, alkaline or enzymatic substances may be present during digestion. With the use of these compounds, the dissection phase can frequently be skipped by initially digesting the entire mussel or fish individual or target organ (Kabir et al., 2023). In tissue digestion, HNO_3 is a typical acid that is employed for a variety of tasks, including the removal of microplastics from marine life. In comparison to other digestive chemicals like HCl, NaOH and H_2O_2, mussel soft tissue can be completely digested in HNO_3 in under 24 hours (Unar et al., 2023). Removal of organic matter from microplastics is a very complicated step without disrupting microplastic particles. Because enzymes, unlike chemical digestion, do not deform or destroy the plastic polymers, they have been utilized in several investigations to hydrolyze or dissolve biological tissues (Luo et al., 2022). Although each enzyme functions at its optimum pH and temperature setting, which must be checked and maintained throughout the experiment, enzymatic digestion is also a time-consuming procedure (Bellasi et al., 2020).

Enzymatic digestion may be a valuable process for removal of organic matter. Most of the studies show that this process was introduced for aquatic samples (Maw et al., 2022). Cole et al. (2014) used proteinase-K for the sample having high planktons and achieved > 97% digestion efficiency. Enzymatic digestion is found to be the most efficient method, as compared to other digestion methods, without destroying microplastic particles. Löder et al. (2017) observed that, for complex samples, consecutive enzymatic digestion combined with H_2O_2 oxidation may provide better results, as compared to using a single oxidation step (Hu et al., 2021).

11.4 QUANTIFICATION AND IDENTIFICATION

Microplastics are quantified by counting them and dividing them into groups according to their size, color and kind of polymer. Microscopes are typically used for optical counting of microplastics, although this method has certain accuracy issues (Liu et al., 2023). Visual counting runs the danger of misunderstanding microplastic content and mistaking non-plastic particles for plastic ones. Given the wide variety of polymers, it is crucial to specify their chemical make-up in order to ensure accuracy. Combining spectroscopy and microscopy may increase the precision of optical counting (Huang et al., 2023).

To minimize the possibility of precisely counting plastics, some studies propose 'the hot needle test' (Dümichen et al., 2017). Zhang et al. (2022) proposed a process for polymer identification in soils. After a density separation with water, the remains present in the supernatant are reviewed by comparing microscope-images taken earlier. After heating the sample at 130 °C for 3–5 seconds, liquefied plastics are recognized as thermoplastics. But some natural polymers also melt at certain temperatures, so the possibility of finding exact polymer time is reduced (An et al., 2023). The various extraction techniques are incorporated into a chromatographic unit for the quantitative and qualitative identification of plastic polymers. While RAMAN spectroscopy and Fourier-transform infrared (FTIR) are particle-based methods for identifying microplastics, these mass-based techniques include pyrolysis, gas chromatography and mass spectrometry, and thermal extraction desorption (Mike et al., 2023) saw the introduction of thermal extraction desorption gas chromatography mass spectrometry (TED GC-MS). This approach takes roughly 2 to 3 hours to complete and does not require any pretreatment. Although TED GC-MS is a rapid approach for analyzing samples, it is a destructive method. Thermo-gravimetric analysis is another technique for identifying microplastics. For improved outcomes, it can be paired with mass spectrometry or differential scanning calorimetry (Ye et al., 2022).

Despite being inexpensive and simple, the approach cannot be recommended as a regular identification method because manual or microscopic counting can result in mistakes and inaccurate figures (Liu et al., 2021). Additionally, because there may be debris in the microplastics, counting the split MPs could be challenging. Microscopy has been used in recent years to optically observe microplastics because it enables error-free identification of plastic polymers; vibrational spectroscopy techniques like RAMAN or FTIR are typically utilized for microplastic characterization (Fang et al., 2023). The Raman spectroscopy uses a scattering method for activating the target molecules with a laser of a specific wavelength (Dobek and Baranowski, 2023). The frequency of the sample's scattered radiation provides a measurement of its elemental composition. According to most of the studies, FTIR is a very suitable method for the analysis of sediment samples (Hanvey et al., 2017). Using the FTIR technique to examine the surface characteristics of materials has long been widely used. In order to produce a disturbance (or vibrations) that may be utilized to study the molecular characteristics and determine the sample's composition, the sample absorbs infrared light at a specific wavelength (Yang X. et al., 2023). Micro-FTIR mapping is the sequential measuring of infrared spectra at different points on the sample's surface. The measurement region is filled with thousands of spectra that are collected to produce a chemical image of the material. FTIR can be coupled with an optical microscope to improve the results accuracy (Rodrigues et al., 2023).

Both RAMAN and FTIR are non-destructive processes. SEM is also used to identify size, shape or morphology of the polymers. Many studies have used SEM for the identification of microplastics from sediment samples. Focused electrons are used to scan the surface of the microplastics (Ali et al., 2023). These electrons engage with the atoms on the surface of the microplastics and provide the required signals, which produce an image indicating the sample's size and morphological properties of the sample when paired with the beam position (Kadja et al., 2023).

Proton nuclear magnetic resonance spectroscopy is a new process for size-independent microplastics analysis developed by Peez et al. (2019); samples containing polyethylene, polystyrene and polyethylene tetrapthalate can be analyzed qualitatively and quantitatively by this method. But this method has some disadvantages, such as it is cost effective and requires quite removal of organic matter. So, this method is not a satisfactory method for analysis of soil samples (Möller et al., 2020).

Hyperspectral imaging (HSI) is one of the novel identifying approaches started for microplastics detection. The technique focuses on measuring reflectance spectra for taking pictures with an image spectrometer or an ultra-sensitive camera. This camera is capable of capturing spectral data as pixels. Each pixel more quickly and precisely detects the chemical substance that is present. These substances exhibit pixelation at high resolution (Faltynkova et al., 2021; Tamin et al., 2023).

11.5 CONCLUSIONS

Microplastics have become a matter of concern these days for the environment and scientific society. It is found in the air we breathe, in the soil where we grow crops and in the water we drink. A large generation of plastic waste is a problem nowadays. The various industries use primary microplastics and generate secondary microplastics indirectly. The waste-plastic products are discarded in soil, which moves to the water bodies. These large plastic particles break down with time and form micro-sized particles; these micro-sized particles are spread in air, water and soil. Microplastics are consumed by different aquatic organisms and enter into the food chain, which further undergo biomagnification. As humans consume fish, these microplastics can bioaccumulate in the human body. The microplastics also act as as carrier for other toxic substance because of their hydrophobic nature. One of the best suitable approach is seen to be eliminating food microplastics at the source. The amount of plastic waste dispersion into the environment will be reduced by using effective solid waste management techniques and infrastructure (Kathryn et al., 2022). It has been reported that microplastics have many adverse effects in organisms such as endocrine disruption, reproductive disorders, genetic disorders and immunotoxicology responses. Identification procedures like SEM, Raman spectroscopy, FTIR, thermo-analytical techniques and hyperspectral imaging can be used for identification of microplastics. The micro-sized particles may break down into nanoparticles, which is also a great problem to deal with in the future.

ACKNOWLEDGMENTS

The authors would like to acknowledge Central University of Haryana for providing suitable space and facility for this work.

Conflict of Interest

The authors declare that they have no known conflict of interest.

BIBLIOGRAPHY

Adegoke, K. A., Adu, F. A., Oyebamiji, A. K., Bamisaye, A., Adigun, R. A., Olasoji, S. O., & Ogunjinmi, O. E. (2023). Microplastics toxicity, detection, and removal from water/wastewater. *Marine Pollution Bulletin, 187*, 114546.

Alberghini, L., Truant, A., Santonicola, S., Colavita, G., & Giaccone, V. (2022). Microplastics in fish and fishery products and risks for human health: A review. *International Journal of Environmental Research and Public Health, 20*(1), 789.

Ali, A., Koloor, S. S. R., Alshehri, A. H., & Arockiarajan, A. (2023). Carbon nanotube characteristics and enhancement effects on the mechanical features of polymer-based materials and structures—A review. *Journal of Materials Research and Technology, 24*, 6495–6521.

Alimba, C. G., & Faggio, C. (2019). Microplastics in the marine environment: Current trends in environmental pollution and mechanisms of toxicological profile. *Environmental Toxicology and Pharmacology, 68*, 61–74.

Amin, R. M., Sohaimi, E. S., Anuar, S. T., & Bachok, Z. (2020). Microplastic ingestion by zooplankton in Terengganu coastal waters, southern South China Sea. *Marine Pollution Bulletin, 150*, 110616.

An, S. Y., Wu, X., Zhao, Y., Liu, T., Yin, R., Ahn, J. H., Walker, L. M., Whitacre, J. F., & Matyjaszewski, K. (2023). Highly conductive polyoxanorbornene-based polymer electrolyte for lithium-metal batteries. *Advance Science, 10*, 2302932. DoI: 10.1002/advs.202302932

Anagnosti, L., Varvaresou, A., Pavlou, P., Protopapa, E., & Carayanni, V. (2021). Worldwide actions against plastic pollution from microbeads and microplastics in cosmetics focusing on European policies. Has the issue been handled effectively? *Marine Pollution Bulletin*, *162*, 111883.

Andeobu, L., Wibowo, S., & Grandhi, S. (2023). Informal E-waste recycling practices and environmental pollution in Africa: What is the way forward? *International Journal of Hygiene and Environmental Health*, *252*, 114192.

Aparna, K. (2019). Microplastics in food chain. *Health Action 2019*, 27–28.

Bellasi, A., Binda, G., Pozzi, A., Galafassi, S., Volta, P., & Bettinetti, R. (2020). Microplastic contamination in freshwater environments: A review, focusing on interactions with sediments and benthic organisms. *Environments*, *7*(4), 30.

Borrelle, S. B., Ringma, J., Law, K. L., Monnahan, C. C., Lebreton, L., McGivern, A., Murphy, E., Jambeck, J., Leonard, G. H., Hilleary, M. A., & Eriksen, M. (2020). Predicted growth in plastic waste exceeds efforts to mitigate plastic pollution. *Science*, *369*(6510), 1515–1518.

Bostan, N., Ilyas, N., Akhtar, N., Mehmood, S., Saman, R. U., Sayyed, R. Z., Shatid, A. A., Alfaifi, M. Y., Elbehairi, S. E. I., & Pandiaraj, S. (2023). Toxicity assessment of microplastic (MPs); a threat to the ecosystem. *Environmental Research*, 116523.

Britannica. (2023). Polyethylene. In *Encyclopedia Britannica*. www.britannica.com/science/polyethylene (Accessed 24 November 2023).

Cai, Y., Yang, T., Mitrano, D. M., Heuberger, M., Hufenus, R., & Nowack, B. (2020). Systematic study of microplastic fiber release from 12 different polyester textiles during washing. *Environmental Science & Technology*, *54*(8), 4847–4855.

Campanale, C., Massarelli, C., Savino, I., Locaputo, V., & Uricchio, V. F. (2020). A detailed review study on potential effects of microplastics and additives of concern on human health. *International Journal of Environmental Research and Public Health*, *17*(4), 1212.

Campillo, A., Almeda, R., Vianello, A., Gómez, M., Martínez, I., Navarro, A., & Herrera, A. (2023). Searching for hotspots of neustonic microplastics in the Canary Islands. *Marine Pollution Bulletin*, *192*, 115057.

Chandra, A., McNamara, K. E., Clissold, R., Tabe, T., & Westoby, R. (2023). Climate-induced non-economic loss and damage: Understanding policy responses, challenges, and future directions in pacific small island developing states. *Climate*, *11*(3), 74. DoI: 10.3390/cli11030074

Chappell, B., Pramanik, A., Basak, A. K., Sarker, P. K., Prakash, C., Debnath, S., & Shankar, S. (2022). Processing household plastics for recycling—A review. *Cleaner Materials*, 100158.

Choudhury, A., Simnani, F. Z., Singh, D., Patel, P., Sinha, A., Nandi, A., Ghosh, A., Saha, U., Kumari, K., Jaganathan, S. K., & Kaushik, N. K. (2023). Atmospheric microplastic and nanoplastic: The toxicological paradigm on the cellular system. *Ecotoxicology and Environmental Safety*, *259*, 115018.

Claessens, M., Van Cauwenberghe, L., Vandegehuchte, M. B., & Janssen, C. R. (2013). New techniques for the detection of microplastics in sediments and field collected organisms. *Marine Pollution Bulletin*, *70*(1–2), 227–233.

Cole, M., Webb, H., Lindeque, P. K., Fileman, E. S., Halsband, C., & Galloway, T. S. (2014). Isolation of microplastics in biota-rich seawater samples and marine organisms. *Scientific Reports*, *4*(1), 4528.

De Frond, H., Cowger, W., Renick, V., Brander, S., Primpke, S., Sukumaran, S., Elkhatib, D., Barnett, S., Navas-Moreno, M., Rickabaugh, K., & Vollnhals, F. (2023). What determines accuracy of chemical identification when using microspectroscopy for the analysis of microplastics? *Chemosphere*, *313*, 137300.

Debraj, D., & Lavanya, M. (2023). Microplastics everywhere: A review on existing methods of extraction. *Science of the Total Environment*, 164878.

Dellisanti, W., Leung, M. M. L., Lam, K. W. K., Wang, Y., Hu, M., Lo, H. S., & Fang, J. K. H. (2023). A short review on the recent method development for extraction and identification of microplastics in mussels and fish, two major groups of seafood. *Marine Pollution Bulletin*, *186*, 114221.

Diggle, A., & Walker, T. R. (2022). Environmental and economic impacts of mismanaged plastics and measures for mitigation. *Environments*, *9*(2), 15.

Dissanayake, P. D., Kim, S., Sarkar, B., Oleszczuk, P., Sang, M. K., Haque, M. N., Ahn, J. H., Bank, M. S., & Ok, Y. S. (2022). Effects of microplastics on the terrestrial environment: A critical review. *Environmental Research*, *209*, 112734.

Dobek, K., & Baranowski, M. (2023). The multiplanar imaging microscope with a laser induced thermal lens: A practical case study. *Optics and Lasers in Engineering*, *164*, 107506.

Dümichen, E., Eisentraut, P., Bannick, C. G., Barthel, A. K., Senz, R., & Braun, U. (2017). Fast identification of microplastics in complex environmental samples by a thermal degradation method. *Chemosphere*, *174*, 572–584.

Faltynkova, A., Johnsen, G., & Wagner, M. (2021). Hyperspectral imaging as an emerging tool to analyze microplastics: A systematic review and recommendations for future development. *Microplastics and Nanoplastics*, *1*(1), 1–19.

Fang, C., Luo, Y., & Naidu, R. (2023). Microplastics and nanoplastics analysis: Options, imaging, advancements and challenges. *TrAC Trends in Analytical Chemistry*, 117158.

Foekema, E. M., De Gruijter, C., Mergia, M. T., van Franeker, J. A., Murk, A. J., & Koelmans, A. A. (2013). Plastic in North sea fish. *Environmental Science & Technology*, *47*(15), 8818–8824.

Ghosh, A., Mannab, M. C., Jhab, S., Singha, A. K., Misrac, S., Srivastavab, R. C., Srivastavab, P. P., Laikb, R., Bhattacharyyad, R., Prasadb, S. S., & Singhb, S. P. (2022). Impact of soil-water contaminants on tropical agriculture, animal and societal environment. *Advances in Agronomy*, *176*, 209.

Gilfedder, B. S., Elagami, H., Boos, J. P., Brehm, J., Schott, M., Witt, L., Laforsch, C., & Frei, S. (2023). Filter feeders are key to small microplastic residence times in stratified lakes: A virtual experiment. *Science of the Total Environment*, *890*, 164293.

Hahladakis, J. N., Velis, C. A., Weber, R., Iacovidou, E., & Purnell, P. (2018). An overview of chemical additives present in plastics: Migration, release, fate and environmental impact during their use, disposal and recycling. *Journal of Hazardous Materials*, *344*, 179–199.

Hanvey, J. S., Lewis, P. J., Lavers, J. L., Crosbie, N. D., Pozo, K., & Clarke, B. O. (2017). A review of analytical techniques for quantifying microplastics in sediments. *Analytical Methods*, *9*(9), 1369–1383.

Haque, F., & Fan, C. (2023). Fate and impacts of microplastics in the environment: Hydrosphere, pedosphere, and atmosphere. *Environments*, *10*(5), 70.

Hassan, F., Prasetya, K. D., Hanun, J. N., Bui, H. M., Rajendran, S., Kataria, N., Khoo, K. S., Wang, Y. F., You, S. J., & Jiang, J. J. (2023). Microplastic contamination in sewage sludge: Abundance, characteristics, and impacts on the environment and human health. *Environmental Technology & Innovation*, 103176.

Hassoun, A., Pasti, L., Chenet, T., Rusanova, P., Smaouif, S., Aït-Kaddourg, A., & Bonod, G. (2023). Detection methods of micro and nanoplastics. *Nano/Micro-Plastics Toxicity on Food Quality and Food Safety*, *103*, 175.

Hernandez-Milian, G., Tsangaris, C., Anestis, A., Fossi, M. C., Baini, M., Caliani, I., Panti, C., Bundone, L., & Panou, A. (2023). Monk seal faeces as a non-invasive technique to monitor the incidence of ingested microplastics and potential presence of plastic additives. *Marine Pollution Bulletin*, *193*, 115227.

Hu, K., Tian, W., Yang, Y., Nie, G., Zhou, P., Wang, Y., Duan, X., & Wang, S. (2021). Microplastics remediation in aqueous systems: Strategies and technologies. *Water Research*, *198*, 117144.

Huang, Y., Kok, J. F., Saito, M., & Muñoz, O. (2023). Single-scattering properties of ellipsoidal dust aerosols constrained by measured dust shape distributions. *Atmospheric Chemistry and Physics*, *23*, 2557–2577. DoI: 10.5194/acp-23-2557-2023.

Hurley, R. R., Lusher, A. L., Olsen, M., & Nizzetto, L. (2018). Validation of a method for extracting microplastics from complex, organic-rich, environmental matrices. *Environmental Science & Technology*, *52*(13), 7409–7417.

Jan-Georg, R., Langer, R., & Giovanni, T. (2022). Bioplastics for a circular economy. *Nature Reviews Materials*, *7*(2), 117–137.

Justino, A. K., Ferreira, G. V., Fauvelle, V., Schmidt, N., Lenoble, V., Pelage, L., Martins, K., Travassos, P., & Lucena-Frédou, F. (2023). From prey to predators: Evidence of microplastic trophic transfer in tuna and large pelagic species in the southwestern Tropical Atlantic. *Environmental Pollution*, *327*, 121532.

Kabir, M. S., Wang, H., Luster-Teasley, S., Zhang, L., & Zhao, R. (2023). Microplastics in landfill leachate: Sources, detection, occurrence, and removal. *Environmental Science and Ecotechnology*, 100256.

Kadja, G. T. M., Azhari, N. J., Culsum, N. T. U., Ledyastuti, M., & Radiman, C. L. (2023). New development in polymeric materials characterization and modelling. In *Reference Module in Materials Science and Materials Engineering*. Elsevier. DoI: 10.1016/B978-0-323-96020-5.00003-0

Kallenbach, E. M. F., Rødland, E. S., Buenaventura, N. T., & Hurley, R. (2022). Microplastics in terrestrial and freshwater environments. In M. S. Bank (eds.), *Microplastic in the Environment: Pattern and Process. Environmental Contamination Remediation and Management*. Springer. DoI: 10.1007/978-3-030-78627-4_4

Kathryn, W., Hardesty, B. D., Vince, J., & Wilcox. C. (2022) Local waste management successfully reduces coastal plastic pollution. *One Earth*, *5*(6), 666–676. ISSN 2590-3322.

Khanashyam, A. C., Shanker, M. A., & Nirmal, N. P. (2023). Nano/micro-plastics: Sources, trophic transfer, toxicity to the animals and humans, regulation, and assessment. *Advances in Food and Nutrition Research*, *103*, 141–174.

Kibria, G., Haroon, A. K., Nugegoda, D., & Siddique, M. A. M. (2023). *Book (2023): Microplastic Pollution and Risks. Toxicity, Ecosystems, Water, Food, Air, and Human Health.*

Kumar, R., & Sharma, P. (2021). Recent developments in extraction, identification, and quantification of microplastics from agricultural soil and groundwater. *Fate and Transport of Subsurface Pollutants*, 125–143.

Kye, H., Kim, J., Ju, S., Lee, J., Lim, C., & Yoon, Y. (2023). Microplastics in water systems: A review of their impacts on the environment and their potential hazards. *Heliyon*, *9*, 14359.

Langenfeld, N., & Bugbee, B. (2023). An improved digestion and analysis procedure for silicon in plant tissue. *PloS One*, *18*, e0289151. DoI: 10.1371/journal.pone.0289151.

Liu, K., Huang, J., Xu, M., Perc, M., & Li, X. (2021). Density saliency for clustered building detection and population capacity estimation. *Neurocomputing*, *458*, 127–140.

Liu, Y., Shi, H., Chen, L., Teng, X., Xue, C., & Li, Z. (2023). An overview of microplastics in oysters: analysis, hazards, and depuration. *Food Chemistry*, 136153.

Löder, M. G. J., Imhof, H. K., Ladehoff, M., Löschel, L. A., Lorenz, C., Mintenig, S., Piehl, S., Primpke, S., Schrank, I., Laforsch, C., & Gerdts, G. (2017). Enzymatic purification of microplastics in environmental samples. *Environmental Science & Technology*, *51*(24), 14283–14292.

Luo, M., Wang, Z., Fang, S., Song, B., Cao, P., Liu, H., & Yang, Y. (2022). Removal and toxic forecast of microplastics treated by electrocoagulation: Influence of dissolved organic matter. *Chemosphere*, *308*, 136309.

Madhav, N. V., Gopinath, K. P., Krishnan, A., Rajendran, N., & Krishnan, A. (2020). A critical review on various trophic transfer routes of microplastics in the context of the Indian coastal ecosystem. *Watershed Ecology and the Environment*, *2*, 25–41.

Mandal, A., Singh, N., Mondal, A., Talib, M., Basu, R., Biswas, M. K., & Darbha, G. K. (2023). The extent of microplastic pollution along the eastern coast of India: Focussing on marine waters, beach sand, and fish. *Marine Pollution Bulletin*, *194*, 115265.

Maw, M. M., Boontanon, N., Fujii, S., & Boontanon, S. K. (2022). Rapid and efficient removal of organic matter from sewage sludge for extraction of microplastics. *Science of the Total Environment*, *853*, 158642.

McCann, J. (2022). Collaborative design principles for smart clothing. In *Smart Clothes and Wearable Technology*. Woodhead Publishing, pp. 283–325. DoI: 10.1016/B978-1-84569-357-2.50024-1

Mian, H. R., Chhipi-Shrestha, G., McCarty, K., Hewage, K., & Sadiq, R. (2022). An estimation of tire and road wear particles emissions in surface water based on a conceptual framework. *Science of the Total Environment*, *848*, 157760.

Mike, W., Schoettl, J., Pruin, L., Fischer, B., Wolf, C., Kube, C., Renner, G., Schram, J., Schmidt, T. C., & Tuerk, J. (2023). Determination of atmospherically deposited microplastics in moss: Method development and performance evaluation. *Green Analytical Chemistry*, *7*, 100078.

Miller, M. E., Hamann, M., & Kroon, F. J. (2020). Bioaccumulation and biomagnification of microplastics in marine organisms: A review and meta-analysis of current data. *PLoS One*, *15*(10), e0240792.

Mohanan, N., Montazer, Z., Sharma, P. K., & Levin, D. B. (2020). Microbial and enzymatic degradation of synthetic plastics. *Frontiers in Microbiology*, *11*, 580709. DoI: 10.3389/fmicb.2020.580709

Möller, J. N., Löder, M. G., & Laforsch, C. (2020). Finding microplastics in soils: A review of analytical methods. *Environmental Science & Technology*, *54*(4), 2078–2090.

Mubin, A. N., Arefin, S., Mia, M. S., Islam, A. R. M. T., Bari, A. M., Islam, M. S., Ali, M. M., Siddique, M. A. B., Rahman, M. S., Senapathi, V., & Idris, A. M. (2023). Managing the invisible threat of microplastics in marine ecosystems: Lessons from coast of the Bay of Bengal. *Science of the Total Environment*, 164224.

Nawab, J., Khan, H., Ghani, J., Zafar, M. I., Khan, S., Toller, S., Fatima, L., & Hamza, A. (2023). New insights into the migration, distribution and accumulation of micro-plastic in marine environment: A critical mechanism review. *Chemosphere*, *330*, 138572.

Nicholson, S. R., Rorrer, J. E., Singh, A., Konev, M. O., Rorrer, N. A., Carpenter, A. C., Jacobsen, A. J., Román-Leshkov, Y., & Beckham, G. T. (2022). The critical role of process analysis in chemical recycling and upcycling of waste plastics. *Annual Review of Chemical and Biomolecular Engineering*, *13*, 301–324.

Nuelle, M. T., Dekiff, J. H., Remy, D., & Fries, E. (2014). A new analytical approach for monitoring microplastics in marine sediments. *Environmental Pollution*, *184*, 161–169.

Okoye, C. O., Addey, C. I., Oderinde, O., Okoro, J. O., Uwamungu, J. Y., Ikechukwu, C. K., Okeke, E. S., Ejeromedoghene, O., & Odii, E. C. (2022). Toxic chemicals and persistent organic pollutants associated with micro-and nanoplastics pollution. *Chemical Engineering Journal Advances*, *11*, 100310.

Pavluk, T., & Bij, D. V. A. (2008). Trophic index and efficiency. In *Encyclopedia of Ecology*, pp. 3602–3608. DoI: 10.1016/B978-0-12-409548-9.00608-4

Peez, N., Janiska, M. C., & Imhof, W. (2019). The first application of quantitative 1 H NMR spectroscopy as a simple and fast method of identification and quantification of microplastic particles (PE, PET, and PS). *Analytical and Bioanalytical Chemistry*, *411*, 823–833.

Plastics Europe. (2019). *Plastics e the Facts 2019*. www.plasticseurope.org/en/resources/publications/1804-plastics-facts-2019

Poerio, T., Piacentini, E., & Mazzei, R. (2019). Membrane processes for microplastic removal. *Molecules*, *24*(22), 4148.

Rahman, N., Shozib, S. H., Akter, Y., Islam, A. R. T., Islam, S., Sohel, S., Kamaraj, C., Rakib, R. J., Idris, A. M., Sarker, A., & Malafaia, G. (2023). Microplastic as an invisible threat to the coral reefs: Sources, toxicity mechanisms, policy intervention, and the way forward. *Journal of Hazardous Materials*, 131522.

Reddy, A. S., & Nair, A. T. (2022). The fate of microplastics in wastewater treatment plants: An overview of source and remediation technologies. *Environmental Technology & Innovation*, *28*, 102815.

Reichle, D. E. (2023). Food chains and trophic level transfers. In David E. Reichle (Eds.), *The Global Carbon Cycle and Climate Change: Scaling Ecological Energetics from Organism to the Biosphere*. Elsevier, pp. 129–151. ISBN 9780443187759

Reineccius, J., Bresien, J., & Waniek, J. J. (2021). Separation of microplastics from mass-limited samples by an effective adsorption technique. *Science of the Total Environment*, *788*, 147881.

Rødland, E. S., Lind, O. C., Reid, M., Heier, L. S., Skogsberg, E., Snilsberg, B., Gryteselv, D., & Meland, S. (2022). Characterization of tire and road wear microplastic particle contamination in a road tunnel: From surface to release. *Journal of Hazardous Materials*, *435*, 129032.

Rodrigues, S., Fonteneau, L., & Esterle, J. (2023). Characterisation of coal using hyperspectral core scanning systems. *International Journal of Coal Geology*, *269*, 104220.

Rogers, K. (2022, 5 April). Microplastics. In *Encyclopedia Britannica*. www.britannica.com/technology/microplastic (Accessed 5 August 2023).

Rose, P. K., Jain, M., Kataria, N., Sahoo, P. K., Garg, V. K., & Yadav, A. (2023). Microplastics in multimedia environment: A systematic review on its fate, transport, quantification, health risk, and remedial measures. *Groundwater for Sustainable Development*, 100889.

Rummel, C. D., Löder, M. G., Fricke, N. F., Lang, T., Griebeler, E. M., Janke, M., & Gerdts, G. (2016). Plastic ingestion by pelagic and demersal fish from the North Sea and Baltic Sea. *Marine Pollution Bulletin*, *102*(1), 134–141.

Sambolino, A., Iniguez, E., Herrera, I., Kaufmann, M., Dinis, A., & Cordeiro, N. (2023). Microplastic ingestion and plastic additive detection in pelagic squid and fish: Implications for bioindicators and plastic tracers in open oceanic food webs. *Science of the Total Environment*, 164952.

Seyyedi, S. R., Kowsari, E., Ramakrishna, S., Gheibi, M., & Chinnappan, A. (2023). Marine plastics, circular economy, and artificial intelligence: A comprehensive review of challenges, solutions, and policies. *Journal of Environmental Management*, *345*, 118591.

Shahul Hamid, F., Bhatti, M. S., Anuar, N., Anuar, N., Mohan, P., & Periathamby, A. (2018). Worldwide distribution and abundance of microplastic: How dire is the situation? *Waste Management & Research*, *36*(10), 873–897.

Tamin, O., Moung, E. G., Dargham, J. A., Yahya, F., & Omatu, S. (2023). A review of hyperspectral imaging-based plastic waste detection state-of-the-arts. *International Journal of Electrical and Computer Engineering*, *13*, 3407–3419.

Thiele, C. J., Hudson, M. D., & Russell, A. E. (2019). Evaluation of existing methods to extract microplastics from bivalve tissue: Adapted KOH digestion protocol improves filtration at single-digit pore size. *Marine Pollution Bulletin*, *142*, 384–393.

Thiele, C. J., Hudson, M. D., Russell, A. E., Saluveer, M., & Sidaoui-Haddad, G. (2021). Microplastics in fish and fishmeal: An emerging environmental challenge? *Scientific Reports*, *11*(1), 2045.

Thomas, D., Schütze, B., Heinze, W. M., & Steinmetz, Z. (2020). Sample preparation techniques for the analysis of microplastics in soil—A review. *Sustainability*, *12*(21), 9074.

Tirkey, A., & Upadhyay, L. S. B. (2021). Microplastics: An overview on separation, identification and characterization of microplastics. *Marine Pollution Bulletin*, *170*, 112604.

Unar, A., Sarfraz, M., Ajarem, J. S., Allam, A. A., Bhatti, U., Chanihoon, G. Q., & Afridi, H. I. (2023). Mitigating marine hazardous contaminants: A sustainable management perspective. *Chemosphere*, 139292.

Van Cauwenberghe, L., & Janssen, C. R. (2014). Microplastics in bivalves cultured for human consumption. *Environmental Pollution (Barking, Essex: 1987)*, *193*, 65–70. DoI: 10.1016/j.envpol.2014.06.010.

Widziewicz-Rzońca, K., Janas, S., Błaszczak, B., Błaszczak, M., Pyta, H., Rogula-Kopiec, P., Mathews, B., & Klejnowski, K. (2023). Advancing the understanding of pm filter mass stability: unveiling the influence of humidity and temperature. *Zeszyty Naukowe Pro Publico Bono*, *1*, 7–26. DoI: 10.5604/01.3001.0053.9741.

Wiesinger, H., Wang, Z., & Hellweg, S. (2021). Deep dive into plastic monomers, additives, and processing aids. *Environmental Science & Technology*, *55*(13), 9339–9351.

Wu, H., Hou, J., & Wang, X. (2023). A review of microplastic pollution in aquaculture: Sources, effects, removal strategies and prospects. *Ecotoxicology and Environmental Safety*, *252*, 114567.

Wypych, G. (2023). Odor in different polymers, handbook of odors in plastic materials. In *Handbook of Odors in Plastic Materials* (Third Edition). Elsevier, Chem Tec Publishing, pp. 125–181. ISBN 9781774670200.

Yan, P., Hao, X., & Zhang, S. (2023). Extraction and decontamination of microplastics from high organic matter soils: A simple, cost-saving and high efficient method. *Journal of Environmental Management*, *344*, 118381.

Yang, J., Monnot, M., Sun, Y., Asia, L., Wong-Wah-Chung, P., Doumenq, P., & Moulin, P. (2023). Microplastics in different water samples (seawater, freshwater, and wastewater): Methodology approach for characterization using micro-FTIR spectroscopy. *Water Research*, *232*, 119711.

Yang, X., Huang, S., Li, D., Xu, H., Zeng, Y., Yang, L., Wang, D., Zhang, N., Cao, J., & Shen, Z. (2023). Water-soluble organic matter with various polarities in PM2. 5 over Xi'an, China: Abundance, functional groups, and light absorption. *Particuology*, *84*, 281–289.

Ye, Y., Yu, K., & Zhao, Y. (2022). The development and application of advanced analytical methods in microplastics contamination detection: A critical review. *Science of the Total Environment*, *818*, 151851.

Yu, X., Ladewig, S., Bao, S., Toline, C. A., Whitmire, S., & Chow, A. T. (2018). Occurrence and distribution of microplastics at selected coastal sites along the south-eastern United States. *Science of the Total Environment*, *613*, 298–305.

Yuan, Z., Nag, R., & Cummins, E. (2022). Human health concerns regarding microplastics in the aquatic environment-from marine to food systems. *Science of the Total Environment*, *823*, 153730.

Zhang, Z., Gao. S.-H., Luo, G., Kang, Y., Zhang, L., Pan, Y., Zhou, X., Fan, L., Liang, B., & Wang, A. (2022). The contamination of microplastics in China's aquatic environment: Occurrence, detection and implications for ecological risk. *Environmental Pollution*, *296*, 118737. DoI: 10.1016/j.envpol.2021.118737

Zhao, J., Liu, L., Zhang, Y., Wang, X., & Wu, F. (2018). A novel way to rapidly monitor microplastics in soil by hyperspectral imaging technology and chemometrics. *Environmental Pollution*, *238*, 121–129.

Zhou, Q., Chen, J., Zhang, D., & Pan, X. (2022). Evaluation of organic matter removal by H_2O_2 from microplastic surface by nano-physicochemical methods. *Green Analytical Chemistry*, *3*, 100035.

Zhu, Z., Hossain, K. B., Wei, H., Jia, R., Gao, X., Jin, H., & Cai, M. (2023). Distribution and sources of microplastics in the Beibu Gulf using in-situ filtration technique. *Marine Pollution Bulletin*, *188*, 114614.

Ziani, K., Ioniţă-Mîndrican, C. B., Mititelu, M., Neacşu, S. M., Negrei, C., Moroşan, E., Drăgănescu, D., & Preda, O. T. (2023). Microplastics: A real global threat for environment and food safety: A state of the art review. *Nutrients*, *15*(3), 617.

Zolotova, N., Kosyreva, A., Dzhalilova, D., Fokichev, N., & Makarova, O. (2022). Harmful effects of the microplastic pollution on animal health: A literature review. *PeerJ*, *10*, e13503.

Index

G

H

I

K

L

M

N

O

P